PRECALCULUS:
ELEMENTARY
FUNCTIONS
AND
RELATIONS

PRECALCULUS: ELEMENTARY FUNCTIONS AND RELATIONS

DONALD R. HORNER, PH.D.
EASTERN WASHINGTON STATE COLLEGE

HOLT, RINEHART AND WINSTON
New York   Chicago   San Francisco   Atlanta
Dallas   Montreal   Toronto   London   Sidney

1 2 3 4 5 6 7 8 9 10

Preface

Precalculus: Elementary Functions and Relations necessarily represents the author's conception of what precalculus preparation ought to be. His classroom efforts are an attempt to help the student appreciate and understand the topics and the role of the topics included within this text. The character of the work is based solely on the awareness that students' appreciation (or lack of appreciation) of the function and its importance as a central concept running through mathematics is one of the prime reasons for academic impotence in the study of calculus.

The allotments of time, space, and emphasis reflect those ideas often not understood by students of the calculus. Accordingly, thought processes are channeled in such a direction as to picture the entire experience as an inquiry into various characteristics of the elementary real functions.

As a result, a number of concepts are learned gradually and do not appear in the table of contents. An example is the continuing experience in graphing functions. The reader may not even be conscious of this learning process. Furthermore, virtually every graphing exercise and every incidence of the examination of an equation has a directly related set of problems in inequalities and their graphs. The author knows of no other similar textual material doing so much with inequalities. Inequalities, however, do not appear as an entry in the table of contents.

Similarly, the totality of the theory of equations given is not reflected in a cursory examination of the table of contents. The work with absolute values is not so shown. The fact that many statements and problems reflect those encountered in the calculus is not apparent there, nor are the uniqueness of the study of the trigonometric functions and the considerable amount of effort given toward identities and conditional equations. Hyperbolic functions afford another example of the great deal of material embedded naturally in this approach.

The general problem lists are intended to perform a definite and important task, a major task being that of complete and meaningful preinstruction. Careful consideration of the role of this area of the text is essential, for it is intended to be a very real part of the learning process.

Review exercises and advanced exercises allow for a great deal of flexibility of interests and tastes. The quizzes at the end of each chapter should check mechanical, theoretical, and philosophical understanding.

Considerable analytical geometry is built into the body of the text and given again in the examination of functions and relations. The amount of geometry used in any one study may be varied considerably, depending on needs.

The book is not simply another integrated algebra and trigonometry text claiming to emphasize functions. It is a functions text exploring applications to algebra and trigonometry. The difference is considerable and of significant import. The fine Math O recommendations of the Mathematical Association of America Committee on the Undergraduate Program in Mathematics are met in a genuine and meaningful investigation of real-valued functions — the only nonartificial method of presenting all precalculus topics in one study.

In a nutshell, the text is a concentrated effort to sell two beliefs: (1) that the unifying element of the material is the real function, and (2) that the student, once convinced of the central theme, will appreciate and comprehend to a more gratifying degree.

Mechanically, the text may be used in a number of ways and in courses of a number of different lengths. The flow chart at the end of these remarks gives a view of the different natural orders. By using one class period per section (excluding review and advanced exercises and quizzes) courses varying from less than 45 periods to 54 periods may be taught. By omitting some topics to be done completely in a following calculus course, this length may be shortened, whereas spending class meetings on the review exercises and appendix material can extend the time to 62 or more lectures. Even longer courses can be taught through particular instructor emphasis, instructor additions, or the use of quizzes and selected advanced exercises. In an extreme case, an 80-period course is feasible. The author's familiarity is with a 50-period investigation.

It is indeed a privilege to have written a text that looks ahead to the prosperity of new programs in mathematics. It is as much a privilege to acknowledge the assistance of Professor Joseph Dorsett of St. Petersburg Junior College and Mr. David Straughan and Mr. James Mulligan of Eastern Washington State College. Again the author takes pride in publicly acknowledging that he need give no thought to fear of clerical or secretarial ineptitudes. His wife Donna does a simply superb job of manuscript composition.

D. R. HORNER

CHENEY, WASHINGTON
FEBRUARY 1969

Contents

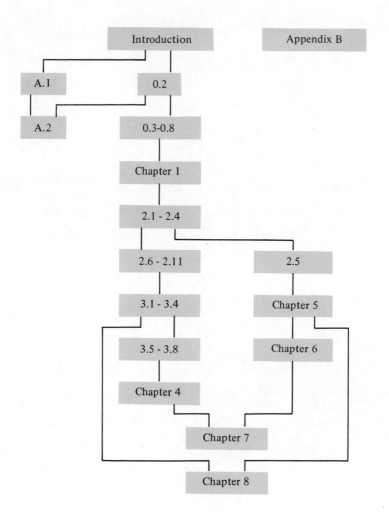

Introduction

Appendix B

A.1

0.2

A.2

0.3-0.8

Chapter 1

2.1 - 2.4

2.6 - 2.11

2.5

3.1 - 3.4

Chapter 5

3.5 - 3.8

Chapter 6

Chapter 4

Chapter 7

Chapter 8

Prefatory Material

INTRODUCTION

In the present chapter we expect to set the stage for virtually all of what is to follow by introducing some symbolism and notation common to contemporary mathematics. This notation and these symbols will help you develop a feel for and an appreciation of the philosophy of mathematical thinking and processes. The discussion of set theory helps you, in a very real way, along this route. Your level of mathematical maturity should swing upward and you can expect to achieve a certain sophistication in the use of mathematical symbolism.

One single most important objective is the acquisition of a well-grounded concept of the function. A good deal of the material of the text, and of this prefatory chapter in particular, is geared toward realizing this hope.

Factual knowledge and mechanical aptitude involving other topics discussed are of great importance but are superseded by the necessity for maturity, sophistication, and a knowledge of the function.

0.1 SYMBOLISM AND SETS

Symbolism in mathematics is both a system of shorthand and a descriptive language. To illustrate some useful symbolism, we introduce a number of topics from set theory. Examine the following

1

(capital letters will be used to represent sets, lowercase letters to signify elements of sets):

$$x \in A$$

The above symbol is read in any of the following ways:

1. x is an element of A.   3. x belongs to A.

2. x is a member of A.   4. x is in A.

In all probability you can think of several further phrases that exhibit this element-set relationship. The symbol $x \notin A$ is read, "x is not an element of A." The meaning of the slash as introduced here is somewhat standard in mathematics. It will be well for you to realize that the terms *element* and *set* have not been defined. Neither will they be defined, for each is an example of the use of *undefined* terms. The mathematician does this to prevent "circular" definitions and to generate a universality for his work.

The set builder symbolism is particularly important in mathematics today. The following is an example of this concept:

$$A = \{x : x \text{ has property } P\}$$

The symbolic statement is read, "A is the set of all elements x such that x has property P." A more vivid example follows.

$$A = \{x : x \text{ is a dog or } x \text{ is a cat}\}$$

The set A is the set of all objects x where x is either a dog or a cat. In particular, A is the set of all dogs and cats. A similar arrangement may be used to list the elements of a set.

$$B = \{p, \$, z, \pi, <\}$$

In this instance B is the set consisting of the objects p, $\$$, z, π, and $<$.

Consider at this time the following relationship (between two sets) as symbolized by either

1. $A \subset B$ or   2. $B \supset A$

We read the sentence: A is a *subset* of B. It (the statement) means that every element of A is an element of B as well. In this situation we might say descriptively that A is *contained* in B or that B *contains A.*

The definition encompasses two distinct statements:

1. If $A \subset B$ then each $x \in A$ satisfies $x \in B$; and

2. If $x \in A$ implies $x \in B$, then $A \subset B$.

Now let us examine some specific examples of the *inclusion relation "$\subset$."*

EXAMPLE 0.1.1. Let the sets A, B, and C be as given below.

$$A = \{1, 2, 3, 4, 5\}$$
$$B = \{0, 1, 3, 5, 9\}$$
$$C = \{1, 3, 4\}$$

Which of the following are true and which are false?

1. $C \subset A$ 2. $A \subset C$ 3. $A \subset B$

4. $B \subset C$ 5. $C \subset B$

Examine first the solution of 1. Is $C \subset A$? To answer this question, we must compare the elements of C with those of A. For C to be a subset of A, we need only show that each element of C is also an element of A. Is this true? Yes. Each of the elements 1, 3, and 4 of C is found also in A. Thus 1 is true.

The remaining four statements are all false. Statement 2 is false, since 2 is an element of A not also contained in C.

Now, $2 \in A$ but $2 \notin B$, whence $A \not\subset B$. Furthermore, since $9 \in B$ but not in A, we see that $B \not\subset A$ either.

Finally, $4 \in C$ but $4 \notin B$, which is to say that $C \not\subset B$. Thus we find correct our prediction that 2–5 are false.

In viewing the inclusion (subset) relationship, you may have raised in your mind the question of the meaning of equality of sets. Intuitively, we expect two sets to be equal when they have precisely the same elements. One definition of equality which is very usable (in a mechanical sense) is: Two sets are *equal* whenever each is a subset of the other. This will be the definition prescribed for use in this text.

Symbolically,

$$A = B \text{ means } A \subset C \text{ and } B \subset A$$

EXAMPLE 0.1.2. Are any of the sets A, B, and C from Example 0.1.1 above found to be equal? The answer is no. This is because no two of the sets contain exactly the same elements. Viewing the question from the standpoint of the subset relation, $A \neq B$ since $A \not\subset B$ (nor is $B \subset A$). Furthermore, $A \neq C$ since $A \not\subset C$. Finally, $B \neq C$ since neither is a subset of the other.

EXERCISES 0.1

1. Let $A = \{1, 2, 3, 4, 5\}$, $B = \{1, 3, 5\}$, $C = \{2, 4\}$, $D = \{0, 1, 3\}$, $E = \{0, 2, 4\}$, and $F = \{0, 1, 2, 3, 4, 5\}$. Answer the following by True or False.

a. $A \subset A$	h. $B \subset B$	o. $C \subset C$	v. $D \subset D$
b. $A \subset B$	i. $B \subset C$	p. $C \subset D$	w. $D \subset E$
c. $A \subset C$	j. $B \subset D$	q. $C \subset E$	x. $D \subset F$
d. $A \subset D$	k. $B \subset E$	r. $C \subset F$	y. $E \subset A$
e. $A \subset E$	l. $B \subset F$	s. $D \subset A$	z. $E \subset B$
f. $A \subset F$	m. $C \subset A$	t. $D \subset B$	
g. $B \subset A$	n. $C \subset B$	u. $D \subset C$	

a'. $E \subset C$	d'. $E \subset F$	g'. $F \subset C$	j'. $F \subset F$
b'. $E \subset D$	e'. $F \subset A$	h'. $F \subset D$	
c'. $E \subset E$	f'. $F \subset B$	i'. $F \subset E$	

2. After having answered the questions in Exercise 0.1.1, you have probably formed an opinion about the truth of $P \subset P$. What conclusion have you drawn? Justify your conclusion.

3. In Exercise 0.1.1 we observe that if we "throw" all the sets between one pair of braces, the resulting set is F. If we combine A and B by this process, the resulting set is the same as A. In some sense we might declare that we are adding sets when we so combine them. In this context add the following pairs of sets.

a. A, B	f. A, A	k. B, B	p. D, E
b. A, C	g. B, C	l. C, D	q. D, F
c. A, D	h. B, D	m. C, E	r. D, D
d. A, E	i. B, E	n. C, F	s. E, F
e. A, F	j. B, F	o. C, C	t. E, E

4. On the basis of Exercise 0.1.3, decide what set results if we add a set P to another set Q where $P \subset Q$. What set results if you add P to itself?

5. In Exercise 0.1.1 various of the sets have some element(s) in common. In each case list the complete set of elements common to each pair of sets.

a. A, B	f. A, A	k. B, B	p. D, E
b. A, C	g. B, C	l. C, D	q. D, F
c. A, D	h. B, D	m. C, E	r. D, D
d. A, E	i. B, E	n. C, F	s. E, F
e. A, F	j. B, F	o. C, C	t. E, E

6. What can you say about the set of elements common to sets P and Q if $P \subset Q$? Need two sets have any elements in common?

7. Relative to Exercises 0.1.3 and 0.1.5 do any of your answers vary if the order of the sets is reversed? Explain.

8. Using the concept of "common elements," we can form yet further sets. In each pair below, describe the set obtained by deleting from the first set elements also found in the second set. The sets are from Exercise 0.1.1.

a.	A, B	f.	A, A	k.	B, B	p.	D, E
b.	A, C	g.	B, C	l.	C, D	q.	D, F
c.	A, D	h.	B, D	m.	C, E	r.	D, D
d.	A, E	i.	B, E	n.	C, F	s.	E, F
e.	A, F	j.	B, F	o.	C, C	t.	E, E

9. From your experience with Exercise 0.1.8 decide what set is formed by deleting from P the elements found also in Q under the condition that $P \subset Q$. If $Q \subset P$? If $P = Q$? If P and Q have no common element? Do your answers from Exercise 0.1.8 change if the order of the pairs is reversed? Explain.

10. In Exercise 0.1.1, the set F contains all the elements from all the sets of the discussion. Describe in each case below the set of elements (in F) not in the given set.

    a. A b. B c. C d. D e. E

How does the process of determining these answers compare with that used in Exercises 0.1.8 and 0.1.9?

0.2 COMBINATIONS OF SETS

The next few paragraphs will be devoted to sets easily constructed as combinations of other sets.

$$A \cup B = \{x : x \in A \quad \text{or} \quad x \in B\}$$
$$A \cap B = \{x : x \in A \quad \text{and} \quad x \in B\}$$
$$A - B = \{x : x \in A \quad \text{but} \quad x \notin B\}$$

The symbol $A \cup B$ represents the *union* of A and B. The union is proclaimed to be the set of all elements belonging to at least one of A and B. The "at least one" is a result of the word "or" (rather, our interpretation of the word "or"). Recall problems 3 and 4 of Exercise 0.1.

The expression $A \cap B$ symbolizes the *intersection* of A and B. The intersection of A and B is the collection of all elements belonging to both sets. (See Exercises 0.1.5 and 0.1.6.)

Finally, $A - B$ is the *difference* of A and B or the *complement* of B with respect to A. The difference is the set of all elements of A not belonging to B. (Review Exercises 0.1.8 and 0.1.9.)

EXAMPLE 0.2.1. Let A, B, and C be as in Example 0.1.1. Then

$$A \cup B = \{0, 1, 2, 3, 4, 5, 9\} \qquad A \cup C = A$$
$$A \cap B = \{1, 3, 5\} \qquad A \cap C = C$$
$$A - B = \{2, 4\} \qquad A - C = \{2, 5\}$$
$$B - A = \{0, 9\}$$

   The union of A and B was computed by listing within one set of braces the elements of A and the elements of B. Their intersection was found by listing the elements found in common. The difference $A - B$ of A and B was formed by deleting from A those elements that belong to B. The other sets are formed analogously.

   How do we describe the set $C - A$? If we delete from C all the elements also found in A, we find that we have deleted all the elements (remember that $C \subset A$). Thus, $C - A$ has no elements. It then appears useful to define: ϕ is the symbol used to denote any set not having elements (*empty set*). Thus, using this concept we write (from Example 0.2.1): $C - A = \phi$.

EXERCISES 0.2

1. Let $A = \{1, 2, 3, 4, 5\}$, $B = \{1, 3, 5\}$, $C = \{2, 4\}$, $D = \{0, 1, 3\}$, $E = \{0, 2, 4\}$, and $F = \{0, 1, 2, 3, 4, 5\}$. Find each of the following sets.

a. $A \cup A$	h. $B \cup B$	o. $C \cup D$	v. $E \cup D$
b. $A \cup B$	i. $B \cup C$	p. $D \cup C$	w. $E \cup E$
c. $B \cup A$	j. $C \cup B$	q. $C \cup E$	x. $E \cup F$
d. $A \cup C$	k. $B \cup D$	r. $C \cup F$	y. $F \cup E$
e. $A \cup D$	l. $B \cup E$	s. $D \cup D$	z. $F \cup F$
f. $A \cup E$	m. $B \cup F$	t. $D \cup E$	
g. $A \cup F$	n. $C \cup C$	u. $D \cup F$	

2. From your experience with Exercise 0.2.1 what do you find $P \cup P$ to be? Justify your conclusion. What is the relation between $P \cup Q$ and $Q \cup P$? Justify this answer also.

3. Find all the sets formed by replacing "$\cup$" by "$\cap$" in Exercise 0.2.1.

4. Answer all the questions from Exercise 0.2.2 after replacing "$\cup$" by "$\cap$."

5. Find all the sets formed by replacing "$\cup$" by "$-$" in Exercise 0.2.1.

6. Answer the questions from Exercise 0.2.2 by replacing "$\cup$" by "$-$."

7. In Exercise 0.1.1, the set F contains all elements under discussion. In this sense, F is the *universal* set of the discussion. For each subset P of F, $F - P = P'$ is called the *complement* of P (relative to F). The expression "relative to F" is dropped when the set F is the universal set of the discussion. The universal set is, for all practical purposes, the set of all elements (remember that this is relative to the individual discussion). Thus we can say that P' is the set of all elements not in P. From Exercise 0.2.1 find the following sets.

    a. A' b. B' c. C' d. D' e. E' f. F' g. ϕ'

8. If we consider sets A and B we can form a set via the diagram of Figure 0.2.1. The two "axes" are labeled A and B respectively and the lattice

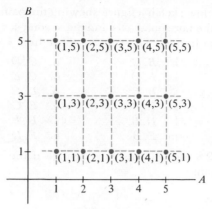

Figure 0.2.1

work of horizontal and vertical lines form fifteen points of intersection. The points are labeled as pairs, the first element being that associated with A and the second with B. For example, the vertical line through the A axis at 2 intersects the horizontal line passing through the 5 on the B axis to form the "pair" $(2, 5)$. The complete list of points so formed is

$(1, 1)$, $(1, 3)$, $(1, 5)$, $(2, 1)$, $(2, 3)$, $(2, 5)$, $(3, 1)$, $(3, 3)$, $(3, 5)$, $(4, 1)$, $(4, 3)$, $(4, 5)$, $(5, 1)$, $(5, 3)$, $(5, 5)$

Figure 0.2.2 shows a similar arrangement with the sets A and B interchanged on the two axes.

a.   List the pairs formed in Figure 0.2.2.

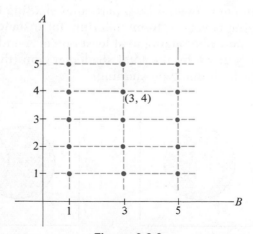

Figure 0.2.2

In each case below sketch a figure showing the pairs formed by placing the first set on the horizontal axis and the second set on the vertical axis. List the pairs so formed. The sets are those from Exercise 0.2.1.

b.	A, C	h.	B, C	n.	C, D	t.	A, A
c.	A, D	i.	B, D	o.	C, E	u.	B, B
d.	A, E	j.	B, E	p.	D, C	v.	C, C
e.	C, A	k.	C, B	q.	E, C	w.	D, D
f.	D, A	l.	D, B	r.	D, E	x.	E, E
g.	E, A	m.	E, B	s.	E, D		

9. In Exercise 0.2.8, do your answers change if the order of the sets is reversed? Explain.

0.3 VENN DIAGRAMS

There are simple diagrams that can be used as an aid to intuition (when discussing sets). These diagrams are called Venn-Euler or simply Venn diagrams. Figures 0.3.1–0.3.5 show all the relationships that can possibly exist between two sets A and B. The sets are taken to be the respective circles together with their interiors.

In Figure 0.3.1 the two sets have an empty intersection (that is, $A \cap B = \phi$). Two such sets are said to be *disjoint*.

Horizontal and vertical shading can be used to illustrate some of the aforementioned combinations of sets. Figures 0.3.6–0.3.10 depict the same conditions as Figures 0.3.1–0.3.5 (respectively) but with the shading added. The set A is shaded horizontally while B has the vertical shading.

In what manner does the shading display certain combinations of sets? Any element covered by a particular shading belongs to the set whose shading covers it. Remembering, for instance, that $A \cup B$ is the set of elements belonging to at least one of A and B, we realize that this union is given by the total shaded region (that is, the total region having at least one type shading).

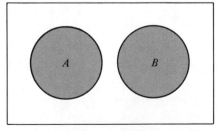

Figure 0.3.1

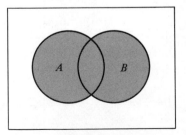

Figure 0.3.2

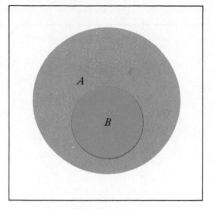

Figure 0.3.3

Figure 0.3.4

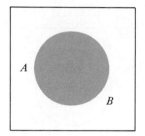

Figure 0.3.5

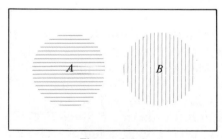

Figure 0.3.6

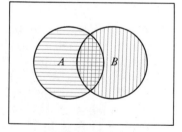

Figure 0.3.7

How is $A \cap B$ described? It is the set endowed with *both* types of shading. Alternatively, the desired intersection is represented by the crosshatched area.

Other combinations of sets are described in an analogous manner.

We now present two examples of proofs of theorems from set theory.

Theorem 0.3.1.   If $A \subset B$ *and* $B \subset C$, *then* $A \subset C$. *(This is called the* transitive *property of* "$\subset$".)

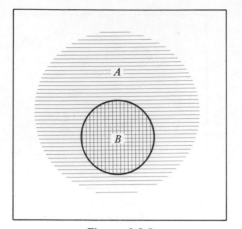

Figure 0.3.8   **Figure 0.3.9**

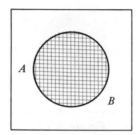

Figure 0.3.10

Proof. What is to be shown? On the hypothesis that $A \subset B$ and $B \subset C$, we are to show that $A \subset C$. How can we accomplish this? The definition of $A \subset C$ declares that we must merely show: each $x \in A$ belongs also to C. The following argument, making use of the hypothesis, is designed to do just this.

Since $A \subset B$, each x in A is in B. Furthermore, since $B \subset C$, each $x \in B$ is in C. Necessarily, each $x \in A$ is found to be an element of C.

Our conclusion? If $x \in A$, then $x \in C$, which is to say that $A \subset C$.

The theorem is seen intuitively in Figure 0.3.11.

Theorem 0.3.2. If $A \subset B$, then $A \cup B = B$.

Proof. To show that $A \cup B = B$, it is sufficient to show:

1. $A \cup B \supset B$ and
2. $B \supset A \cup B$.

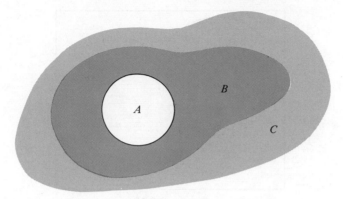

Figure 0.3.11

The proof of statement 1 is left as an exercise for the student, whence let us proceed to the proof of statement 2.

Let x be in $A \cup B$. It follows from the definition of $A \cup B$ that x is in A or x is in B. Since, however, $A \subset B$, each $x \in A$ is likewise in B. Thus, $x \in B$ whenever $x \in A \cup B$. This illustrates the fact that $B \supset A \cup B$.

The two inclusion statements 1 and 2 above demonstrate the equality of B and $A \cup B$.

From Figure 0.3.9 it can be seen intuitively that $A \cup B = B$ if $A \subset B$. B represents the total of the shaded region.

In the exercises below you are asked to prove some uncomplicated statements from set theory. Advanced problems of this nature are given in the advanced section at the end of the chapter.

EXERCISES 0.3

1. The following are in relation to Figure 0.3.12. Describe each set in terms of the shadings. [*Note:* Parentheses are used to describe an order in performing the set combinations. For example, to compute $(P \cup Q) - S$, compute first $P \cup Q$ and then compute the difference of $P \cup Q$ and S.]

a. A	f. $B \cup C$	k. $A \cap B \cap C$	p. $B - C$
b. B	g. $A \cup B \cup C$	l. $A - B$	q. $C - B$
c. C	h. $A \cap B$	m. $B - A$	r. $(A \cup B) - C$
d. $A \cup B$	i. $A \cap C$	n. $C - A$	s. $(A \cup C) - B$
e. $A \cup C$	j. $B \cap C$	o. $A - C$	t. $(B \cup C) - A$

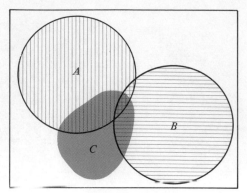

Figure 0.3.12

Let the interior of the entire rectangle of Figure 0.3.12 represent the universal set (denoted here by X). Describe (in terms of the shadings) the following sets.

u. A'	w. C'	y. $(B \cup C)'$	a'. $(A \cap B)'$
v. B'	x. $(A \cup B)'$	z. $(A \cup C)'$	b'. $(A \cap C)'$
			c'. $(B \cap C)'$

2. Show that $A \subset A \cup B, A \supset A \cap B$, and $A - B \subset A$.

3. Mr. D.'s advisees consist of sixteen students enrolled in algebra, twenty in geometry, and twelve in history. Six of this group are in both algebra and geometry, four in both geometry and history, five in both algebra and history, and three are enrolled in all three courses. Assuming that each of his advisees is enrolled in at least one of the classes, how many advisees does Mr. D have? Use a carefully devised Venn diagram to help you.

0.4 CARTESIAN PRODUCTS

In the preceding section we viewed several means of combining two sets A and B. We wish now to combine sets in yet another way. (See Exercises 0.2.8 and 0.2.9.)

The *Cartesian product* $A \times B$ of nonempty sets A and B is the collection of all symbols of the form (a, b) where $a \in A$ and $b \in B$. The elements (a, b) of $A \times B$ are called *ordered pairs*. The element (a, b) and the element (c, d) are said to be *equal* if and only if a and c represent the same element of A and b and d represent the same element from B. Thus we give significance to the order of appearance of the first and second terms of (a, b).

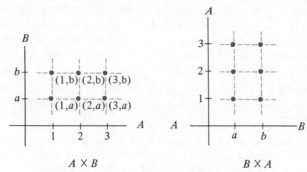

Figure 0.4.1   Sketches of Cartesian products.

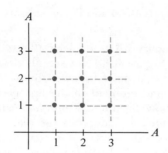

Figure 0.4.2   A sketch of $A \times A$.

EXAMPLE 0.4.1   Let $A = \{1, 2, 3\}$ and $B = \{a, b\}$. Then $A \times B$ is the collection $\{(1, a), (1, b), (2, a), (2, b), (3, a), (3, b)\}$ while the Cartesian product $B \times A$ is determined to be $\{(a, 1), (a, 2), (a, 3), (b, 1), (b, 2), (b, 3)\}$. As we might guess, $B \times A$ can be formed by reversing the position of the elements in each ordered pair of $A \times B$.

A sketch of $A \times B$ and $B \times A$ occurs in Figure 0.4.1 and a sketch of $A \times A$ is formed in Figure 0.4.2.

The Cartesian product is to play an important role in much of what we do from this point.

EXERCISES 0.4

1. Let $A = \{1, 2, 3, 4\}$, $B = \{0, 5, 6\}$, $C = \{1, 3, 4\}$, and $D = \{0, 1, 3\}$. Write out the following Cartesian products and sketch each.

a. $A \times A$	e. $A \times B$	i. $B \times C$	m. $C \times D$
b. $B \times B$	f. $A \times C$	j. $B \times D$	n. $D \times A$
c. $C \times C$	g. $A \times D$	k. $C \times A$	o. $D \times B$
d. $D \times D$	h. $B \times A$	l. $C \times B$	p. $D \times C$

2. Consider $A \times A$ of Example 0.4.1 and Figure 0.4.2. Let $\delta_A = \{(a, a) : a \in A\}$ be the set of all pairs in $A \times A$ having both elements the same. Draw a sketch of δ_A from Figure 0.4.2 and tell why you think δ_A is called the *diagonal* of $A \times A$. Sketch the diagonals of $A \times A$, $B \times B$, $C \times C$, and $D \times D$ from Exercise 0.4.1.

3. In Exercise 0.4.1 is it true that $A \times B = B \times A$? Is any pair of sets found there for which this "reversal of axes" fails to change the Cartesian product?

4. Let A be as in Exercise 0.4.1 with $P \subset A \times A$ given by $P = \{(1, 1), (1, 2), (2, 1)\}$. Observe that if any pair from P has the order of its elements reversed, the resulting element is also in P. For example, $(1, 2)$ becomes $(2, 1)$ under this process. However, $(2, 1) \subset P$. List all subsets of $K \times K$ where $K = \{a, b\}$ and determine which have this property.

5. Let K be as in Exercise 0.4.4 and list all subsets of $K \times K$ containing the diagonal δ_K.

6. Let A be as in Exercise 0.4.1 and let $P \subset A \times A$ be $P = \{(1, 1), (2, 3), (3, 4), (2, 4)\}$. In Figure 0.4.3 we see all possible pairs of elements from P listed. The two pairs followed by a "$\checkmark$" are precisely those satisfying: The second element of the first pair is the first element of the second pair. Now forming an ordered pair in each such case by utilizing the "outside" elements we see that $(1, 1)$, $(1, 1)$ gives $(1, 1)$ while $(2, 3)(3, 4)$ yields $(2, 4)$. The newly formed elements $(1, 1)$ and $(2, 4)$ already lie in P. List all subsets P of $K \times K$ (Exercise 0.4.4) where this "matching" process does not yield a pair outside P. Show that δ_K is one such subset of $K \times K$.

(1,1) (1,1) $\checkmark$	(3,4) (1,1)
(1,1) (2,3)	(3,4) (2,3)
(1,1) (3,4)	(3,4) (2,4)
(1,1) (2,4)	(3,4) (3,4)
(2,3) (1,1)	(2,4) (1,1)
(2,3) (2,3)	(2,4) (2,3)
(2,3) (3,4) $\checkmark$	(2,4) (3,4)
(2,3) (2,4)	(2,4) (2,4)

Figure 0.4.3

0.5 RELATIONS

A *relation* from A to B (or, on $A \times B$) is a subset of $A \times B$. The definition of relation is quite abstract and quite simple but is totally consistent with our usual (nonmathematical) notion of what the word relation should mean.

EXAMPLE 0.5.1. Let A be the set of all people residing in the United States and B the set of all the states of the United States together with Washington, D.C. Let F be the set.

$$\{(a,b): a \in A,\ b \in B,\ \text{and}\ a \text{ resides in } b\}$$

Is F a relation on $A \times B$? We answer in the affirmative, since if $(a,b) \in F$, $(a,b) \in A \times B$. That is, $F \subset A \times B$.

Note that $(a,b) \in F$ implies that a resides in B. Thus we might call F the relation "resides in" and denote $(a,b) \in F$ by aFb. Thus aFb is read: a resides in b. If $(a,b) \in A \times B$ but $(a,b) \notin F$, we know that a does *not* reside in B. However, $(a,b) \notin F$ would clearly be denoted by $aℲb$ and read: a does not reside in b. Consequently, we see that our concept of the relation is not at odds with our prior experience.

If F is a relation on $A \times A$, we say simply: F is a *relation on A*. A relation on A is, therefore, simply a collection of ordered pairs of elements of A. It is in such a circumstance that the adjective "ordered" in the term ordered pair becomes important. Suppose that a and b are elements of A. Is (a, b) the same as (equal to) (b, a)? Unless b and a are the same (equal), we must answer negatively. For example, $(12, 1)$ is not the same pair as $(1, 12)$. The order of appearance of the elements in the pair is as important as the elements themselves.

The importance of order in a pair can be significant in a practical or physical way. For example, at a theatrical world premiere a row twelve, seat one ticket might be $20 while a row one, seat twelve ticket could cost you $200.

Exercises 0.4.4–0.4.6 worked with and hinted at three important characteristics that can be held by relations. One of these characteristics is whether or not the relation contains the diagonal. To make sense, we must obviously be talking about relations on a set A.

If F is a relation on $A(A \times A)$ we say that F is *reflexive* if it contains the diagonal. It ought to be clear that some relations contain the diagonal while others do not. Thus some are reflexive while others are not.

What does it mean for $F \subset A \times A$ to contain the diagonal? It means that F must contain all elements of the form (a, a) (that is, the pairs having first and second terms the same). In other words, F is reflexive if it contains all elements (a, a) belonging to A.

EXAMPLE 0.5.2. Let A be the set of all people with F, G, and H the relations on $A \times A$ given by

$$F = \{(a, b): a, b \in A, \quad \text{and} \quad a \text{ is as tall as } b\}$$
$$G = \{(a, b): a, b \in A, \quad \text{and} \quad a \text{ is taller than } b\}$$
$$H = \{(a, b): a, b \in A, \quad \text{and} \quad a \text{ is exactly as tall as } b\}$$

Do any of F, G, and H contain δ_A? That is, are any among F, G, and H reflexive? Let us examine the relations individually.

F contains δ_A since if $a \in A$ then $(a, a) \in F$ (that is, if a is a person, a is as tall as a). However, a is taller than a cannot hold (for any person a), whence G does not contain δ_A. Now H clearly contains δ_A. Thus F and H are reflexive while G is not.

EXAMPLE 0.5.3. Let K be the relation on $A \times A$ described by

$$K = \{(a, b) \in A \times A: a \text{ is as tall as } b \text{ and both } a \text{ and } b$$
$$\text{have red hair}\}$$

Is K reflexive? The answer is no, since if a has black hair $(a, a) \notin K$.

A second characteristic of interest is that of symmetry. A relation F on A $(A \times A)$ is *symmetric* if reversing the order of the elements of pairs from F results only in pairs from F. We could say quite descriptively that F is symmetric means that whenever $(a, b) \in F$, $(b, a) \in F$ also. Observe that (b, a) is (can be) formed by reversing the elements from (a, b).

EXAMPLE 0.5.4. Relation F from Example 0.5.2 is not symmetric since the fact that a is as tall as b is no guarantee that b is as tall as a (a might be taller than b). Similarly, G from Example 0.5.2 is not symmetric.

However, H from that example is symmetric. We therefore must conclude that the properties of reflexivity and symmetry are independent. [F is symmetric and not reflexive while H is reflexive and symmetric. Furthermore, $L = \{(a, b) \in A \times A: a \text{ is exactly as tall as } b \text{ and } a \text{ and } b \text{ have different color hair}\}$ is symmetric but not reflexive.] K from Example 0.5.3 is not symmetric.

The third concept is that of the idea of "matching" pairs. If F is a relation on $A \times A$ and if (a, b) and (b, c) in F [*Note:* (a, b) and (b, c) are "matching" pairs—the b forces the match] guarantees that (a, c) is in F [(a, c) is formed from the "outside" elements of (a, b) and (b, c)], we say that F is *transitive*.

EXAMPLE 0.5.5. From Example 0.5.2 we find that F is transitive (If a is as tall as b and b is as tall as c, then a is as tall as c. Alternatively, if $(a, b) \in F$ and $(b, c) \in F$, then $(a, c) \in F$.), G is transitive (argue as in the case for F), and H is transitive.

Similarly, K from Example 0.5.3 is transitive but L from Example 0.5.4 is not $[(a, b)$ and $(b, c) \in L$ may not mean $(a, c) \in L$ since a and c may have the same color hair.].

Figure 0.5.1 sums up our examples and their characteristics of satisfying or not satisfying the "R-S-T" properties. Figure 0.5.2 shows all possible combinations of yes and no that can be generated. Our examples have shown cases 1, 3, 6, and 7. The following examples show cases 2, 4, and 8.

Relation	R	S	T
F	yes	no	yes
G	no	no	yes
H	yes	yes	yes
K	no	no	yes
L	no	yes	no

Figure 0.5.1

	R	S	T
1	yes	yes	yes
2	yes	yes	no
3	yes	no	yes
4	yes	no	no
5	no	yes	yes
6	no	yes	no
7	no	no	yes
8	no	no	no

Figure 0.5.2 Cases for a relation on A.

EXAMPLE 0.5.6. Let A be the set of all people with M the relation.

$$M = \{(a, b) \in A \times A : b \text{ is taller than } a \text{ and } a \text{ and } b$$
$$\text{have different color hair}\}$$

Now M satisfies condition 8 since it is neither reflexive, symmetric, nor transitive. Furthermore, since $F \supset \delta_A$, $M \cup F \supset \delta_A$, and

$$M \cup F = \{(a, b) \in A \times A : (1) \ a \text{ is as tall as } b \text{ or } (2) \ b$$
$$\text{is taller than } a \text{ while } a \text{ and } b \text{ have differ-}$$
$$\text{ent color hair}\}$$

Thus the relation $M \cup F$ is reflexive. It is, however, not symmetric and not transitive. Thus, $M \cup F$ satisfies case 4 of Figure 0.5.2.

EXAMPLE 0.5.7. Let A be the set of all people and N the relation.

$N = \{(a, b) \in A \times A:$ a is exactly as tall as b and has
different color hair than b or a and b have the
same color eyes$\}$

N is not transitive since there are persons a, b, and c satisfying: a has red hair and blue eyes, while b has brown hair with green eyes, a and b have same height, and c has green eyes while being shorter than b. That is, $(a, b) \in N$ and $(b, c) \in N$ but $(a, c) \notin N$. Case 2 of Figure 0.5.2 is satisfied.

Exercise 0.5.9 shows case 5 of Figure 0.5.2.

In general, we may (because of our examples) believe that these three properties $(R\text{-}S\text{-}T)$ of relations are mutually independent.

A relation satisfying all three conditions (H satisfies all three) is very special in mathematics and is called an *equivalence* relation. The diagonal δ_A of $A \times A$ and $A \times A$ itself are always equivalence relations associated with the set A.

EXERCISES 0.5

1. Show why the relation L from Example 0.5.4 is symmetric and not reflexive.

2. Show why the relation M from Example 0.5.6 is not reflexive, not symmetric, and not transitive.

3. Show why $M \cup F$ from Example 0.5.6 is reflexive but not symmetric and not transitive.

4. Show N in Example 0.5.7 is reflexive and symmetric.

5. In each case below, determine if the collection is a relation on any Cartesian product from Exercise 0.4.1. If it is, list the appropriate Cartesian products. Tell which relations are (1) reflexive, (2) symmetric, and (3) transitive. Tell which if any are equivalence relations.

    a. $\{(1, 0), (3, 5), (2, 6)\}$
    b. $\{(1, 5), (6, 0), (5, 3)\}$
    c. $\{(1, 1), (3, 3), (4, 4), (1, 4)\}$
    d. $\{(1, 0), (0, 3), (3, 0), (0, 0)\}$
    e. $\{(1, 0), (3, 0), (4, 0)\}$
    f. $\{(1, 3), (1, 4), (3, 1), (3, 4)\}$
    g. $\{(0, 0), (0, 1), (0, 3), (5, 0), (5, 1), (5, 3), (6, 0), (6, 1), (6, 3)\}$
    h. $\{(1, 0), (0, 5)\}$

    i. $\{(1,1)\}$

    j. $\{(0,0), (0,5), (0,6)\}$

    k. $\{(1,3), (5,6), (2,3)\}$

    l. $\{(1,1), (2,2), (3,3), (4,4)\}$

6. Make up a relation with each of the following specifications. (Do not use any given in the text.)

    a. Reflexive but not symmetric and not transitive.
    b. Symmetric but not reflexive and not transitive.
    c. Transitive but not reflexive and not symmetric.
    d. Reflexive and symmetric but not transitive.
    e. Reflexive and transitive but not symmetric.

7. Let A be the collection of all subsets of some set X.

    a. Show that "$\subset$" is a reflexive and transitive relation on A.
    b. Show that "$=$" is an equivalence relation on A.

8. Show that the diagonal of A is an equivalence relation on A. Show furthermore that each equivalence relation on A contains the diagonal.

9. Show that if $F \subset A \times A$, F is symmetric, and F is transitive, and the set of first terms of ordered pairs in F is A, then F is reflexive. Give an example of a symmetric and transitive but not reflexive relation.

10. Let A be the set of all (plane) triangles and let F be the relation on A induced by the phrase: is congruent to (can be superimposed upon). Describe F as a collection of ordered pairs and attribute to F the *RST* properties that hold.

0.6 REVIEW EXERCISES

1. Let $A = \{a,b,c,d,e\}$, $B = \{a,c,e\}$, $C = \{1,2,3,a,c,e\}$, and $D = \{1,2,3\}$.

    a. List all inclusion relations (that is, statements of the form $P \subset Q$) that hold for the sets A, B, C, and D.
    b. Are any two of the sets equal?
    c. Are any two of the sets disjoint?

2. From Exercise 0.6.1 compute:

    a. The "smallest" universal set X for the discussion

b. $A \cup B$	f. $A \cap B$	j. $A - B$	n. $C - A$
c. $A \cup C$	g. $A \cap C$	k. $A - C$	o. $C - D$
d. $B \cup C$	h. $B \cap C$	l. $B - A$	p. $D - B$
e. $C \cup D$	i. $C \cap D$	m. $B - D$	

With X as in a, compute:

    q. A' r. B' s. C' t. D'

3. For B from Exercise 0.6.1 compute $B \times B$ and δ_B.

4. Write a complete sentence describing each indicated set.

    a. $A = \{x : x \text{ is a trigonometry student}\}$
    b. $B = \{x : x \text{ is an algebra student}\}$ e. $E = A - B$
    c. $C = A \cup B$ f. $F = B - A$
    d. $D = A \cap B$ g. $G = (A - B) \cup (B - A)$

5. Draw Venn diagrams for each situation in Exercise 0.6.1 shading A vertically and B horizontally. Describe each of the sets in Exercise 0.6.1 in terms of shadings.

6. Answer the following by True or False.

    a. $A \cup B \subset A$ h. If $A \cap B = \phi$, A and B are disjoint
    b. $A \subset A \cup B$ i $A \cup A = A$
    c. $A \subset A \cap B$ j. $A \cap A = A$
    d. $A \cap B \subset A$ k. $A - A = A$
    e. $A - B \subset A$ l. If $A \cap B = \phi$, $A - B = A$
    f. $\phi \subset A$ m. $(A \cap B)' = A' \cup B'$
    g. $A \cap A' = \phi$ n. $(A \cup B)' = A' \cap B'$

7. Write the Cartesian product $B \times A$ of the sets $A = \{\heartsuit, \clubsuit, \spadesuit, \blacklozenge\}$ and $B = \{a, 2, 3, 4, 5, 6, 7, 8, 9, 10, j, q, k\}$. The Cartesian product is familiar as a set used in various familiar forms of recreation. What is the set? Draw a sketch of the Cartesian products $A \times B$ and $B \times A$.

8. Give an example of a reflexive relation; symmetric relation; transitive relation; equivalence relation.

0.7 QUIZ

Attempt to work this quiz (without the book or any such aid) in one hour. Check the back of the book for the correct answers.

1. Write a complete sentence that interprets the symbolism in each case.

    a. $x \in A$ c. $A = \{x : x \text{ is a carrot}\}$
    b. $x \notin A$

2. Give definitions for each:

    a. $X \subset Y$ b. $X = Y$

3. In view of 2, answer each of the following where $A = \{p, q, 1, 3, 2/3\}$, $B = \{a, 1, 3\}$, and $C = \{1, 3\}$.

    a. Is $C \subset A$?   c. Is $A \subset C$ or $B \subset C$?
    b. Is $C \subset B$?   d. Are any two sets equal?

4. Let A, B, and C be as in 3. Compute each of the following sets.

    a. $A \cup C$ f. $A - C$
    b. $A \cup B$ g. $C - A$
    c. $A \cap B$ h. $B \times C$
    d. $B \cap C$ i. $C \times C$
    e. $A - B$ j. δ_A

5. List all (C is as in Problem 3):

   a. Reflexive relations on C.   c. Transitive relations on C.
   b. Symmetric relations on C.   d. Equivalence relations on C.

0.8 ADVANCED EXERCISES

Prove that the following set relations are true.

1. $A \subset A \cup B$
2. $A \supset A \cap B$
3. $(A - B) \cap B = \phi$
4. $(A - B) \subset A$
5. $(A - B) \cap (B - A) = \phi$
6. $\phi \subset A$ for any set A
7. $A - (B \cup C) = (A - B) \cap (A - C)$
8. $A - (B \cap C) = (A - B) \cup (A - C)$
9. $A \cap (B \cup C) = (A \cap B) \cup (A \cap C)$
10. $A \cup (B \cap C) = (A \cup B) \cap (A \cup C)$
11. $A \cap B = \phi$ implies $A - B = A$
12. $(A - B) \cup B = A$ implies $B \subset A$
13. $B \subset A$ implies $(A - B) \cup B = A$
14. $A \cap B = \phi$ implies $(A \cup B) - B = A$
15. $(A \cup B) - B = A$ implies $A \cap B = \phi$

CHAPTER 1

The Function

1.1 INTRODUCTION TO FUNCTIONS

As mentioned at the very outset, the formulation of a concept of the function is a primary desire. Of the several means whereby we may introduce this idea, we will use two, one because of its intuitive ramifications and the other because of its value in the sense of graphing and abstraction.

Let each of X and Y be a set. A *function f* from X into Y, written $f: X \rightarrow Y$, is a rule which relates to each x in X a unique y in Y. Now y will be denoted by $f(x)$; that is, $y = f(x)$.

There are several facts to observe in connection with a function $f: X \rightarrow Y$.

1. If $x \in X$, there is a $y \in Y$ with $y = f(x)$. Moreover, there is only one such y.

2. There may be some $y \in Y$ with $y \neq f(x)$ for any x in X. Intuitively, Y may not be entirely "used up" by the function.

3. The function may mate more than one x to the same y. That is to say, it may be that $x_1 \neq x_2$ (x_1 and x_2 are different) but $f(x_1) = f(x_2)$.

The following examples will illustrate.

EXAMPLE 1.1.1. Let $X = \{0, 1, 2, 3, 4\}$ and $Y = \{a, b, c, d, e\}$. Define f by the following rule:

$$
\begin{array}{lll}
& f & \\
0 \rightarrow a & \text{that is,} & f(0) = a \\
1 \rightarrow b & \text{that is,} & f(1) = b \\
2 \rightarrow c & \text{that is,} & f(2) = c \\
3 \rightarrow a & \text{that is,} & f(3) = a \\
4 \rightarrow b & \text{that is,} & f(4) = b
\end{array}
$$

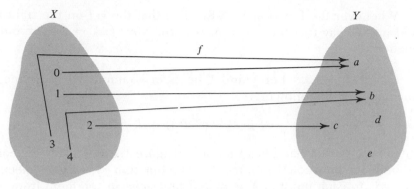

Figure 1.1.1 A function $f: X \to Y$.

Figure 1.1.1 is a sketch representing the manner in which f "carries" the elements of X over to elements of Y. Observe that there is *one and only one* arrow emanating from each element of X. Each arrow terminates at an element of Y. The rule f may be thought of as being the set of arrows, if you please. Is f a function from X into Y? The answer is yes, because:

1.   Every $x \in X$ is mated to some y in Y. That is, for every x there is an $f(x)$.
2.   For no x in X do there exist two choices for $f(x)$.

Thus we conclude that to every $x \in X$, f mates one element y from Y. The definition of a function is satisfied.

EXAMPLE 1.1.2.   Let X be as in Example 1.1.1 with $Z = \{a, b, c\}$. Let f be the same rule given in the example. Again, f is a function ($f: X \to Z$). Note, however, that although Y was not entirely used up (d and e are not used), Z is. (See Figure 1.1.2.)

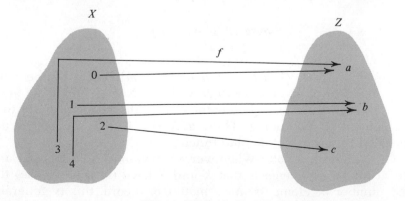

Figure 1.1.2 $f: X \to Z$.

Whenever the function is defined so that the set on the right is used up, we say that the function is *onto*. Now $f: X \to Z$ is an onto function.

EXAMPLE 1.1.3. Let X and Y be as in Example 1.1.1. Define a function g by the following:

$$g(0) = a, \ g(1) = b, \ g(2) = c, \ g(3) = d, \quad \text{and} \quad g(4) = e$$

You should verify that g is a function. (Figure 1.1.3 displays the correspondence between the elements.) The function g has a very special property: to each point of Y is mated *precisely one* element from X. Figure 1.1.3 shows that *no more than one* arrow terminates at any given element of Y. Figure 1.1.2 shows this *not* to be the case for f. As an example (from Figure 1.1.2), two arrows terminate at b.

Whenever no element of Y is used more than once, the function is said to be $1:1$. Clearly, g is $1:1$. In this situation, g happens to be onto also. However, a $1:1$ function need *not* be onto.

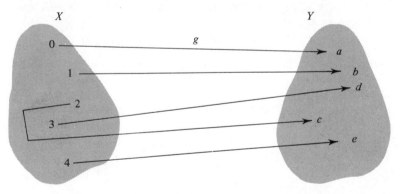

Figure 1.1.3 $g: X \to Y$.

EXAMPLE 1.1.4. Let $P = \{a, b, c, d, e, l, m\}$. Let g be the same correspondence as given immediately above. Now $g: X \to P$ is a function (see Figure 1.1.4). No element of P is the image of more than one element of X under g. Thus, $g: X \to P$ is $1:1$. Since P is not exhausted by g, g is not an onto function.

One interesting note: Whenever a function $h: X \to Y$ is $1:1$ and onto, it intuitively suggests that X and Y have (in some sense) the same number of elements. As a matter of record, this is generally the technique used to define such a condition. See Exercise 1.8.7.

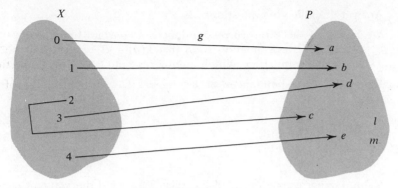

Figure 1.1.4 $g\colon X \to P$.

EXERCISE 1.1

1. Let $X = \{0, 1, 2, 3, 4, 5\}$, $A = \{0, 1, 2\}$, $Y = \{a, b, c, d, e\}$, $B = \{a, c, e\}$.
    Let f, g, and h be the correspondences listed below.

f	g	h
$0 \to a$	$0 \to a$	$0 \to a$
$1 \to c$	$1 \to a$	$1 \to c$
$2 \to e$	$2 \to a$	$2 \to e$
$3 \to b$	$3 \to a$	
$4 \to d$	$4 \to a$	
$5 \to a$	$5 \to a$	

    a. Give reasons to justify the statement that f is a function in the sense
        $f\colon X \to Y$.
    b. Is f onto? 1:1?
    c. Why is $g\colon X \to Y$ a function? Note that g is an example of a *constant*
        function (every element of X is mated with the same element of Y).
    d. Is $h\colon A \to Y$ a 1:1 function? Onto?
    e. Is $h\colon A \to B$ a 1:1 function? Onto?

2. Follow the instructions and answer the questions of Exercise 1.1.1
    where $X = \{0, -1, 1, 2, -2, -3, 3, 4, -4\}$, $A = \{0, 1, 2, 3, 4\}$, $Y =$
    $\{a, b, c, d, e, i, j, k, m, n\}$, $B = \{a, c, d, i, k, n\}$ and f, g, and h are the
    following correspondences.

f	g	h
$0 \to a$	$0 \to a$	$0 \to a$
$1 \to b$	$1 \to b$	$1 \to b$
$-1 \to c$	$-1 \to b$	$2 \to c$
$2 \to d$	$2 \to d$	$3 \to i$
$-2 \to e$	$-2 \to d$	$4 \to k$
$3 \to i$	$3 \to i$	
$-3 \to j$	$-3 \to i$	
$4 \to k$	$4 \to k$	
$-4 \to n$	$-4 \to k$	

Answer the following questions also.

a.   Is any of the three correspondences a constant function?
b.   If $k: P \rightarrow Q$ is a function such that $k(a) = k(-a)$ for each $a \in P$, k is an *even* function. Which among f, g, and h is an even function?

3.   Using the correspondences of Exercise 1.1.1, find the value of the following:

a. $f(0)$ e. $g(1)$
b. $f(2)$ f. $g(3)$
c. $f(4)$ g. $h(1)$
d. $f(5)$ h. $h(2)$

4.   Using the correspondences of Exercise 1.1.2, find the value of each of the following.

a. $f(0)$ e. $g(1)$ i. $h(0)$
b. $f(-1)$ f. $g(2)$ j. $h(2)$
c. $f(3)$ g. $g(-3)$ k. $h(3)$
d. $f(-4)$ h. $g(4)$ l. $h(4)$

1.2   INVERSE FUNCTIONS

Upon examining Figure 1.1.3, we should note that if the arrows are reversed, a function from Y to X results (Figure 1.2.1). The fact that reversing the arrows leads to a new function holds only for functions which are both 1:1 and onto. Figure 1.2.2 shows the situation occurring when the arrows of Figure 1.1.1 are reversed.

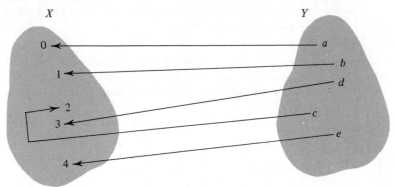

Figure 1.2.1   Reversing the arrows.

Two observable things happen. First, there are elements in Y from which no arrow emanates. Second, two arrows originate from both a and b. Each situation is indicative of the fact that the described correspondence is not a function.

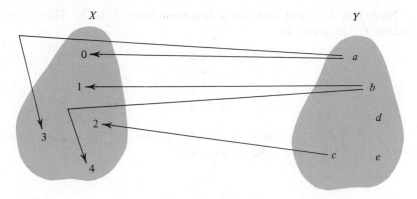

Figure 1.2.2 No function.

Figure 1.2.3 illustrates the reversal of the arrows of Figure 1.1.2. At least one arrow radiates from each element of Z, but, as before, two arrows start at each of a and b. Thus the new correspondence does *not* yield a function according to our definition.

If a function $h: X \to Y$ is 1:1 and onto, the correspondence h^{-1} given by $h^{-1}(y) = x$ where $h(x) = y$ is a function called the *inverse function* for h. This is a mathematical description of the act of reversing the arrows. The idea of an inverse function will be used frequently throughout the text, whence some specific examples of inverse functions are in order.

EXAMPLE 1.2.1. Let $X = \{0, 1, 2, 3\}$ and $Y = \{r, s, t, u\}$ with f defined by

$$
\begin{array}{c}
f \\
0 \to s \\
1 \to u \\
2 \to t \\
3 \to r
\end{array}
$$

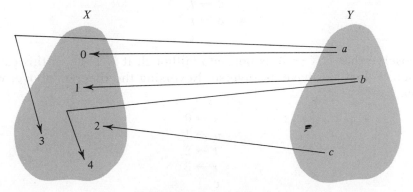

Figure 1.2.3 No function.

Now f is $1:1$ and onto as a function from X to Y. The inverse function f^{-1} is given by

$$f^{-1}$$
$$s \to 0$$
$$u \to 1$$
$$t \to 2$$
$$r \to 3$$

EXAMPLE 1.2.2. Using X as in Example 1.2.1 and letting $Z = \{r, s, t\}$, define h by the following.

$$h$$
$$0 \to r$$
$$1 \to s$$
$$2 \to t$$
$$3 \to r$$

Does h have an inverse? No, h is not $1:1$ (it is onto, however). In trying to reverse the diagram for h, we have

$$h^{-1}$$
$$r \to 0$$
$$s \to 1$$
$$t \to 2$$
$$r \to 3$$

Thus h^{-1} takes r to both 0 and 3, whence it is not a function.

EXAMPLE 1.2.3. Let X be as in Example 1.2.1 and let $W = \{r, s, t, u, v\}$. Also let f be the correspondence of that example. That is,

$$f$$
$$0 \to s$$
$$1 \to u$$
$$2 \to t$$
$$3 \to r$$
$$v$$

Observe that $f: X \to W$ is not onto (although it is $1:1$). In this case, then, f does not have an inverse. Reversing the diagram above, we see

$$f^{-1}$$
$$s \to 0$$
$$u \to 1$$
$$t \to 2$$
$$r \to 3$$
$$v$$

Since f^{-1} does not pair v with an element of X, f^{-1} is not a function (W must be "used up").

In retrospect, we see that 1:1 and onto functions are precisely those having inverses. Furthermore, inverse functions are of necessity 1:1 and onto. If the 1:1 and onto function carries x to y, the inverse mates y back to x [that is, $f^{-1}(f(x)) = x$, while $f(f^{-1}(y)) = y$].

EXERCISE 1.2

1. Tell which of the following functions have inverses and which do not. Describe each inverse function and if a function does not have an inverse, tell why.

a. f
$$a \to 1$$
$$b \to 2$$
$$c \to 3$$
$$d \to 4$$
$$e \to 5$$

c. h
$$a \to 1$$
$$b \to 2$$
$$c \quad 3$$
$$d \quad 4$$
$$e \quad 5$$

e. u
$$a \to 1$$
$$b \quad 2$$
$$c \quad 3$$
$$d \quad 4$$
$$e \quad 5$$
$$\quad 6$$

b. g
$$a \to 1$$
$$b \to 2$$
$$c \quad 3$$
$$d \to 4$$
$$e \to 5$$

d. t
$$a \to 1$$
$$b \quad 2$$
$$c \quad 3$$
$$d \to 4$$
$$e \to 5$$

f. k
$$a \quad 1$$
$$b \quad 2$$
$$c \quad 3$$
$$d \quad 4$$
$$e \quad 5$$

2. Answer the following relative to (1) Exercise 1.1.1, (2) Exercise 1.1.2.

   a. Does f have an inverse? Why or why not?
   b. Does g have an inverse? Why or why not?
   c. Does $h: A \to B$ have an inverse? If so, what is the correspondence?
   d. Does $h: A \to Y$ have an inverse? If so, what is the correspondence?

3. If f is a function and has an inverse function f^{-1}, what is $f(f^{-1}(x))$ and what is $f^{-1}(f(y))$?

1.3 MORE ABOUT FUNCTIONS

Some further terminology and symbolism will prove useful and descriptive. Given a function $f: X \to Y$, X is called the *domain* of f while Y is a *codomain*. Examples of domains and codomains are given in the chart in Figure 1.3.1.

Example	Function	Domain	Codomain
1.1.1	f	X	Y
1.1.2	f	X	Z
1.1.3	g	X	Y
1.1.4	g	X	P
1.2.1	f	X	Y
1.2.2	h	X	Z
1.2.3	f	X	W

Figure 1.3.1 Domains and codomains.

We have noted earlier that the codomain may not be entirely used up by the function. It seems, however, that our interest should center around those elements that are involved in the matings. Consequently we take this opportunity to give a name to that subset of the codomain Y "used up" under the correspondence. We call that set the *range* of f and write it as $f[X]$. Symbolically, for $f: X \to Y$,

$$f[X] = \{y \in Y : y = f(x) \quad \text{for some} \quad x \in X\}$$

The ranges for the functions in the six previous examples are listed in Figure 1.3.2.

We are aware that a function is onto if and only if its range and codomain are one and the same. That is, $k: X \to Y$ is onto if and only

Example	Function	Range
1.1.1	f	$\{a,b,c\}$
1.1.2	f	Z
1.1.3	g	Y
1.1.4	g	$\{a,b,c,d,e\}$
1.2.1	f	Y
1.2.2	h	Z
1.2.3	f	$\{s,t,u,r\}$

Figure 1.3.2 Ranges.

if $k[X] = Y$. We could have taken this for the definition of an onto function.

We can generalize the concept of the domain and range. Given that $f: X \to Y$ is a function with $A \subset X$ while $B \subset Y$, define the following descriptive sets:

Image of $A = f[A] = \{y \in Y : y = f(a)$ for some $a \in A\}$
Inverse image of $B = f^{-1}[B] = \{x \in X : f(x) \in B\}$

EXAMPLE 1.3.1. Let $X = \{0, 1, 2, 3, 4, 5\}$, $A = \{0, 1, 2\}$, $Y = \{a, b, c, d\}$ and $B = \{a, c, d\}$. Consider the correspondence f given below.

$$
\begin{array}{c}
f \\
0 \to a \\
1 \to b \\
2 \to c \\
3 \to d \\
4 \to d \\
5 \to d
\end{array}
$$

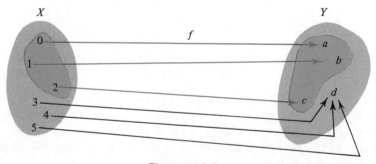

Figure 1.3.3

Figure 1.3.3 shows the situation described by the correspondence f. Now, $f[A]$ is that part of Y used up as f "takes" the elements of A over to Y. The figure shows this set to be $\{a, b, c\}$, that is, $f[A] = \{a, b, c\}$.

Now examine the set B. What elements of X are carried into B via the correspondence f? Figure 1.3.4 shows this set to be $\{0, 2, 3, 4, 5\}$. According to the definition, $f^{-1}[B] = \{0, 2, 3, 4, 5\}$.

In this usage, the f^{-1} is *not* to be construed as denoting an inverse function but rather consider $f^{-1}[B]$ as a symbol representing a subset of the domain of the function f. In fact, f^{-1} may not even exist (as a function).

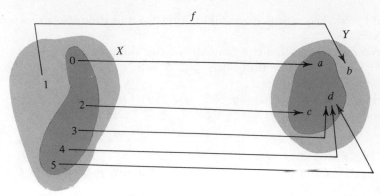

Figure 1.3.4

Two functions f and g are said to be *equal* if they have the same domain and represent the same correspondence [that is, $f(x) = g(x)$ for each x in the common domain].

EXAMPLE 1.3.2. $X = \{0, 1, 2, 3\}$, $Y = \{a, b, c\}$, and $W = \{a, b\}$. Define f, g, h, and k as below.

$f: X \to Y$	$g: X \to W$	$h: X \to Y$	$k: Y \to W$
f	g	h	k
$0 \to a$	$0 \to a$	$0 \to a$	$a \to a$
$1 \to b$	$1 \to b$	$1 \to b$	$b \to b$
$2 \to b$	$2 \to b$	$2 \to c$	$c \to a$
$3 \to a$	$3 \to a$	$3 \to a$	

Now f, g, and h have the same domain. Is $f = g$? $f = h$? $g = h$?

It is true that $f = g$ since the correspondences are the same. However $f(2) \neq h(2)$, whence $f \neq h$. Similarly, $g \neq h$.

None of the three (f, g, and h) is equal to k, since the domain of k is Y and not X.

Suppose that $f: X \to Y$ and $g: W \to Z$ are functions with $f[X] \subset W$. Then we can define a function $h: X \to Z$ by using f and g. The function so obtained is known as the *composition* of g with f and is written $h = g \circ f$. The composition function $g \circ f$ is defined by $(g \circ f)(x) = g(f(x))$.

Figure 1.3.5 illustrates the construction of the composition. Selecting any x from X, we automatically acquire a y in Y ($y = f(x)$). Having obtained y, we are directed (by g) to a unique element z ($z = g(y)$) in the set Z. Thus the choosing of any x in X results in the mating of x to a single z in Z. This then shows that $g \circ f$ is a function.

That the function exists is true is *solely because the range of f is contained in the domain of g*. This is important! The following

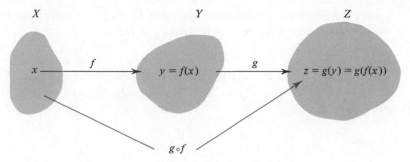

Figure 1.3.5 Composition of functions.

example shows the construction of a composition of two functions.

EXAMPLE 1.3.3. Let $X = \{0, 1, 2, 3, 4, 5\}$, $Y = \{6, 7, 8, 9\}$ and $Z = \{10, 11, 12, 13\}$. Define $f: X \to Y$ and $g: Y \to Z$ by the following.

$$
\begin{array}{ll}
f(0) = 6 & g(6) = 10 \\
f(1) = 7 & g(7) = 11 \\
f(2) = 8 & g(8) = 12 \\
f(3) = 9 & g(9) = 10 \\
f(4) = 6 & \\
f(5) = 7 &
\end{array}
$$

Hence, $g \circ f$ is the correspondence below.

$$
\begin{array}{l}
g \circ f(0) = g(f(0)) = g(6) = 10 \\
g \circ f(1) = g(f(1)) = g(7) = 11 \\
g \circ f(2) = g(f(2)) = g(8) = 12 \\
g \circ f(3) = g(f(3)) = g(9) = 10 \\
g \circ f(4) = g(f(4)) = g(6) = 10 \\
g \circ f(5) = g(f(5)) = g(7) = 11
\end{array}
$$

(See Figure 1.3.6.)

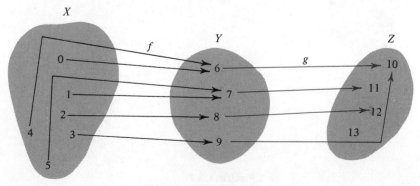

Figure 1.3.6 $g \circ f$.

Again, the item of importance insuring that the composition is a function is that the range of f is a subset of the domain of g.

EXAMPLE 1.3.4. To illustrate the last remark above, let $X = \{1, 2, 3, 4, 5\}$, $Y = \{a, b, c\}$, and $Z = \{0, -1, -2\}$ with f and g the following functions.

$$
\begin{array}{cc}
f & g \\
1 \to a & 1 \to 0 \\
2 \to b & 2 \to -1 \\
3 \to c & 3 \to -2 \\
4 \to a & 4 \to 0 \\
5 \to b & 5 \to 0
\end{array}
$$

What is $g \circ f$? The answer is that $g \circ f$ does not make sense. Why? Try to calculate $g \circ f(1)$ for example. If $g \circ f(1)$ can be found, $g \circ f(1) = g(f(1)) = g(a)$. However, *a is not in the domain of g* and $g(a)$ is therefore not defined (see Figure 1.3.7). This points out the necessity that f's range be a subset of g's domain.

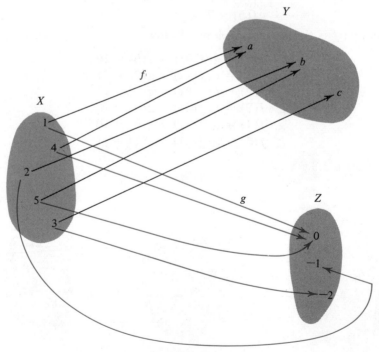

Figure 1.3.7

EXERCISES 1.3

1. Let $X = \{0, 1, 2, 3, 4\}$, $Y = \{a, b, c, d\}$, and $Z = \{1/2, 1/3, 1/4, 1/5\}$, with f and g as below.

$$
\begin{array}{cc}
f & g \\
0 \to a & a \to 1/2 \\
1 \to b & b \to 1/3 \\
2 \to c & c \to 1/4 \\
3 \to a & d \to 1/5 \\
4 \to d &
\end{array}
$$

   a. What is the correspondence $g \circ f$?

   b. Does $f \circ g$ make sense? Why or why not?

2. From Exercise 1.3.1 find each of the following.

a. $f(0)$	d. $f(4)$	g. $g(c)$	j. $g \circ f(1)$
b. $f(1)$	e. $g(a)$	h. $g(d)$	k. $g \circ f(3)$
c. $f(3)$	f. $g(b)$	i. $g \circ f(0)$	l. $g \circ f(4)$

3. Using X, Y, and Z as in Exercise 1.3.1, define f, g, h, and k as below.

$$
\begin{array}{cccc}
f: X \to Y & g: X \to Y & h: Y \to Z & k: Z \to Y \\
0 \to a & 0 \to c & a \to 1/2 & 1/2 \to a \\
1 \to b & 1 \to d & b \to 1/3 & 1/3 \to c \\
2 \to c & 2 \to a & c \to 1/4 & 1/4 \to d \\
3 \to d & 3 \to b & d \to 1/3 & 1/5 \to a \\
4 \to a & 4 \to b & &
\end{array}
$$

Which of the following make sense?

a. $f \circ g$	e. $g \circ h$	i. $h \circ k$	m. $f \circ f$
b. $f \circ h$	f. $g \circ k$	j. $k \circ f$	n. $g \circ g$
c. $f \circ k$	g. $h \circ f$	k. $k \circ g$	o. $h \circ h$
d. $g \circ f$	h. $h \circ g$	l. $k \circ h$	p. $k \circ k$

Show (via diagrams) the compositions that make sense. What must be true for $f \circ f$ to make sense?

4. Prove that if f is 1:1 and onto, f has an inverse. If f fails to be 1:1 and onto, f has no inverse. If f has an inverse, the inverse is 1:1 and onto (use $f: X \to Y$). If f has an inverse, what are the domain and range for f^{-1}?

5. Give the domain and range for each function in Exercises 1.3.1, 1.3.2, and 1.3.3.

1.4 FUNCTIONS, RELATIONS, AND ORDERED PAIRS

    In the previous section the concept of a function was described in terms of the words relate, rule, correspond, mate, and the like. The

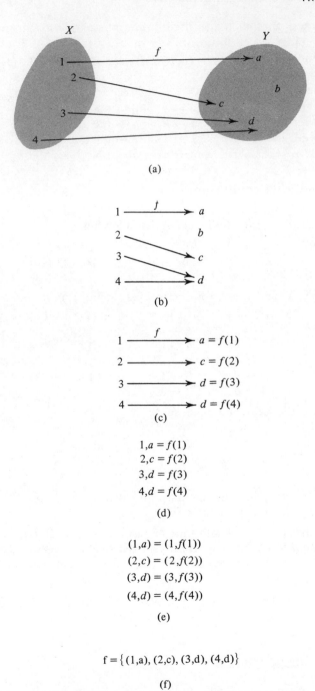

Figure 1.4.1 Various means of describing a function.

basic idea was that each x in the domain was paired with some unique element $f(x)$ in the codomain. Figure 1.4.1 illustrates various ways of picturing one such function.

The final diagram of Figure 1.4.1 shows the *function as a collection of ordered pairs* $(x, f(x))$. The first element in each pair is the domain element while the second element is its image under the correspondence. The function f is seen to be described in detail by listing all such ordered pairs. Examples 1.1.1–1.1.3 of Section 1.1 can be restated (the functions can be redesignated) in the following way:

Examples 1.1.1 and 1.1.2:

$$f = \{(0, a), (1, b), (2, c), (3, a), (4, b)\}$$

Example 1.1.3:

$$g = \{(0, a), (1, b), (2, c), (3, d), (4, e)\}$$

In this notation the set of all first elements from the ordered pairs is the domain while the set of all second elements forms the range of the function. The more general codomain as given before does not (usually) appear explicitly in the new setting. An exception to this statement occurs when the function being discussed is onto.

Thus we can define a *function* as a collection of ordered pairs such that each element which is a first element in some ordered pair is a first element in exactly one ordered pair. That is, if (x, y) and (x, z) are ordered pairs in some function f, then y must be the same as (equal to) z. Two different ordered pairs cannot possibly have the same first element. (Two ordered pairs are the same if and only if they have the same first terms and the same second terms.)

In this new sense, the concept of a function appears to be very near that of the relation. If $f: A \rightarrow B$ is a function, f (in ordered-pair notation) satisfies $f \subset A \times B$. However, subsets of Cartesian products were called *relations*. Hence, every function in the ordered-pair sense is a relation. In fact, we may define a function $f: A \rightarrow B$ as a relation on $A \times B$ satisfying: for each $a \in A$, a is the first element in exactly one ordered pair from f.

Is each relation a function? The answer is negative because of the special condition a function must satisfy. The following example illustrates two relations on $A \times B$, neither of which is a function from A to B.

EXAMPLE 1.4.1. Let $A = \{1, 2, 3\}$ and $B = \{a, b, c, d\}$. Define

$$F = \{(1, a), (1, b), (2, a), (3, c)\} \quad \text{and} \quad G = \{(1, a), (2, b)\}$$

Now, F and G are relations on $A \times B$. Are they functions from A to B?

F is not a function from A to B since $1 \in A$ is the first element of *two* ordered pairs on F. [Is $F(1) = a$ or is $F(1) = b$?] G is not a function from A to B since $3 \in A$ but 3 is not the first element of *any* ordered pair in G.

In the example, $G: \{1, 2\} \to B$ is a function since $G \subset \{1, 2\} \times B$ and both 1 and 2 are first elements of a single ordered pair from G.

We have previously drawn sketches of Cartesian products. Since relations and functions are subsets of Cartesian products, we can draw graphs (sketches) of these subsets.

Figure 1.4.2 shows a graph of $A \times B$ for A and B as in Example 1.4.1. The *points* of the product are the intersections of the dotted horizontal and vertical lines. In Figure 1.4.3 (a) and (b), dots are placed at various points of the Cartesian product $A \times B$. The dots in (a) represent the points of the relation F while the dots in (b) represent those of G.

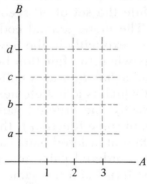

Figure 1.4.2 $A \times B$.

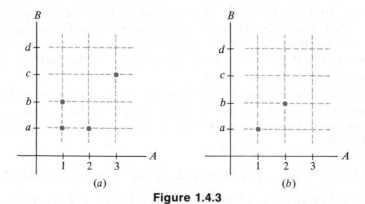

(a)   (b)

Figure 1.4.3

The horizontal axis contains the domain elements (set of possible first elements of the pairs) while the vertical axis displays the codomain (that is, the collection of possible second elements).

Intuitively we see that a relation is a function if and only if there is exactly one dot (point) of the relation on each vertical line through a domain element. F is not a function, since the vertical line through 1 contains *two* points of F. G is not a function since the vertical line through 3 contains *no* point of G. Figure 1.4.4 shows, however, that G is a function from $\{1, 2\}$ to B.

Graphs of the functions from Examples 1.1.1–1.1.3 are shown in Figures 1.4.5–1.4.7 respectively (see the beginning of this section

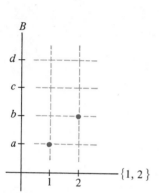

Figure 1.4.4 $G \subset \{1, 2\} \times B.$

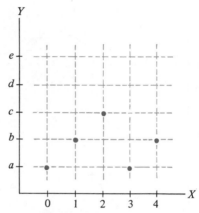

Figure 1.4.5 $f : X \to Y.$

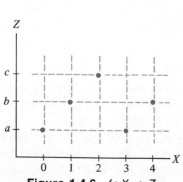

Figure 1.4.6 $f : X \to Z.$

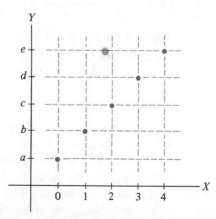

Figure 1.4.7 $g : X \to Y.$

for the ordered-pair notation for these functions). The function g
from Example 1.1.4 is graphed in Figure 1.4.8.

A function is onto if every codomain element is used. Graphically, this means that a function is an onto function if *each* horizontal
line through a codomain element contains a point of the function.
Similarly, a function is 1:1 if *no* horizontal line contains more than
one point of the function. Consequently, a function is both 1:1 and
onto if *every* horizontal line through a codomain element contains
exactly one point of the function.

From Figures 1.4.5–1.4.8 we see that $f: X \rightarrow Z$ and $g: X \rightarrow Y$ are
onto functions while $g: X \rightarrow Y$ and $G: X \rightarrow P$ are 1:1. Only $g: X \rightarrow Y$
is both 1:1 and onto (see Figure 1.4.7).

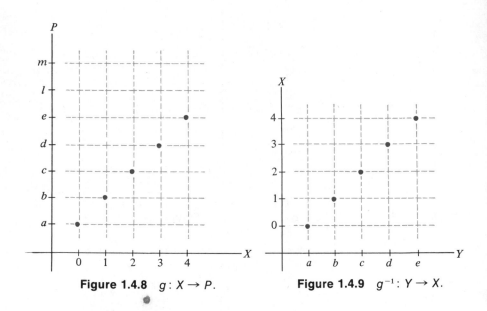

Figure 1.4.8 $g: X \rightarrow P$.   Figure 1.4.9 $g^{-1}: Y \rightarrow X$.

We might also note that the range can be formed by tracing each
horizontal line containing a point of the function. The points on the
vertical axis lying on such horizontal lines form the range.

If a function has an inverse, how is its inverse formed (relative
to the idea of ordered pairs)? The inverse is found by reversing the
order of the terms in each ordered pair of the function. The inverse
of the function in Figure 1.4.7 is shown in Figure 1.4.9.

EXERCISES 1.4

1. Graph each function:

   a. f b. $g(a) = \alpha$ c. $H = \begin{cases} (a, \alpha),\ (b, \alpha),\ (c, \alpha), \\ (d, \alpha),\ (e, \alpha), \\ (l, \alpha),\ (m, \alpha) \end{cases}$

   $a \to \alpha$ $g(b) = \beta$
   $b \to \beta$ $g(c) = \gamma$
   $c \to \gamma$ $g(d) = \phi$
   $d \to \delta$ $g(e) = \mu$ d. $T = \begin{cases} (a, \alpha),\ (b, \beta), \\ (c, \gamma),\ (d, \delta) \end{cases}$
   $e \to \alpha$ $g(l) = \eta$
   $l \to \beta$ $g(m) = \delta$
   $m \to \gamma$

   e. $S = \{(e, \eta),\ (l, \gamma),\ (b, \beta),\ (c, \gamma)\}$

   f. u

   $a \to \alpha$
   $b \to \gamma$
   $c \to \delta$
   $d \to \beta$
   $e \to \mu$

2. State the domain and range of each function in Exercise 1.4.1.

3. Which functions (if any) from Exercise 1.4.1 are 1:1?

4. Which functions in Exercise 1.4.1 are onto? Why?

5. Putting together Exercises 1.4.3 and 1.4.4, decide which functions from Exercise 1.4.1 have inverses. Graph each inverse.

6. Consider each function in Exercise 1.4.1 as having codomain $Y = \{\alpha, \beta, \gamma, \delta, \eta, \mu, \phi\}$. Sketch each function as a subset of $\{a, b, c, d, e, l, m\} \times Y$. Which of the functions are functions under this concept (that is, where Y is considered to be the codomain)? Which are onto?

1.5 SOME SPECIAL FUNCTIONS

Now that the function concept has been introduced, let us examine some new and associated ideas. Suppose a function $f: X \to Y$ is given and $A \subset X$. The function f also defines a function from A into Y and this function is called the *restriction* of f to A; it is written $f|A$. To say explicitly that $g = f|A$ means $g: A \to Y$ is such that $g(a) = f(a)$ for each $a \in A$. Conversely, we might say that f is an *extension* of the function g.

In the ordered pair notation for functions, $f|A$ is the set of ordered pairs from f having elements of A as their first elements.

EXAMPLE 1.5.1. Let $X = \{0, 1, 2, 3, 4, 5\}$, $A = \{0, 1, 2\}$ and $Y = \{a, b, c, d\}$ with f given by

$$f(0) = a$$
$$f(1) = b$$
$$f(2) = c$$
$$f(3) = d$$
$$f(4) = d$$
$$f(5) = d$$

(See Figure 1.5.1.) The restriction of f to A, $f|A$, consists of those arrows originating from the subset A of X. Thus $f|A$ is as pictured in Figure 1.5.1 and is given by

$$0 \to a \quad \text{that is,} \quad f|A(0) = a$$
$$1 \to b \quad \text{that is,} \quad f|A(1) = b$$
$$2 \to c \quad \text{that is,} \quad f|A(2) = c$$

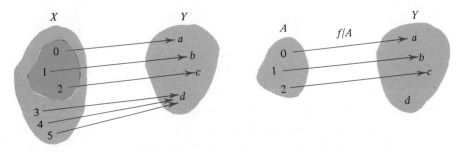

Figure 1.5.1 $f \mid A$.

In the ordered-pair notation, $f = \{(0, a),\ (1, b),\ (2, c),\ (3, d),\ (4, d),\ (5, d)\}$ while $f|A = \{(0, a),\ (1, b),\ (2, c)\}$.

EXAMPLE 1.5.2. Let $A = \{a, b, c, d, 1, 2, 3, 4\}$ with $f: A \to A$ given by $f(x) = x$. Let B denote the set $\{1, 2, 3, 4\}$. What is $f|B$?

First of all, what is f itself? We see that f is that function carrying each element of A to itself. (Here, x is used to signify an arbitrary element of A.) A graph of f is seen in Figure 1.5.2. Now, $f|B$ is precisely that set of ordered pairs having elements from B as the first term. That is, $f|B = \{(1, 1),\ (2, 2),\ (3, 3),\ (4, 4)\}$. Figure 1.5.3 shows a graph of $f|B$.

Functions of the sort given in Example 1.5.2 are very important ones. Functions $f: A \to A$ given by $f(x) = x$ are known as *identity* functions. The name is self-explanatory; the identity function takes

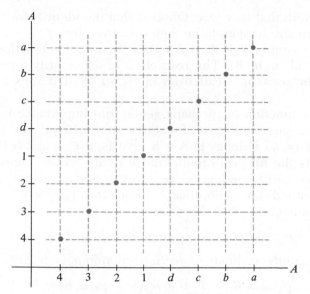

Figure 1.5.2 The identity function on A.

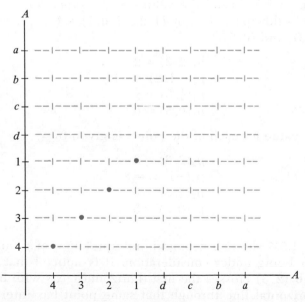

Figure 1.5.3 The identity function on A restricted to $\{4, 3, 2, 1\}$.

each x to itself: that is, $x \rightarrow x$. Observe that the identity functions are $1:1$ and onto and happen to be their own inverses.

In the example, $f|B : B \rightarrow A$ is like an identity function in that $f(x) = x$ for all x (in B). The restriction of an identity function to a subset of the domain is called an *injection* (of that subset into the parent set).

Another function of (perhaps geometric) importance is the projection function. Projection functions are defined on Cartesian products. If (a, b) belongs to $A \times B$ (that is, $a \in A$ and $b \in B$), we say that a is the *first coordinate* and b is the *second coordinate* of (a, b). The correspondences $(a, b) \rightarrow a$ and $(a, b) \rightarrow b$ denote functions called the (coordinate) projection functions. The *first projection* function p_1 is defined by

$$p_1 : A \times B \rightarrow A \quad \text{is given by} \quad p_1(a, b) = a$$

while the *second* coordinate *projection* function p_2 is described thusly:

$$p_2 : A \times B \rightarrow B \quad \text{is given by} \quad p_2(a, b) = b$$

The projection functions simply "pick out" the appropriate coordinate values.

EXAMPLE 1.5.3. Find the values of the projection functions at each of the following points of $\{1, 2, -1, 0, \frac{1}{2}\} \times \{3, 5, 2, -6\}$: $(2, 3)$, $(-1, 2)$, $(\frac{1}{2}, 3)$, and $(0, 5)$.

$$p_1(2, 3) = 2$$
$$p_1(-1, 2) = -1$$
$$p_1(\tfrac{1}{2}, 3) = \tfrac{1}{2}$$
$$p_1(0, 5) = 0$$

Now the p_2 value for each of the points is given by

$$p_2(2, 3) = 3$$
$$p_2(-1, 2) = 2$$
$$p_2(\tfrac{1}{2}, 3) = 3$$
$$p_2(0, 5) = 5$$

Figure 1.5.4 shows a geometric interpretation of p_1 and p_2, the point $(2, 3)$ being under consideration. It is noticed that a vertical line through $(2, 3)$ crosses the first (horizontal) axis with intercept 2 while a horizontal line through that same point has intercept 3 on the second (vertical) axis. The numbers 2 and 3, then, are the values of $p_1(2, 3)$ and $p_2(2, 3)$ respectively.

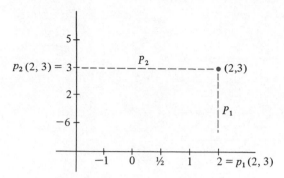

Figure 1.5.4 The projections.

EXERCISES 1.5

1. Consider Exercise 1.1.1. Is $h = f|A$?

2. Consider Exercise 1.1.3. Is $h = f|A$?

3. Describe $g|A$ in Exercises 1.1.1 and 1.1.2.

4. a. Given that $f = \{(0,0), (1,2), (3,4), (5,6), (\frac{1}{2}, 7), (\frac{1}{4}, 8)\}$, find $f|A$ where $A = \{0, 3, \frac{1}{2}\}$.
   b. Does f^{-1} exist?
   c. If your answer to b. is affirmative, what is f^{-1}?
   d. What is $(f|A)^{-1}$?

5. Find each of the following:

   a. $p_1(6, 4)$ d. $p_1(1, 2)$ g. $p_2(1, 3)$ j. $p_2(0, \frac{1}{2})$
   b. $p_1(2, 5)$ e. $p_1(1, 3)$ h. $p_2(2, 3)$ k. $p_2(-1, -1)$
   c. $p_1(\frac{1}{2}, 3)$ f. $p_1(1, 6)$,   i. $p_2(-1, 3)$ l. $p_2(2, 2)$

6. Let $A = \{a, b, c, 1\}$ and $B = \{\#, *, ?\}$. Define $f: A \to B$ by $f = \{(a, \#), (b, *), (c, ?), (1, ?)\}$ and $g: A \to B$ by $g(x) = (x, f(x))$. Sketch a graph of each of f and g (g is a subset of $A \times (A \times B)$ and this product is shown in Figure 1.5.5). Compute each of the following where i_A and i_B are the identity functions on A and B respectively.

   a. $p_1 \circ g(a)$ e. $p_1 \circ g(c)$ h. $i_B \circ f(b)$
   b. $p_2 \circ g(a)$ f. $p_2 \circ g(1)$ i. $i_A \circ p_1 \circ g(1)$
   c. $p_1 \circ g(b)$ g. $i_B \circ f(a)$ j. $i_B \circ p_2 \circ g(c)$
   d. $p_2 \circ g(b)$

   What can you say about $i_B \circ f$, $f \circ i_A$, $p_1 \circ g$, $p_2 \circ g$, $i_A \circ p_1 \circ g$, and $i_B \circ p_2 \circ g$? Sketch $p_1 \circ g$ and $p_2 \circ g$.

7. In Exercise 1.5.6 elements of $A \times (A \times B)$ are of the form $(a, (b, c))$ where $a \in A$, $b \in A$, and $c \in B$. We can define $A \times B \times C = \{(a, b, c) : a \in A, b \in B, \text{ and } c \in C\}$, a set of ordered triples. We, of course, wish to have $(a, b, c) = (d, e, f)$ if and only if $a = d$, $b = e$, and $c = f$. Show

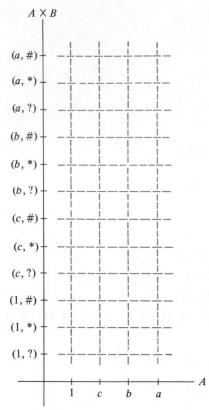

Figure 1.5.5 $A \times (A \times B)$.

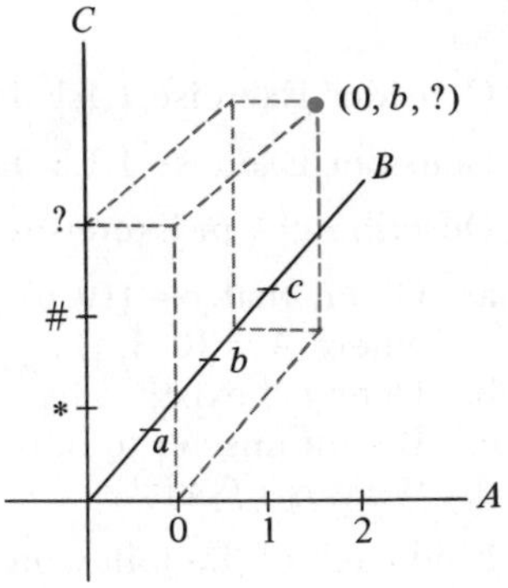

Figure 1.5.6 $A \times B \times C$.

that there is a 1:1 onto function from $A \times (A \times B)$ onto $A \times A \times B$.
(Pick an obvious function.) Figure 1.5.6 shows a two-dimensional model
(graph) of $A \times B \times C$ where $A = \{0, 1, 2\}$, $B = \{a, b, c\}$, and $C = \{*, \#, ?\}$
with the point $(0, b, ?)$ located.

1.6 REVIEW EXERCISES

1. Let $X = \{0, 1, 2, 3, 4, 5, 6\}$ and $Y = \{a, b, c, d, e\}$.
   a. Can there exist a function $f: X \to Y$ so that f is 1:1? Why or why not?
   b. Can there exist a function $g: Y \to X$ such that g is a 1:1 function?
   c. Can there exist a function $h: X \to Y$ with h an onto function?
   d. Can there exist a function $k: Y \to X$ with k onto?
   e. Can there exist a 1:1 and onto function $t: X \to Y$?
   f. Can there exist a 1:1 and onto function $s: Y \to X$?
   g. If the answer to any of parts a.–f. is yes, construct an example of
      such a function.

Let $A = \{3, 5, 6\}$ and $B = \{a, b, e\}$. Let u and v be correspondences as given by

$$
\begin{array}{cc}
u & v \\
0 \to a & 0 \to c \\
1 \to c & 1 \to b \\
2 \to e & 2 \to e \\
3 \to b & 3 \to c \\
4 \to c & 4 \to b \\
5 \to a & 5 \to e \\
6 \to e & 6 \to a
\end{array}
$$

h. Is each of u and v a function? If so, give the domain and range of such functions.
i. What is $u[X]$? $v[X]$? $u[A]$? $v[A]$? $u^{-1}[Y]$? $v^{-1}[Y]$? $u^{-1}[B]$? $v^{-1}[B]$? $u^{-1}[u[X]]$? $v^{-1}[v[X]]$? $u^{-1}[u[A]]$? $v^{-1}[v[A]]$? $u[u^{-1}[B]]$? $v[v^{-1}[B]]$?
j. Describe the correspondences $u|A$, $v|A$.
k. Is $u|A = v|A$? Why or why not?
l. Is u 1:1 and onto? v?
m. Is $u|A$ onto? $v|A$?
n. Is $u|A$ 1:1 and onto? $v|A$?
o. Write u and v as collections of ordered pairs.
p. Write $u|A$ and $v|A$ as collections of ordered pairs.
q. If any of $u, v, u|A$, or $v|A$ is 1:1, write a corresponding inverse for each such, using ordered-pair notation.

Find the value of each of the following.

r. $u(0)$ u. $u(5)$ w. $v(3)$ y. $v(2)$
s. $u(1)$ v. $u(6)$ x. $v(4)$ z. $v(6)$
t. $u(2)$

2. Let $f(x) = 3x$ represent a functional relationship (for each number x, f "triples" x). Let $g(x) = x/3$ (that is, g divides each number by 3). Now g is a function also.
   a. Show that $g \circ f$ and $f \circ g$ are both the identity function. That is, show that $g \circ f(x) = x = f \circ g(x)$ for all numbers x.
   b. For what numbers x does $f(x) = g(x)$?

Compute the following values.

c. $f(0)$ g. $g(1)$ k. $g(3)$ o. $f \circ g(-1)$
d. $f(1)$ h. $g(-1)$ l. $g(-3)$ p. $g \circ f(0)$
e. $f(-1)$ i. $f(1/3)$ m. $g \circ f(0)$ q. $g \circ f(1)$
f. $g(0)$ j. $f(-1/3)$ n. $f \circ g(1)$ r. $g \circ f(-1)$

1.7 QUIZ

Work this self-test in one hour and then correct your work.

1. Let f, g, h, and k be defined as below and let

$$A = \{\text{Oscar, James, Frank}\}$$
$$B = \{\text{Cad, Rolls}\}$$
$$C = \{1, 2, a, b\}$$

f	$g = \{(\text{Rolls}, 1), (\text{Cad}, 2)$
Oscar $\to$ Rolls	$h = \{(1, \text{Oscar}), (2, \text{Oscar}), (a, \text{Oscar}),$
James $\to$ Cad	$\quad\quad (b, \text{James}), (1, \text{James}) \}$
Frank $\to$ Rolls	$k = \{(1, \text{Oscar}), (2, \text{James}), (3, \text{Frank})\}$

   a. Graph each of the relations f, g, h, and k.
   b. Which of the relations are functions between sets among A, B, and C?
   c. List the domain and range of each function from b.
   d. Do any of the nonfunctions from a. become functions if either the domain or codomain is changed? If so, which ones?
   e. Are any of the functions 1:1? List all such.
   f. Are any of the functions onto? List all such.
   g. Can there exist 1:1 and onto functions between any pair of sets among A, B, and C? Why or why not?
   h. Are any of the functions composable? Describe such compositions completely.
   i. Graph the identity function on A. What relation is this?
   j. Let $D = \{\text{Oscar, James}\}$. What is $f|D$?
   k. What is $p_1(f(\text{Oscar}), g(\text{Cad}))$? $p_2(f(\text{Oscar}), g(\text{Cad}))$?

1.8 ADVANCED EXERCISES

1. Let $i: X \to X$ be the identity function X. Show that i is *bijective* (1:1 and onto) and that $i^{-1} = i$.

2. Given $f: X \to Y$ and $g: Y \to Z$, show that:
   a. $g \circ f$ is 1:1 implies f is 1:1.
   b. $g \circ f$ is onto shows g is onto.

3. Prove $f: X \to Y$ has an inverse implies that f is 1:1 and onto. Moreover, f^{-1} is 1:1 and onto when it exists.

4. If $f: X \to Y$ and $g: Y \to Z$, show each of the following.
   a. $A \subset B \subset X$ implies $f[A] \subset f[B]$.
   b. A and $B \subset X$ show $f[A \cup B] = f[A] \cup f[B]$.
   c. $A \subset B \subset Y$ implies $f^{-1}[A] \subset f^{-1}[B]$.
   d. A and $B \subset Y$ show $f^{-1}[A \cup B] = f^{-1}[A] \cup f^{-1}[B]$.
   e. With the hypothesis of b., $f[A \cap B] \subset f[A] \cap f[B]$.
   f. With the hypothesis of d., $f^{-1}[A \cap B] = f^{-1}[A] \cap f^{-1}[B]$.
   g. $A \subset Y$ implies $f^{-1}[Y - A] = f^{-1}[Y] - f^{-1}[A] = X - f^{-1}[A]$.
   h. $A \subset Z$ shows $(g \circ f)^{-1}[A] = f^{-1}[g^{-1}[A]]$.

5. Sets X and Y are said to be *equivalent,* written $X \sim Y$, if and only if there is a 1:1 and onto function $f: X \to Y$. Show that
   a. $X \sim X$ (There is an obvious 1:1 and onto function.)
   b. $X \sim Y$ implies $Y \sim X$ (Use an inverse function.)
   c. $X \sim Y$ and $Y \sim Z$ show $X \sim Z$. (Composition of functions.)

   We realize that "$\sim$" is an equivalence relation on sets. Let A be the relation given by $A = \{(X, Y): X, Y$ are sets and $X \sim Y\}$. A is the relation "induced" by "$\sim$."

6. Let $N = \{1, 2, 3, \ldots\}$ with F the relation on N given by: $(a, b) \in F$ means each of a and b have the same remainder upon division by 7. An example of an element in F is $(7, 14)$ since both have a remainder of zero upon division by 7. The element $(1, 9)$ of $N \times N$ is not in F since 1 has a remainder of 1 upon division by 7 and 9 has a remainder of 2.

   a. Is F an equivalence relation on N? Verify.

   F can be divided into *equivalence classes* (seven such) by grouping together all equivalent pairs (that is, pairs having a common remainder for their elements). For instance,

   Let $\boxed{0} = \{(x, y) \in F:$ the remainder upon division of x by 7 is 0$\}$
   Let $\boxed{1} = \{(x, y) \in F:$ the remainder upon division of x by 7 is 1$\}$

   Continue this process for $\boxed{2}$, $\boxed{3}$, $\boxed{4}$, $\boxed{5}$, and $\boxed{6}$.

   b. Show that $F = \boxed{0} \cup \boxed{1} \cup \boxed{2} \cup \boxed{3} \cup \boxed{4} \cup \boxed{5} \cup \boxed{6}$
   Let $\boxed{x}$ denote the equivalence class to which (x, x) belongs.
   c. Show that $\boxed{a - b} = \boxed{0}$ if $\boxed{a} = \boxed{b}$.
   Define: $\boxed{a} + \boxed{b} = \boxed{a + b}$ where $\boxed{a + b}$ is the equivalence class to which $(a + b, a + b)$ belongs. For example:

   $\boxed{3} + \boxed{6} = \boxed{9} = \boxed{2}$ since $(9, 9) \in \boxed{2}$, 9 having a remainder of 2 upon division by 7. Furthermore, we can define $\boxed{a} \cdot \boxed{b} = \boxed{ab}$ as the equivalence class to which (ab, ab) belongs.

   $$\boxed{3} \cdot \boxed{3} = \boxed{9} = \boxed{2}$$

   d. Make complete addition and multiplication tables for this Modulo 7 system.

   Define
   $$\boxed{a} - \boxed{b} = \boxed{c}$$
   where
   $$\boxed{b} + \boxed{c} = \boxed{a} \qquad \text{(if such a } \boxed{c} \text{ exists)}$$
   Next define $\boxed{a} \div \boxed{b} = \boxed{d}$ where $\boxed{b} \cdot \boxed{d} = \boxed{a}$, provided a *unique* such $\boxed{d}$ exists. The following illustrates.
   $$\boxed{3} - \boxed{6} = \boxed{4}$$
   since
   $$\boxed{6} + \boxed{4} = \boxed{10} = \boxed{3}$$

Furthermore;

$$\boxed{3} \div \boxed{6} = \boxed{4}$$

since

$$\boxed{6} \cdot \boxed{4} = \boxed{24} = \boxed{3}$$

e. Make subtraction tables to show that subtraction is always defined.

f. Division is well defined except that $\boxed{a} \div \boxed{0}$ is not valid for precisely the same reasons we cannot divide by zero in the set of real numbers. Make a table for division to show that division exists except for this special case. Perform the following operations.

g. $(\boxed{3} + \boxed{5})(\boxed{4} + \boxed{5}) - \boxed{3}$

h, $(\boxed{3} - \boxed{5})\boxed{4} + \boxed{2}$

i. $(\boxed{3}\ \boxed{5} - \boxed{2}) \cdot \boxed{6} - \boxed{4}$

7. Go through the entire development of a system like that derived in Exercise 1.8.6 except that the remainder be that remainder upon division by 5 instead of 7.

CHAPTER 2

The Plane and Real Functions

2.1 A MODEL OF THE REAL NUMBERS

In an earlier problem we viewed functions f and g described by the equations $f(x) = 3x$ and $g(x) = x/3$. It was remarked that x in each case represented a number. There are many types of entities the mathematician calls "numbers." Our main concern centers around those called *real* numbers. In particular, we will have an intense interest in functions whose domains and ranges lie in the collection of real numbers.

Since these numbers have been the central item of study of virtually all high school preparation, it will not be our expressed purpose to take up an extensive review and make up a background in the properties of these numbers. (Appendix B does, however, give more information.) Some operations with the numbers and subclassifications of the numbers will be viewed in order to fix certain ideas not generally a part of everyone's usable knowledge.

In searching for a suitable model (picture, graph) for the real numbers, we find that a generalization of the concept of a ruler is entirely satisfactory. Since primary school days we have been asked to consider the relative sizes of these numbers and in performing our task we have made use of the symbol "$<$." Figure 2.1.1 illustrates our model of R, making explicit use of the relation "$<$."

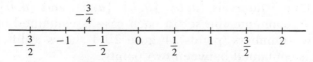

Figure 2.1.1 The real line.

64358

The real numbers will be visualized as elements of a (horizontal) straight line. Each point on the line has a number designation and each number designates a point on the line (that is, there is a 1:1 and onto function between R and the points of the line). The numbers are "ordered" via the concept of "$<$."

If a is less than b ($a < b$), a will be found to the left of b in our sketch. The positive numbers (those greater than zero) are to the right of zero while the negative numbers (those less than zero) lie to the left of zero. The positioning of numbers will be taken to be uniform in the sense that the calibrations on a ruler are uniform.

In Figure 2.1.2 we see the real line pictured but with only numbers of the form 1, 2, 3, . . . and 0 and $-1, -2, -3, . . .$ labeled. These numbers constitute the set Z of *integers*. Geometrically we feel that they "run through" the real numbers. That is, if a is any real number, then some integer n lies to the right of a ($n > a$) and some integer m lies to the left of a ($m < a$). We will also have occasion to denote the *positive integers* by the set N.

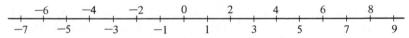

Figure 2.1.2 The integers in R.

The concept of relative size leads to a determination of various types of subsets of R, each being known as an *interval* of one sort or another. A number of intervals are illustrated in Figure 2.1.3. We repeat here the definitions shown graphically ($a \leq b$ means a is less than b or a is equal to b).

$$(a, b) = \{x \in R: a < x < b\}$$
$$[a, b] = \{x \in R: a \leq x \leq b\}$$
$$[a, b) = \{x \in R: a \leq x < b\}$$
$$(a, b] = \{x \in R: a < x \leq b\}$$
$$(a, \infty) = \{x \in R: a < x\}$$
$$[a, \infty) = \{x \in R: a \leq x\}$$
$$(-\infty, b) = \{x \in R: x < b\}$$
$$(-\infty, b] = \{x \in R: x \leq b\}$$

In this context we might write $R = (-\infty, \infty)$. The interval notation is very descriptive.

The "finite" intervals (a, b), $[a, b]$, $[a, b)$, and $(a, b]$ all have one thing in common: the set in each case is contained or trapped between two numbers (points). Figure 2.1.4 shows a different type of set that is contained between two points.

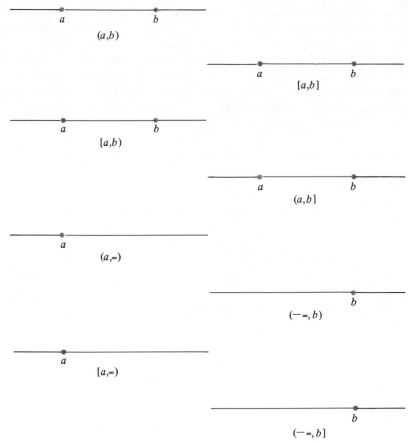

Figure 2.1.3   Intervals.

If A is a subset of R and A is contained between two numbers (that is, there are points u and l where $u \leqslant a \leqslant l$ for each $a \in A$), we say that A is *bounded*. All numbers having the property of l above are called *upper bounds* of A while numbers like u (numbers less than or equal to each element of A) are *lower bounds* of A. If A has an upper bound (a lower bound), A is said to be *bounded above* (*bounded below*).

We see that being bounded is equivalent to being bounded both above and below. Being bounded above or below is not sufficient

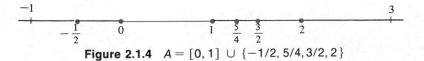

Figure 2.1.4 $A = [0, 1] \cup \{-1/2, 5/4, 3/2, 2\}$

to guarantee boundedness $[(a, \infty)$ is bounded below by a but is not bounded above while $(-\infty, b)$ is bounded above and fails to be bounded].

The real numbers have the property that if a subset A has an upperbound, it has a *least* (smallest) *upper bound* denoted by $l.u.b.A$. Geometrically, if A has an upper bound there is an upper bound l such that if k lies to the left of l, k is not an upper bound. Alternatively, if $k < l$, there is a member $a \in A$ satisfying $k < a \leq l$. See Figure 2.1.5.

Similarly, if A is bounded below, A has a *greatest* (largest) *lower bound* denoted by $g.l.b A$. Figure 2.1.6 illustrates this idea.

The property "each bounded subset has a greatest lower bound and a least upper bound" is known as the *completeness* property.

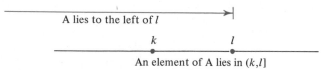

An element of A lies in $(k,l]$

Figure 2.1.5   Geometric consideration of l.u.b.A.

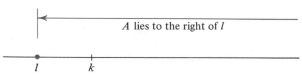

An element of A lies in $[l, k)$

Figure 2.1.6   Geometric consideration of g.l.b.A.

EXAMPLE 2.1.1.   Let $A = \{1, \ 1/2, \ 1/3, \ 1/4, \ \ldots\}$, $B = \{x \in R : a < x < b\}$, and $C = \{x \in R : x \text{ is negative}\}$. Determine which among A, B, C, and N have upper and lower bounds and find the greatest lower bounds and least upper bounds where they exist.

Now 2 is an upper bound for A and C. In fact, 1 is an upper bound for A and 0 is an upper bound for C. The number 1 is l.u.b.A and 0 is l.u.b.C. (Why?). Furthermore, b is l.u.b.B, the set of all numbers between a and b. N has no upper bound.

C has no lower bound (C contains the negative integers), g.l.b.N = 1, g.l.b.A = 0, and g.l.b.B = a.

The completeness property for the real numbers gives rise to a function

$$\text{l.u.b.}: \mathcal{A} \to R$$

and a function

$$\text{g.l.b.}: \mathcal{B} \to R$$

where $\mathscr{A}$ is the set of all subsets of R bounded above and $\mathscr{B}$ is the set of all subsets of R bounded below. The correspondences are obvious.

EXERCISES 2.1

1. Show that the upper (lower) bounds and least upper (greatest lower) bounds are as specified in Example 2.1.1.

2. Sketch a graph of the following intervals.

    a. $(0, 1)$ d. $(0, 1]$ g. $[0, \infty)$
    b. $[0, 1]$ e. $(0, \infty)$ h. $(-\infty, 0]$
    c. $[0, 1)$ f. $(-\infty, 0)$ i. $(-\infty, \infty)$

3. For each of the sets in parts a.–d. of Exercise 2.1.2 give the l.u.b. and g.l.b. From your answers, do you conclude that the l.u.b. (g.l.b.) of a set *must* be in the set? *can* be in the set?

4. Let $A = \{(n+1)/n : n \in N\} = \{2, 3/2, 4/3, 5/4, \ldots\}$. Find g.l.b. A and l.u.b. A.

5. Let $A = \{(n+1)/n : n \in Z - \{0\}\}$. (See Exercise 2.1.4 above.) What are g.l.b.A and l.u.b.A?

6. Describe each of the following sets.
    a. $(0, 1) \cup (1/2, 2)$
    b. $(0, 1) \cap (1/2, 2)$
    c. $(0, 1) - (1/2, 2)$

7. Why do you think that, if $A \subset B$, l.u.b.$A \leqslant$ l.u.b.B and g.l.b.$A \geqslant$ g.l.b.B where all g.l.b.'s and l.u.b.'s exist?

2.2   A MODEL OF $R \times R$

In Section 2.1 we stated that a goodly portion of our work will involve functions between subsets of the real numbers. In the ordered-pair sense, the functions are subsets of $R \times R = R^2$, the *Euclidean plane*. Figure 2.2.1 shows a model of R^2, a model very similar to our models for other Cartesian products but still somewhat different.

Points (ordered pairs of real numbers) are located as before. A number of example points are shown in Figure 2.2.2. In Figure 2.2.3 we see that points on the first (horizontal) axis have the form $(a, 0)$ while those on the second (vertical) axis look like $(0, b)$. We consider the axes as integral parts of the space (product), *not* as isolated factors simply determining the position of the ordered pairs. Our means of picturing the Cartesian product gives rise to easy descriptions of two types of sets.

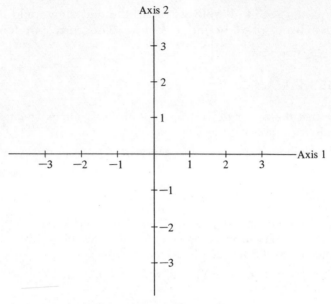

Figure 2.2.1 $R \times R$.

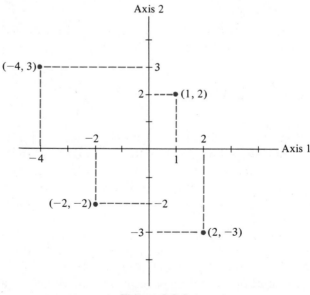

Figure 2.2.2

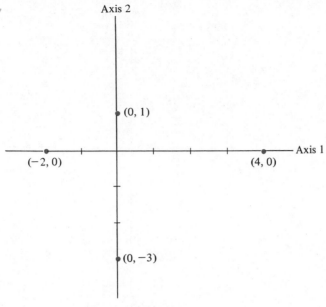

Figure 2.2.3

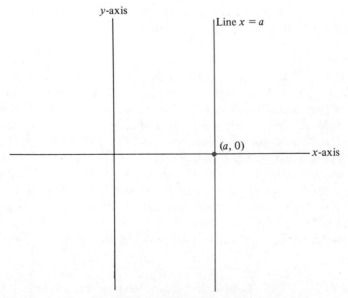

Figure 2.2.4 A vertical line.

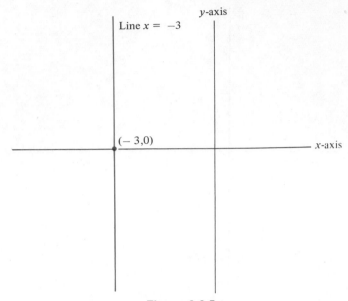

Figure 2.2.5

What are vertical lines? What characteristic is predominant? A vertical line intersects the horizontal axis at some point $(a, 0)$. The points on that vertical line all have the same first coordinate. Thus the vertical line passing through $(a, 0)$ is the set: $\{(x, y) \in R \times R: x = a\}$. We say that the "equation" of the vertical line is $x = a$. (See Figure 2.2.4.)

EXAMPLE 2.2.1. Graph the vertical line given by $x = -3$.

The line is $\{(x, y) \in R^2: x = -3\}$. A point is on the line if and only if its first coordinate is -3. The value of the second coordinate has no role in determining whether the point is on the line. The second coordinate merely describes which point of the line is at hand. Figure 2.2.5 shows this line.

Now horizontal lines should share an analogous characteristic. A horizontal line must pass through the y-axis (vertical axis) at some point $(0, b)$. The line is then the set of all points having second coordinate b. That is, the line is the set $\{(x, y) \in R \times R: y = b\}$. Accordingly, we say that the equation of the line is $y = b$. (See Figure 2.2.6.)

EXAMPLE 2.2.2. Graph the line having the equation $y = 2/3$.

The line is the set $\{(x, y) \in R \times R: y = \frac{2}{3}\}$. Whether (x, y) is on this line is independent of (does not depend in any way upon)

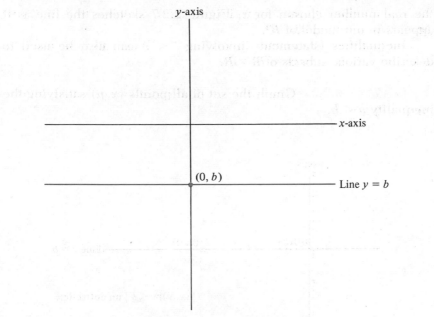

Figure 2.2.6 A horizontal line.

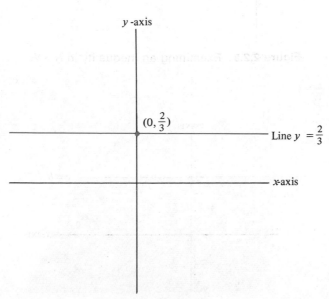

Figure 2.2.7

the real number chosen for x. Figure 2.2.7 sketches the line as it appears in our model of R^2.

Inequalities (statements involving "$<$") can also be used to describe various subsets of $R \times R$.

EXAMPLE 2.2.3. Graph the set of all points (x, y) satisfying the inequality $y < b$.

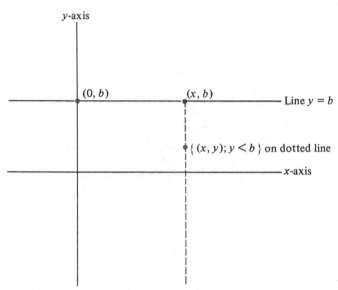

Figure 2.2.8 Examining an inequality in $R \times R$.

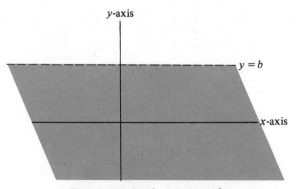

Figure 2.2.9 $\{(x, y)\ y < b\}$

Observe that the value of x is not a part of the description. A point (x, y) satisfying $y < b$ is dependent only upon the value of the second coordinate.

The equation $y = b$ represents a horizontal line. If (x, y) is a point on this line, (x, y) does not satisfy the inequality given above. The statement $y = b$ is satisfied, not $y < b$. Thus for any particular x, the points (x, z) having second coordinate $z < b$ lie "below" (x, y). (See Figure 2.2.8.) Consequently, $\{(x, y) \in R \times R : y < b\}$ is the set shaded in Figure 2.2.9. This set is known as the *solution set* for the inequality $y < b$.

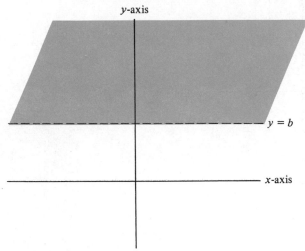

Figure 2.2.10 $\{(x, y) : y > b\}$

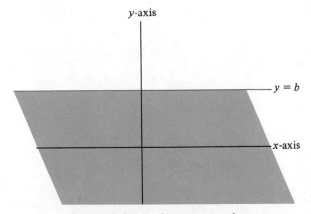

Figure 2.2.11 $\{(x, y) : y \leq b\}$.

The solution set for the inequality $y > b$ is shown in Figure 2.2.10 and that for $y \leqslant b$ is shown in Figure 2.2.11. The "=" in the last inequality adds the "boundary" (the line $y = b$).

We similarly form inequalities $x \leqslant a$, $x < a$, and $x \geqslant a$ and their solution sets as seen in Figure 2.2.12.

EXAMPLE 2.2.4. Graph the set $\{(x, y) : x > a \text{ and } y < b\}$.

The solution set must satisfy both inequalities. That is to say, the desired solution set is the intersection of two separate solution

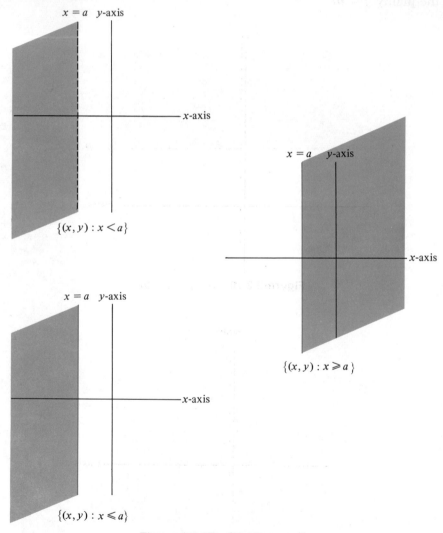

Figure 2.2.12 $\{(x, y) ; x \leqslant a\}$.

sets. The crosshatched area of Figure 2.2.13 (recall the Venn-diagram concept) displays the solution set.

EXAMPLE 2.2.5.   Graph the solution set for $2 < y < 4$.

Two inequalities are implied: $2 < y$ and $y < 4$. The solution set is then determined by the *intersection* of the individual solution sets. (See Figure 2.2.14.)

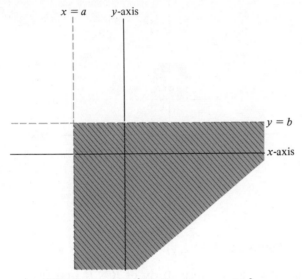

Figure 2.2.13 $\{(x, y): x > a, y < b\}$.

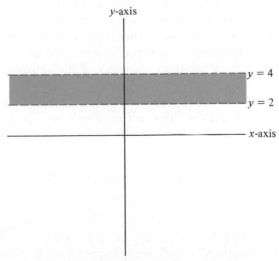

Figure 2.2.14 $\{(x, y): 2 < y < 4\}$.

EXAMPLE 2.2.6. Graph the set $\{(x, y) : -1 < x < 1$ and $2 < y < 4\}$.

The four inequalities have separate solution sets and the intersection shown in Figure 2.2.15 results.

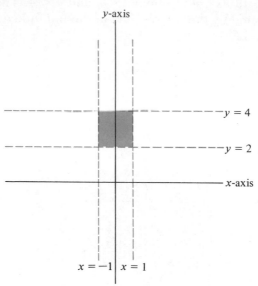

Figure 2.2.15 $\{(x, y): -1 < x < 1, 2 < y < 4\}$.

We originally started our study of $R \times R$ because of a desire to be able to graph functions between sets of real numbers. Two such functions will be examined presently. A function with range in R is called a *real-valued function* while if f is a function between two sets of real numbers, f is a *real-valued function of one real variable* or simply a *real function*. The term *real variable* is used to signify that the x in $f(x)$ is an arbitrary element of the domain of f and is a real number.

One example of a real function that "doesn't do much" is the function that one might say "doesn't do anything." It is quite important, however, and can be used as the basis for a significant portion of our study to come.

EXAMPLE 2.2.7. The identity function $i_R : R \to R$ is graphed in Figure 2.2.16:

$$i_R = \{(x, x) : x \in R\}$$

We see that i_R includes points such as $(1, 1)$, $(0, 0)$, $(2, 2)$, $(-1, -1)$, and so forth. By locating "a few" sample points we feel that (the graph of) i_R is a straight line connecting them. At this stage our conclusion is just a belief, nothing more.

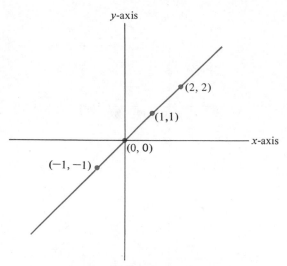

Figure 2.2.16 Graph of i_R.

In general our graphing of functions will be handled in this manner: We will sketch a sufficient number of points to obtain an "idea" of what the curve looks like and then "rough in" the rest of the graph. Although this sounds crude (indeed, sketches are crude in many respects) we will find that our excursion through a study of various functions will give us an increasing background from which to draw conclusions about the "shapes" of graphs. We will find our abilities and accuracy advance readily.

EXAMPLE 2.2.8. Let $f: R \to R$ be given by $f(x) = 6$ for each $x \in R$. That is, the value of f at any point of the domain is 6. We have said that such a function is *constant*.

$$f = \{(x, 6) : x \in R\}$$

the set of all ordered pairs having second coordinate 6. Earlier we saw that this was represented by the equation $y = 6$. That is, f is a horizontal line passing through $(0, 6)$. (See Figure 2.2.17.)

In general, if $g: R \to R$ is a constant, say $g(x) = k$, the graph of g is the horizontal line through $(0, k)$. (See Figure 2.2.18.)

EXAMPLE 2.2.9. Let [ ]: $R \to R$ be a function given by: [ ](x) (or $[x]$) is the largest integer not greater than x. Graph [ ].

Since the integer described is unique in each case, [ ] is indeed a function. For example, [ ]$(2) = [2] = 2$, $[1/2] = 0$, $[-1] = -1$, and $[-2/3] = -1$. In fact, if n is any integer and $n \leqslant x < n + 1$, $[x] = n$. Thus, [ ] is constant on each interval of the form $[n, n + 1)$. Figure

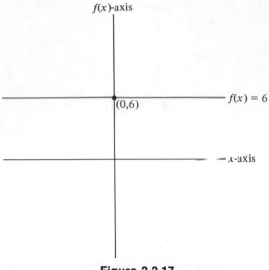

Figure 2.2.17

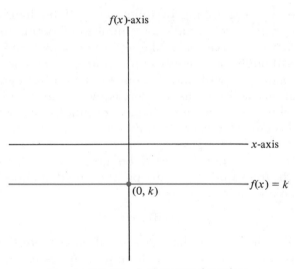

Figure 2.2.18 A constant function.

2.2.19 is a result of this analysis and the knowledge of the graphs of constant functions.

The concept of inequalities as seen earlier can be applied to functions also.

EXAMPLE 2.2.10. Sketch a graph of $\{(x, y): y \geq [x]\}$.

Figure 2.2.19 shows a sketch of $y = [x]$. The desired solution set includes these points together with all those "above" these points

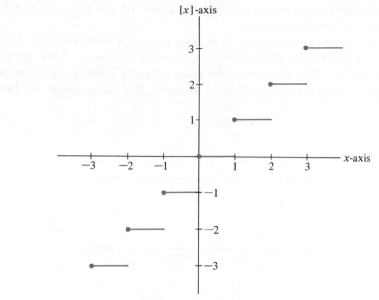

Figure 2.2.19 Graph of $[x]$.

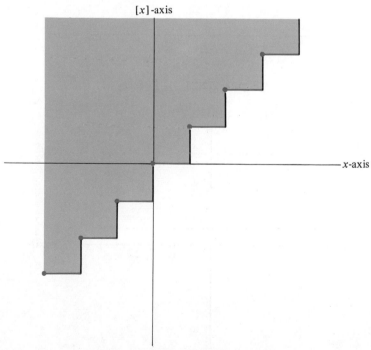

Figure 2.2.20 $\{(x, y) : y \geq [x]\}$.

$((x, a)$ above (x, y) means $a > y$, whence if $y = [x]$, $a > [x])$.
Figure 2.2.20 shows a sketch.

We have described a set A as bounded if it lies between two
numbers. We describe a function $f: A \to R$ as *bounded* if the range
is a bounded set. We recall that the range is $p_2[f]$, the projection of
the function onto the vertical axis. Because the range must lie be-

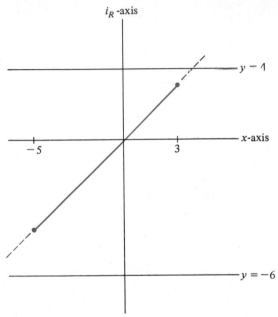

Figure 2.2.21 $i_R|[-5, 3]$.

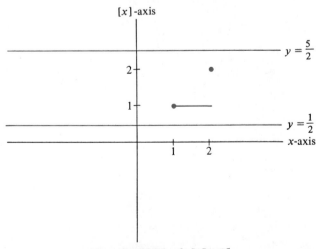

Figure 2.2.22 $[x]|[1, 2]$.

tween two numbers, say u and l, the graph of the function must lie between the lines $y = u$ and $y = l$. (Why?)

Of the three real functions studied above, only the constant function is bounded. However, each of the functions is bounded when restricted to bounded subsets of the domain. Figures 2.2.21 and 2.2.22 show two such restrictions and associated "bounding lines."

EXERCISES 2.2

1. Evaluate:
   a. $[-\frac{2}{3}]$ c. $[0]$ e. $[1]$
   b. $[-51]$ d. $[\frac{1}{2}]$ Is $[\ ]$ even? odd?

2. What is $[x]$ where $x \in [1, 2)$? For each integer n, what is $[x]$ where $x \in [n-1, n)$? What function is $[\ ]|N$? Sketch $[x]$ on N as a subset of $R \times R$. Does $[\ ]$ have an inverse? $[\ ]|N$?

3. Describe $p_2[[\ ]]$ where $[\ ]$ is taken to be the function in ordered-pair notation.

4. Graph each solution set:
   a. $\{(x, y): y < -2\}$ f. $\{(x, y): x \leq 2, y \leq 3\}$
   b. $\{(x, y): y \leq -2\}$ g. $\{(x, y): x < -1, y \geq 2\}$
   c. $\{(x, y): x > 5\}$ h. $\{(x, y): y < [x]\}$
   d. $\{(x, y): x \geq 5\}$ i. $\{(x, y): y > x\}$
   e. $\{(x, y): x > 3, y < 2\}$

5. Sketch each of the following functions.
   a. $f: R \to R$ where $f(x) = 1$ if x is an integer, $f(x) = 0$ otherwise.
   b. $f: N \to R$ given by $f(x) = x$ if x is even, $f(x) = 0$ if x is odd.
   c. $f: R \to R$ defined by $f(x) = [x] - 1$.
   d. $f: R \to R$ where $f(x) = [x] - x$.

6. In each case in Exercise 2.2.5 sketch:
   a. $\{(x, y) < f(x)\}$ c. $\{(x, y): y \leq f(x)\}$
   b. $\{(x, y): y > f(x)\}$ d. $\{(x, y): y \geq f(x)\}$

7. Is every straight line a function?

8. Is the solution set of any inequality a function?

2.3 BINARY OPERATIONS

Our knowledge of the real numbers is not limited to the numbers as entities. We are aware also of the existence of operations such as addition, multiplication, subtraction, and division. The properties of these operations as they pertain to the set R gives rise to what is known as the algebraic structure of a *complete ordered field*. The properties of completeness and order have been mentioned in connection with "<" and upper and lower bounds. It is our purpose here to discuss addition and multiplication as functions. We will then generate new real functions from old, making use of these operations.

The concept of addition, known to all of us, is really the concept of a special function having very special properties. Addition (+) is a function defined by: $+: R \times R \to R$ where $+(a, b) = a + b$. Since, for each pair of elements from R, the sum is a unique real number, $+$ is a well-defined function. (Functions from $A \times A$ into A are often called *binary operations* on A and A is said to be *closed* relative to the operation.)

Multiplication ($\cdot$) is another binary operation on R defined by: $\cdot: R \times R \to R$ where $\cdot(a, b) = a \cdot b = ab$.

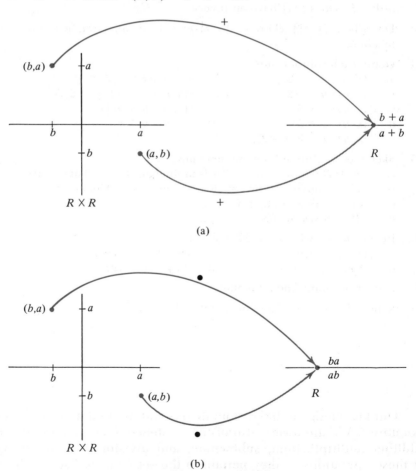

(a)

(b)

Figure 2.3.1 Commutativity.

Figure 2.3.1 shows diagrams of the functions $+$ and $\cdot$. Each diagram indicates that the function maps both (a, b) and (b, a) (generally different domain elements) to the same element of the range. Symbolically,

$$+ (a, b) = + (b, a)$$

and

$$\cdot (a, b) = \cdot (b, a)$$

or

$$a + b = b + a \qquad \text{and} \qquad ab = ba$$

Binary operations indifferent to the order within the pairs on which they operate are said to be *commutative*. Thus addition and multiplication are commutative operations on R.

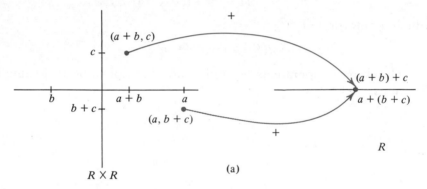

(a)

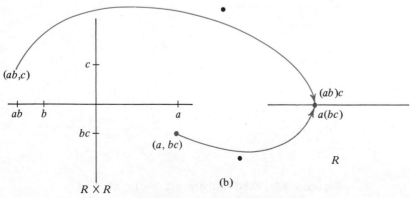

(b)

Figure 2.3.2   Associativity.

Figure 2.3.2 illustrates the fundamental properties:

$$+(+(a,b),c) = +(a,+(b,c))$$

and

$$\cdot(\cdot(a,b),c) = \cdot(a,\cdot(b,c))$$

In more usual terms,

$$(a+b)+c = a+(b+c)$$

and

$$(ab)c = a(bc)$$

Such operations are said to be *associative*.

Figure 2.3.3 shows the following *distributive* property:

$$\cdot(a,+(b,c)) = +(\cdot(a,b),\cdot(a,c))$$

which is interpreted

$$a(b+c) = ab + ac$$

The binary operations $+$ and $\cdot$ are examples of real-valued functions.

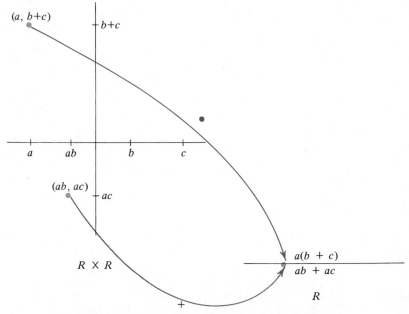

Figure 2.3.3 Distributivity, $a(b+c) = ab+ac$.

The binary operations are *not* real functions of a single real variable. The binary operations are real-valued *functions of two real variables.* [In $+(a, b)$ both a and b are real variables in that for each arbitrary domain element (a, b), both a and b are real numbers.]

Let $f: R \times R \times R \to R$ be given by $f(a, b, c) = a + b + c$, $R \times R \times R$ being the set of ordered triples of real numbers. (See Exercise 1.5.7.) Since the domain element is a triple of real numbers, we say that f is a real-valued function of three (real) variables.

The operation of addition gives rise to a $1:1$ and onto real function. For each $x \in R$, there is a unique real number $-x$ (the *negative* of x or the *additive inverse* for x) where $x + (-x) = 0$. Thus $n: R \to R$ given by $n(x) = -x$ is a $1:1$ onto function and $-(-x) = x$ whence $x = n(-x)$ and n is its own inverse.

Similarly, for each nonzero real number x there is exactly one number $1/x$ satisfying $x(1/x) = 1$. The number $1/x$ is the *reciprocal* or *multiplicative inverse* for x. The function $r: R - \{0\} \to R - \{0\}$ given by $r(x) = 1/x$ is $1:1$ and onto and $r^{-1} = r$ (that is, $1/(1/x) = x$).

The operations also give rise to two further operations (subtraction and division). These will be examined in the exercises.

EXERCISES 2.3

1. Let $n: R \to R$ be the function defined by $n(x) = -x$ as given in this section. Find
   a. $n(0)$ e. $n(-1)$
   b. $n(1)$ f. $n(-2)$
   c. $n(2)$ g. $n(-3)$
   d. $n(3)$

2. Sketch a graph of n from Exercise 2.3.1.

3. Sketch (n is as in Exercises 2.3.1 and 2.3.2)
   a. $\{(x, y): y < -x\}$
   b. $\{(x, y): y \leqslant -x\}$
   c. $\{(x, y): y > -x\}$
   d. $\{(x, y): y \geqslant -x\}$

4. Sketch $r: R - \{0\} \to R - \{0\}$ given by $r(x) = 1/x$. Compute:

   a. $r(1)$ e. $r(1/3)$ i. $r(-1/3)$
   b. $r(2)$ f. $r(1/4)$ j. $r(-1/4)$
   c. $r(3)$ g. $r(1/5)$ k. $r(-1/5)$
   d. $r(1/2)$ h. $r(-1/2)$ l. $r(0)$

5. Show that n in Exercise 2.3.1 and r in Exercise 2.3.4 are their own inverses (that is, $n^{-1} = n$ and $r^{-1} = r$, which is to say that $-(-x) = x$ and $1/(1/x) = x$).

6. Define $a - b = a + (-b)$ and show that $m: R \times R \to R$ given by $m(a, b) = a - b$ is a binary operation. Moreover, m is onto (write each number x as the difference of two real numbers).

7. Define $a \div b = a(1/b)$ if $b \neq 0$. Show that $q: R \times R - A \to R$ (A is the horizontal axis) given by $q(a, b) = a \div b$ is a function.

8. Define

$$| \ |: R \to R \text{ by } : | \ |(x) = |x| = \begin{bmatrix} x & \text{if} & x \geq 0 \\ \text{The additive inverse of } x \text{ if } x < 0 \end{bmatrix}$$

Since x cannot be both negative and nonnegative, $| \ |$ is a function.

   a. Sketch $| \ |$.
   b. Show that $|x| = |-x|$.
   c. What is $| \ ||P$ where $P = \{x \in R : x \geq 0\}$?
   d. What is $| \ ||(R - P)$?
   e. Is $| \ |$ even? odd? (A function f is odd if $f(-x) = -f(x)$ for all x in the domain.)

9. Let $f: A \to R$ and $g: A \to R$ be real-valued functions and define:

$$\begin{aligned}
(f + g): A \to R & \quad \text{by} \quad (f + g)(a) = f(a) + g(a) \\
(f - g): A \to R & \quad \text{by} \quad (f - g)(a) = f(a) - g(a) \\
fg: A \to R & \quad \text{by} \quad (fg)(a) = f(a)g(a) \\
f/g: B \to R & \quad \text{by} \quad (f/g)(a) = f(a)/g(a) \\
& \quad \text{where} \quad B = \{a \in A : g(a) \neq 0\}
\end{aligned}$$

If k is a real number, define:

$$kf: A \to R \quad \text{by} \quad (kf)(a) = k(f(a))$$

The functions defined are called (respectively) the *sum, difference, product,* and *quotient* of f and g and a *scalar product* of f.

Let $n, r, m, q, | \ |$, and $[\ ]$ be as given earlier in the text and exercises. Compute the following.

   a. $(n + r)(1)$
   b. $(n - r)(2)$
   c. $nr(5)$
   d. $(n/r)(4)$
   e. $(r/n)(-1)$
   f. $(n + | \ |)(1)$
   g. $(n + | \ |)(-1)$
   h. $(n \div | \ |)(1)$
   i. $(n + [\ ])(0)$
   j. $(n - [\ ])(1)$

   k. $(n + [\ ])(2)$
   l. $(n [\ ])(1)$
   m. $(n [\ ])(-2)$
   n. $(n/[\ ])(1/2)$
   o. $(n/[\ ])(-5/3)$
   p. $([\ ]r)(1)$
   q. $([\ ]r)(2/3)$
   r. $([\ ]/r)(-3)$
   s. $(r/[\ ])(-2/3)$
   t. $(m + q)(2, 3)$

   u. $(m - q)(1/2, 2/3)$
   v. $(mq)(3, 1)$
   w. $(m/q)(1/2, 1)$
   x. $(q/m)(3, 3)$
   y. $(3n)(-3/2)$
   z. $(5r)(1/4)$
   a'. $(-2| \ |)(-3)$
   b'. $([\ ]/2)(-4)$
   c'. $(5m)(9, 7)$
   d'. $(2q)(1, 2)$

10. Why in Exercise 2.3.9 can m and q not be combined with $n, r, | \ |$, and $[\ ]$ by addition, multiplication, and other operations as defined there?

11. Graph (see Exercise 2.3.9):

    a. i_R
    b. $i_R + n$
    c. $n - i_R$
    d. nr
    e. $i_R + [\ ]$
    f. $i_R + |\ |$
    g. $[\ ]^2 = [\ ][\ ]$
    h. $i_R^2 = i_R i_R$
    i. $i_R^3 = i_R(i_R^2)$
    j. $i_R^{-2} = 1/i_R^2$
    k. $i_R^{-3} = 1/i_R^3$

12. From Exercise 2.3.11 evaluate:

    a. $2i_R^3(3)$
    b. $-3i_R^2(-1)$
    c. $5i_R^{-3}(-2)$
    d. $(2/3)\, i_R^{-2}(-4)$

13. Sketch the following:

    a. $\{(x, y) : y \leqslant x^2 = i_R^2(x)\}$
    b. $\{(x, y) : y \geqslant x^3 = i_R^3(x)\}$
    c. $\{(x, y) : y \leqslant r(x) = 1/x, x \neq 0\}$
    d. $\{(x, y) : y > i^{-3}(x), x \neq 0\}$

2.4 DISTANCE AND LOCUS PROBLEMS IN R^2

Since the model of R^2 resembles the plane examined in Euclidean geometry, we might feel that analysis of the sort from geometry would yield fruitful results.

First we will equip ourselves with a distance concept that is compatible with the Pythagorean theorem.

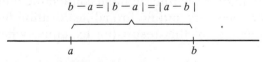

$$b - a = |b - a| = |a - b|$$

Figure 2.4.1 Distance in R.

The distance between the numbers a and b in Figure 2.4.1 is $b - a = |b - a| = |a - b| = ((a - b)^2)^{1/2}$. Thus in Figure 2.4.2, lengths of the two sides of the (right) triangle are $((x - u)^2)^{1/2}$ and $((y - v)^2)^{1/2}$.

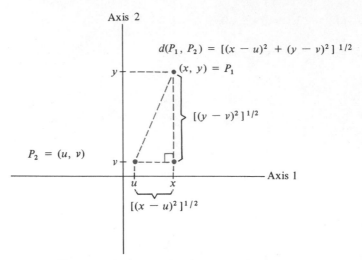

Figure 2.4.2   An interpretation of distance.

The Pythagorean theorem from geometry would declare that the distance $d(P_1, P_2)$ then ought to satisfy

$$[d(P_1, P_2)]^2 = \{[(x - u)^2]^{1/2}\}^2 + [(y - v)^2]^{1/2}\}^2$$
$$= (x - u)^2 + (y - v)^2$$

or that

$$d(P_1, P_2) = [(x - u)^2 + (y - v)^2]^{1/2}$$

Our desire to be compatible with the Pythagorean Theorem leads us to define $d(P_1, P_2)$ as above.

Assuming that the triangle in Figure 2.4.3 is a right triangle, we deduce from the Pythagorean theorem that $c^2 = a^2 + b^2$. The statements $a = 5$ and $b = 12$ mean that $c = 13$. Now by the distance formula, $d(P_1, P_2) = [(6 - 1)^2 + (13 - 1)^2]^{1/2} = 169^{1/2} = 13$. Thus in R^2 the Pythagorean theorem and the distance concept seem to be compatible (the distance between two points is apparently what we would like the length of the line segment connecting them to be). This condition is not necessary and no formal justification has really been given the correctness of this result (for example, what is a "right" triangle in R^2?).

Another typical problem we encounter is that of providing an analytic description for some geometrically defined locus. Consider the "usual" definition of a circle. A *circle* C of center $P_0 = (h, k)$ and radius $r > 0$ is the set of all points $P = (x, y)$, where $d(P_0, P) = r$. (See Figure 2.4.4.) Symbolically,

$$C = \{P = (x, y) : d(P_0, P) = r\}$$

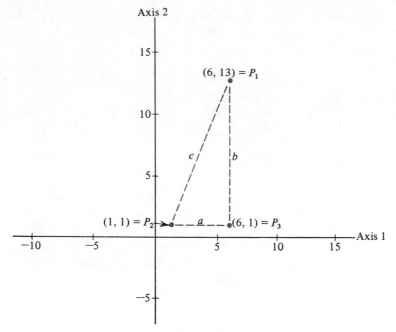

Figure 2.4.3

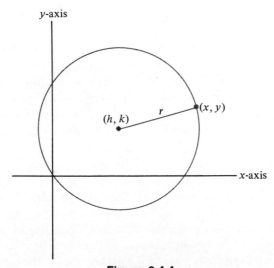

Figure 2.4.4

One problem is that of finding an equation (involving x and y) which describes the set of points on this circle. Such an equation has already been given twice. It is

$$d(P_0, P) = r$$

Alternatively,

$$((x - h)^2 + (y - k)^2)^{1/2} = r$$

or

$$(x - h)^2 + (y - k)^2 = r^2$$

It is not difficult to show conversely that each point satisfying the above equation is on the circle C. The beauty of this last form of the equation is that at a single glance we know the center and radius of the circle. These two items completely characterize or determine the circle. The last form of the equation is called the *standard form* for a circle in the plane.

EXAMPLE 2.4.1.   Analyze the following equation:

$$(x - 3)^2 + (y - 2)^2 = 36$$

This must evidently be a circle of center $(3, 2)$ and radius 6.

EXAMPLE 2.4.2.   One circle which will be used extensively is given by

$$x^2 + y^2 = 1$$

and is known as the *unit circle*. The name is derived from the fact that the circle has unit radius ($r = 1$) and center $(0, 0)$.

Given a circle we (in geometry) attach a positive real number as the "distance around the circle." This number is called the *circumference* and is $2\pi r$, where r is the radius of the circle and π is an irrational number slightly larger than 3 (see Exercise 2.11.2).

If we place one end of a string on any point of the circle and proceed to wrap the string around the circle, the shortest length of string needed to return to the original point is $2\pi r$.

What is the circumference of the unit circle? It must be 2π, since $r = 1$ in this instance.

An *arc* of a circle is considered to be all of the circle contained "between" two of its points. Actually, two arcs are determined by any pair of points on the circle. Since an arc can be thought of as being a portion of the circle, we intuitively expect to determine a positive number for the length of the arc. This number is fixed by that portion of the circumference described by the arc. That is, if the arc is one-fourth of a circle of radius r, the arc's length is $(1/4)(2\pi r) = \pi r/2$.

Remember, all the discussion concerning arc length is on an intuitive basis. There is no elementary method of measuring arc length. That such a number should even exist is a result of the completeness property of the real numbers. (See Exercise 2.11.2.) For the purposes of this material, intuitive reckoning will be relied upon for the acquisition and existence of numbers given as lengths of particular arcs.

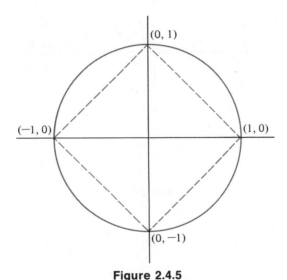

Figure 2.4.5

EXAMPLE 2.4.3. What is the length of the arc (see Figure 2.4.5) between $(1, 0)$ and $(0, 1)$ where the arc is described by traversing the circle in a counterclockwise direction from $(1, 0)$ to $(0, 1)$?

Several observations lead to an answer. First, the circle is the unit circle so that the circumference is 2π. Second, $d((1, 0), (0, 1))$ $= d((0, 1), (-1, 0)) = d((-1, 0), (0, -1)) = d((0, -1), (1, 0))$. We recall from geometry that *chords* (line segments connecting two points on a circle) of equal length determine equal arcs. The chords in this case are the line segments drawn in the figure. Apparently the axes partition the circle into four arcs of equal length. Thus the length of the arc in question ought to be $(1/4)(2\pi) = \pi/2$.

EXAMPLE 2.4.4. Consider the problem of finding the length of the shorter arc from $(1, 0)$ to $(1/\sqrt{2}, 1/\sqrt{2})$. Simple calculation shows that $d((1, 0), (1/\sqrt{2}, 1/\sqrt{2})) = d((1/\sqrt{2}, 1/\sqrt{2}), (0, 1))$, whereby the arc under question has length one-half that of the arc discussed in Example 2.4.3. Thus the length of our arc under examination is $(1/2)(\pi/2) = \pi/4$ (see Figure 2.4.6).

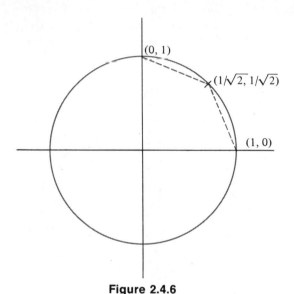

Figure 2.4.6

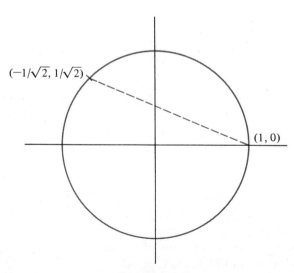

Figure 2.4.7

EXAMPLE 2.4.5. What is the length of the arc traversed counter-clockwise in going from $(1,0)$ to $(-1/\sqrt{2}, 1/\sqrt{2})$? Figure 2.4.7 depicts this arc. Now the length of the shorter arc from $(0,1)$ to $(-1/\sqrt{2}, 1/\sqrt{2})$ is equal to the length of the arc mentioned above (the shorter arc from $(1,0)$ to $(1/\sqrt{2}, 1/\sqrt{2})$). This arc then must have length $\pi/4$. However, that part of the arc from $(1,0)$ to $(0,1)$ has length $\pi/2$, giving a sum total length of $3\pi/4$ for the desired arc.

EXERCISES 2.4

1. Find the distance between each of the pairs of points in R.

    a. 0 and 5   b. 0 and -5 c. -5 and 5

2. Find the distance between the following pairs of points.

    a. $(0,0)$ and $(3,4)$ e. $(5,-4)$ and $(-1,2)$
    b. $(0,0)$ and $(-2,3)$ f. $(-1,-1)$ and $(-2,-2)$
    c. $(-1,2)$ and $(0,3)$ g. $(1,4)$ and $(4,1)$
    d. $(3,5)$ and $(1,2)$ h. $(1,0)$ and $(5,0)$

3. Each set of three points can be considered as the set of vertices of a triangle. Tell whether each of the following triples of points form a right triangle (according to the converse of the Pythagorean theorem), an isosceles triangle (two sides of equal length), or an equilateral triangle (all sides of equal length).

    a. $(-1,2)$, $(7,2)$, and $(7,17)$
    b. $(0,12)$, $(1,2)$, and $(5,10)$
    c. $(5,12)$, $(5,8)$, and $(9,12)$
    d. $(-1,0)$, $(1,0)$, and $(0,3)$
    e. $(0,0)$, $(1,0)$, and $(0,1)$
    f. $(0,0)$, $(1,0)$, and $(1/2, \sqrt{3}/2)$

4. Write the standard form of the equation of each of the following circles.

    a. Center $(1,1)$ and radius 10
    b. Center $(-1,-1)$ and radius 4
    c. Center $(1,-1)$ and radius 3
    d. Center $(0,0)$ and radius 5
    e. Center $(4,-2)$ and radius 2
    f. Center $(1,1)$ passing through $(1,4)$
    g. Center $(2,3)$ passing through $(6,6)$

5. Write an equation (as was done for the circle) depicting the set of points (x,y) which are equidistant from the two points $(1,3)$ and $(-1,2)$.

6. Follow the instructions for Exercise 2.4.5 where the points (x,y) have the property that the distance from (x,y) to $(1,1)$ added to the distance from (x,y) to $(-1,1)$ is 4.

7. Using your intuition and experiences in geometry, decide what distance along the unit circle a point travels in moving from the first point

to the second in each case (movement from the first point to the second to be taken in a counterclockwise direction).

a. $(1,0)$ to $(0,1)$ d. $(1,0)$ to $(\sqrt{3}/2, 1/2)$
b. $(1,0)$ to $(-1,0)$ e. $(1,0)$ to $(1/\sqrt{2}, 1/\sqrt{2})$
c. $(1,0)$ to $(0,-1)$ f. $(1,0)$ to $(1/2, \sqrt{3}/2)$

Hint for (d): $d((1,0), (\sqrt{3}/2, 1/2)) = d((\sqrt{3}/2, 1/2), (1/2, \sqrt{3}/2)) = d((1/2, \sqrt{3}/2), (0,1))$.

g. $(1,0)$ to $(-1/2, \sqrt{3}/2)$ l. $(1,0)$ to $(-1/2, -\sqrt{3}/2)$
h. $(1,0)$ to $(-1/\sqrt{2}, 1/\sqrt{2})$ m. $(1,0)$ to $(1/2, -\sqrt{3}/2)$
i. $(1,0)$ to $(-\sqrt{3}/2, 1/2)$ n. $(1,0)$ to $(1/\sqrt{2}, -1/\sqrt{2})$
j. $(1,0)$ to $(-\sqrt{3}/2, -1/2)$ o. $(1,0)$ to $(\sqrt{3}/2, -1/2)$
k. $(1,0)$ to $(-1/\sqrt{2}, -1/\sqrt{2})$

See Figure 2.4.8.

8. Redo Exercise 2.4.7 by taking a clockwise direction of traversal.

9. Consider the circle $x^2 + y^2 = 25$. This circle is centered at the origin and has radius 5. Find the lengths of the arcs between the following pairs of points on the circle (counterclockwise traversal). See Figure 2.4.8 for an analogous situation.

a. $(5,0)$ to $(0,5)$ i. $(5,0)$ to $(-5\sqrt{3}/2, 5/2)$
b. $(5,0)$ to $(-5,0)$ j. $(5,0)$ to $(-5\sqrt{3}/2, -5/2)$
c. $(5,0)$ to $(0,-5)$ k. $(5,0)$ to $(-5/\sqrt{2}, -5/\sqrt{2})$
d. $(5,0)$ to $(5\sqrt{3}/2, 5/2)$ l. $(5,0)$ to $(-5/2, -5\sqrt{3}/2)$
e. $(5,0)$ to $(5/\sqrt{2}, 5/\sqrt{2})$ m. $(5,0)$ to $(5/2, -5\sqrt{3}/2)$
f. $(5,0)$ to $(5/2, 5\sqrt{3}/2)$ n. $(5,0)$ to $(5/\sqrt{2}, -5/\sqrt{2})$
g. $(5,0)$ to $(-5/2, 5\sqrt{3}/2)$ o. $(5,0)$ to $(5\sqrt{3}/2, -5/2)$
h. $(5,0)$ to $(-5/\sqrt{2}, 5/\sqrt{2})$

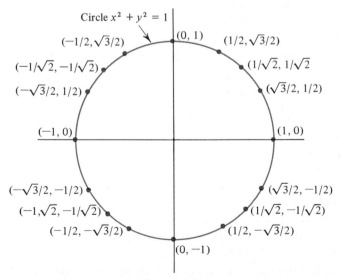

Figure 2.4.8

How do these arc lengths compare with those in Exercise 2.4.7?

10. Redo Exercise 2.4.9 by taking clockwise traversals.

11. Do Exercises 2.4.9 and 2.4.10 have any relation to Exercises 2.4.7 and 2.4.8? If so, what relationship is noted?

12. Observe that $\{(x, y): (x - h)^2 + (y - k)^2 < 25\}$ is the set of all points less than a distance five from (h, k). The set $\{(x, y): (x - h)^2 + (y - k)^2 = 25\}$ is the set of all points a distance (exactly) five from (h, k). The set above determined by the inequality is shown in Figure 2.4.9. Sketch the following subsets of the plane.

   a. $\{(x, y): x^2 + y^2 < 1\}$ c. $\{(x, y): (x - 1)^2 + (y - 2)^2 \leq 4\}$
   b. $\{(x, y): x^2 + y^2 > 1\}$ d. $\{(x, y): (x - 1)^2 + (y - 2)^2 > 4\}$

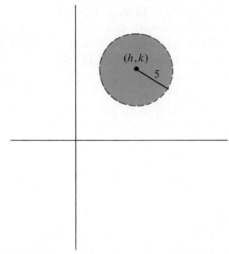

Figure 2.4.9

2.5 THE FUNCTION α

The following statement will be made without proof, the justification being beyond the level of this course (see Exercise 2.11.2). The given properties 1–6 below should be viewed as proved theorems.

Let C be the unit circle. There is an onto function $\alpha\colon R \to C$ with the following properties.

1. $\alpha(0) = (1, 0)$.

2. α restricted to the interval $[0, 2\pi)$ is 1:1 and onto ($\alpha|[0, 2\pi)$ is 1:1 and $\alpha[[0, 2\pi)] = C$).

3. $0 < a < e < 2\pi$ implies that $\alpha(a)$ is a point on the arc from

(1, 0) to $\alpha(e)$, where the arc is taken to be that one traversed by counterclockwise movement from (1, 0) to $\alpha(e)$.

4. $\alpha(a + 2\pi) = \alpha(a)$.

5. $\alpha(-a) = (c, -d)$ where $\alpha(a) = (c, d)$.

6. $d(\alpha(a), \alpha(b)) = d(\alpha(a - b), \alpha(0))$.

Furthermore, the following values are some "special values" found under the correspondence. [The term "special value" will be used to refer to (1) the matings listed below, (2) the special-domain elements used below, or (3) the special-image points below.]

$$\alpha(\pi/6) = (\sqrt{3}/2, 1/2)$$
$$\alpha(\pi/4) = (1/\sqrt{2}, 1/\sqrt{2})$$
$$\alpha(\pi/3) = (1/2, \sqrt{3}/2)$$
$$\alpha(\pi/2) = (0, 1)$$
$$\alpha(2\pi/3) = (-1/2, \sqrt{3}/2)$$
$$\alpha(3\pi/4) = (-1/\sqrt{2}, 1/\sqrt{2})$$
$$\alpha(5\pi/6) = (-\sqrt{3}/2, 1/2)$$
$$\alpha(\pi) = (-1, 0)$$
$$\alpha(7\pi/6) = (-\sqrt{3}/2, -1/2)$$
$$\alpha(5\pi/4) = (-1/\sqrt{2}, -1/\sqrt{2})$$
$$\alpha(4\pi/3) = (-1/2, -\sqrt{3}/2)$$
$$\alpha(3\pi/2) = (0, -1)$$
$$\alpha(5\pi/3) = (1/2, -\sqrt{3}/2)$$
$$\alpha(7\pi/4) = (1/\sqrt{2}, -1/\sqrt{2})$$
$$\alpha(11\pi/6) = (\sqrt{3}/2, -1/2)$$

Compare these special values with your solutions to Exercise 2.4.7.

The function α is called the *arc length function:* intuitively, $\alpha(a)$ is the point on the unit circle with the property that the arc from (1, 0) to $\alpha(a)$ has length $|a|$. (If a is positive, the arc is that traversed in a counterclockwise direction, while for negative a, the arc is described by a clockwise traversal.) Necessarily, our concept of an arc must be generalized to include those generated by moving in such a way as to traverse the circle more than once.

In some sense, property 1 shows that (1, 0) is a "starting" point on the unit circle.

Property 2 declares that α is an onto function; that is, each point (x, y) on the unit circle is $\alpha(a)$ for some real number a. Moreover, we can find a real number $a \in [0, 2\pi)$ which has this property. That this is true is of significance. (Recall that $a \in [0, 2\pi)$ implies the inequality $0 \le a < 2\pi$.) Furthermore, if $\alpha(a) = (x, y)$ and $a \in [0, 2\pi)$, a is the *only such* element from $[0, 2\pi)$ having this particular property. In other words, α is 1:1 on the given restricted domain.

Does α order the points on the unit circle in any way? Property 3 shows that the ordering of the real numbers by "≤" orders these points in some manner. We could say that if $0 < a < e < 2\pi$, $\alpha(a)$ "precedes" $\alpha(e)$ in the sense that upon traversing the unit circle in a counterclockwise direction from $(1, 0)$, we reach $\alpha(a)$ before we reach $\alpha(e)$. This is not necessarily true unless the condition $0 < a < e < 2\pi$ is fulfilled.

EXAMPLE 2.5.1. As an example, $0 < \pi/4 < 3\pi/4 < 2\pi$. Figure 2.5.1 shows that upon traversing the unit circle C in a counterclockwise direction, $\alpha(\pi/4) = (1/\sqrt{2}, 1/\sqrt{2})$ is reached prior to $\alpha(3\pi/4) = (-1/\sqrt{2}, 1/\sqrt{2})$.

Property 4 says several things. First, α is not a 1:1 function on its entire domain. Adding 2π to the argument of the function results in the same image point. That is, given the real number x, we can find $\alpha(x)$ (theoretically). However, $\alpha(x) = \alpha(x + 2\pi)$ according to property 4.

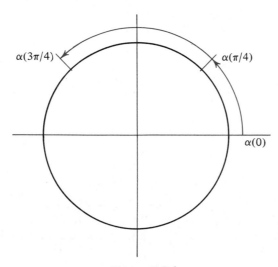

Figure 2.5.1

EXAMPLE 2.5.2. Since $\alpha(0) = (1, 0)$, we also know that $\alpha(2\pi) = (1, 0)$ $(2\pi = 0 + 2\pi)$. Then we also know $(1, 0) = \alpha(6\pi) = \alpha(8\pi)$, and so forth.

The fact that $\alpha|[0, 2\pi)$ is 1:1 and onto is important (α itself does not share the former of the two properties).

Theorem 2.5.1. *If n is any integer, $\alpha(a + 2n\pi) = \alpha(a)$.*

Proof. If $n = 0$, $\alpha(a + 2n\pi) = \alpha(a + 0) = \alpha(a)$ and the theorem is trivially satisfied.

For positive n, we need to use mathematical induction. If $n = 1$, $\alpha(a + 2n\pi) = \alpha(a + 2\pi) = \alpha(a)$ by property 4. Thus the theorem holds for $n = 1$.

If the theorem holds for $n = k$ [that is, $\alpha(a + 2k\pi) = \alpha(a)$] then $\alpha(a + 2(n + 1)\pi) = \alpha(a + 2(k + 1)\pi) = \alpha((a + 2\pi) + 2k\pi) = \alpha(a + 2\pi) = \alpha(a)$. Consequently, the theorem holds for $n = k + 1$ and necessarily for all positive integers n.

If n is a negative integer, $-n$ is a positive integer and we have $\alpha(a + 2n\pi) = \alpha((a + 2n\pi) + 2(-n)\pi) = \alpha(a + 2n\pi - 2n\pi) = \alpha(a)$. The proof is complete.

The theorem states that adding an even multiple of π to the argument of α does not change the image point.

More than the simple examples above, property 4 discloses that to know $\alpha(a)$ for each $a \in [0, 2\pi)$ is sufficient to know the correspondence for the entire domain. That is, if $b \in R$ we can find $\alpha(b)$ knowing only the correspondence α on $[0, 2\pi)$.

Proof? There is an integer n so that

$$2n\pi \leqslant b < 2(n + 1)\pi$$

(b lies between two successive multiples of 2π).

Then

$$0 \leqslant b - 2n\pi < 2\pi$$

Alternatively,

$$(b - 2n\pi) \in [0, 2\pi)$$

Since $\alpha(b) = \alpha((b - 2n\pi) + 2n\pi) = \alpha(b - 2n\pi)$, it is concluded that $\alpha(b)$ is known. (It was given that we knew the correspondence for the restricted domain $[0, 2\pi)$, whence $\alpha(b - 2n\pi)$ is known).

EXAMPLE 2.5.3. As an example of the calculation of an image point of α (using 4), find $\alpha(5\pi)$. We write 5π as $\pi + 4\pi$. Necessarily, $\alpha(5\pi) = \alpha(\pi + 4\pi) = \alpha(\pi) = (-1, 0)$. The intuitive concept of traversing an arc length of 5π is seen in Figure 2.5.2.

EXAMPLE 2.5.4. Look also at the example $\alpha(-11\pi/6)$. Now $-11\pi/6 = \pi/6 - 2\pi$. Using 4 we see that $\alpha(-11\pi/6) = \alpha(\pi/6 - 2\pi) = \alpha(\pi/6) = (\sqrt{3}/2, 1/2)$.

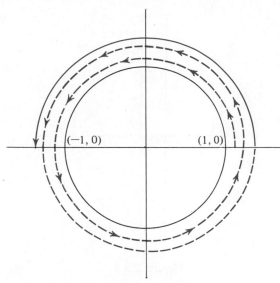

Figure 2.5.2

EXAMPLE 2.5.5. As a third example, find $\alpha(13\pi/6)$. Since $13\pi/6 = \pi/6 + 2\pi$, $\alpha(13\pi/6) = \alpha(\pi/6) = (\sqrt{3}/2, 1/2)$.

In conjunction with property 4, a further note about functions is in order. Let $X \subset R$. If $f: X \to Y$ is a function and there is a nonzero real number p so that $f(x + p) = f(x)$ for all x in the domain, we say that f is *periodic* and that p is a *period* of f. If a smallest positive period exists, it will be called the *primitive period*.

According to property 4 and what was discussed above, the arc length function α is of period (has a period) 2π. It is also true that this is the primitive period of α.

Theorem 2.5.2. *Suppose* f *is a function with periods* p *and* q. *Show that* p + q *is also a period of* f.

Proof. Let x be in the domain of f. Then $f(x + p) = f(x) = f(x + q)$ since each of p and q is a period. Now, $f(x + (p + q)) = f((x + p) + q) = f(x + p) = f(x)$ using first the fact that q is a period and then the fact that p is a period. The technique used in this proof is helpful in the proof of Exercise 2.11.4.

Let $a \in R$. For purposes of simplicity, suppose $0 \leqslant a \leqslant \pi/2$. (See Figure 2.5.3.) Now, $\alpha(a)$ is some point on the unit circle, say (c, d). By property 5, $\alpha(-a)$ is the point $(c, -d)$. It is intuitively clear that the length of the arc from $(1, 0)$ to $\alpha(a)$ taken in a counterclockwise direction is the same as the length of the arc from $(1, 0)$ to $\alpha(-a)$ taken in a clockwise direction $(d(\alpha(a), \alpha(0)) = d(\alpha(-a), \alpha(0)))$. The length of each arc is a, $a \geqslant 0$.

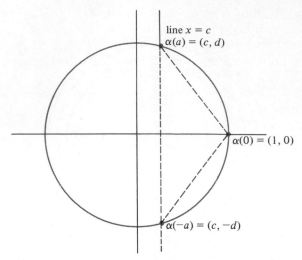

Figure 2.5.3

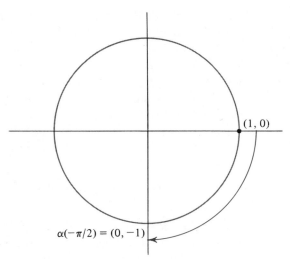

Figure 2.5.4 Arc of negative length.

EXAMPLE 2.5.6. As a concrete example of the message of 5, find $\alpha(-\pi/2)$. The special cases under α include $\alpha(\pi/2)$. Since $\alpha(\pi/2) = (0,1)$, it follows that $\alpha(-\pi/2) = (0,-1)$. Intuitively, we should feel that the negative sign in $-\pi/2$ indicates a clockwise traversal (from $(1,0)$) of arc length $\pi/2$. (See Figure 2.5.4.) The "$-$" in $\alpha(-a)$ indicates traversal opposite to that taken in the case $\alpha(a)$.

Property 6 may seem more complicated than the others. $\alpha(a)$ and $\alpha(b)$ are points on the unit circle and, as such, we can calculate the distance between them. This distance is equal to the distance from $\alpha(0) = (1, 0)$ to $\alpha(a - b)$. An example will illustrate.

EXAMPLE 2.5.7. Consider the real numbers π and $\pi/2$. $\alpha(\pi) = (-1, 0)$ and $\alpha(\pi/2) = (0, 1)$ so that $d(\alpha(\pi), \alpha(\pi/2)) = \sqrt{2}$. On the other hand, $\pi - \pi/2 = \pi/2$, giving $\alpha(\pi - \pi/2) = \alpha(\pi/2) = (0, 1)$. But, $d(\alpha(\pi - \pi/2), \alpha(0)) = d((0, 1), (1, 0)) = \sqrt{2}$ as anticipated from property 6.

Figure 2.5.5 illustrates property 6. The chords are of the same length. Geometrically, we feel that the "spanned" arcs are of equal length also.

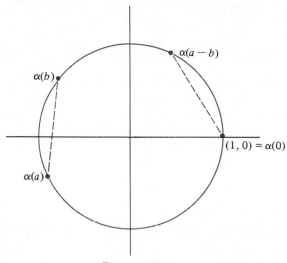

Figure 2.5.5

EXAMPLE 2.5.8. Compute $\alpha(\pi/12)$. We know that since $\pi/12 = \pi/4 - \pi/6$,

$$d(\alpha(\pi/12), (1, 0)) = d(\alpha(\pi/4), \alpha(\pi/6))$$
$$= d((1/\sqrt{2}, 1/\sqrt{2}), (\sqrt{3}/2, 1/2))$$
$$= [(1/\sqrt{2} - \sqrt{3}/2)^2 + (1/\sqrt{2} - 1/2)^2]^{1/2}$$
$$= [2 - (\sqrt{3} + 1)/\sqrt{2}]^{1/2}$$

or

$$[d(\alpha(\pi/12), (1, 0))]^2 = 2 - (\sqrt{3} + 1)/\sqrt{2}$$

However, letting $\alpha(\pi/12) = (x, y)$ we see that $(x^2 + y^2 = 1)$.

$$[d(\alpha(\pi/12), (1, 0))]^2 = (x-1)^2 + (y-0)^2$$
$$= x^2 - 2x + 1 + y^2$$
$$= x^2 - 2x + 1 + 1 - x^2$$
$$= 2 - 2x$$

Consequently,

$$2 - (\sqrt{3} + 1)/\sqrt{2} = 2 - 2x$$

or

$$2x = (\sqrt{3} + 1)/\sqrt{2}$$
$$x = (\sqrt{3} + 1)/2\sqrt{2}$$

Since $\alpha(\pi/12)$ is in the first quadrant,

$$y = (1 - x^2)^{1/2} = (\sqrt{3} - 1)/2\sqrt{2},$$

whence

$$\alpha(\pi/12) = ((\sqrt{3} + 1)/2\sqrt{2}, \ (\sqrt{3} - 1)/2\sqrt{2})$$

The student should familiarize himself with all the special values listed for the correspondence α. As we have mentioned, techniques more advanced than those appearing in this text are needed to compute all points $\alpha(a)$. Because of this the special values listed will be used for the most part.

EXERCISES 2.5

1. Find each of the following. (*Hint:* Use properties 1 and 4 of α.)

   a. $\alpha(0)$ d. $\alpha(-2\pi)$
   b. $\alpha(2\pi)$ e. $\alpha(-4\pi)$
   c. $\alpha(4\pi)$

2. Suppose $a \in [0, 2\pi)$ and suppose $\alpha(a) = \alpha(b)$. Complete the following to make true statements (use property 2 of α).

   a. If $b \in [0, 2\pi)$, then . . .
   b. If $(x, y) \in C$, there is an $a \in [0, 2\pi)$ such that . . .

3. Use property 3 of α to determine the quadrant in which each of the following is located.

   a. $\alpha(5\pi/24)$ e. $\alpha(23\pi/12)$
   b. $\alpha(18\pi/11)$ f. $\alpha(25\pi/51)$
   c. $\alpha(17\pi/16)$ g. $\alpha(24\pi/53)$
   d. $\alpha(18\pi/23)$ h. $\alpha(101\pi/100)$

4. Find each point using property 5 of α.

   a. $\alpha(-\pi/6)$ e. $\alpha(-3\pi/2)$
   b. $\alpha(-\pi/2)$ f. $\alpha(-4\pi/3)$
   c. $\alpha(-\pi)$ g. $\alpha(-11\pi/6)$
   d. $\alpha(-3\pi/4)$

5. Locate the quadrant of $\alpha(-a)$ if $\alpha(a)$ is in

   a. quadrant I   c. quadrant III
   b. quadrant II   d. quadrant IV

6. Give the point designated in each of the following cases.

   a. $\alpha(6\pi)$ g. $\alpha(29\pi/6)$ l. $\alpha(721\pi/6)$
   b. $\alpha(-7\pi/4)$ h. $\alpha(-19\pi/6)$ m. $\alpha(15\pi/4)$
   c. $\alpha(31\pi/3)$ i. $\alpha(37\pi/4)$ n. $\alpha(-3\pi/4)$
   d. $\alpha(-3\pi/2)$ j. $\alpha(-\pi/3)$ o. $\alpha(35\pi/6)$
   e. $\alpha(32\pi/3)$ k. $\alpha(203\pi/2)$ p. $\alpha(26\pi/3)$
   f. $\alpha(-17\pi)$

7. Describe the following sets in terms of the points of the unit circle.

   a. $\alpha[[0, \pi/2]]$ d. $\alpha[[\pi/2, \pi]]$ g. $\alpha[[\pi, 3\pi/2]]$
   b. $\alpha[[0, \pi]]$ e. $\alpha[[\pi/2, 3\pi/2]]$ h. $\alpha[[\pi, 2\pi]]$
   c. $\alpha[[0, 3\pi/2]]$ f. $\alpha[[\pi/2, 2\pi]]$ i. $\alpha[[3\pi/2, 2\pi]]$

8. Describe an arc from the first point to the second point in each case as the image of some interval under α. Find four such intervals in each case listed.

   a. $(1/\sqrt{2}, 1/\sqrt{2})$ to $(-1/\sqrt{2}, 1/\sqrt{2})$. (*Hint:* This arc can be given as $\alpha[[\pi/4, 3\pi/4]]$.) Find some other acceptable intervals.
   b. $(1/2, \sqrt{3}/2)$ to $(-1/\sqrt{2}, 1/\sqrt{2})$
   c. $(-1/2, -\sqrt{3}/2)$ to $(1/2, \sqrt{3}/2)$
   d. $(-\sqrt{3}/2, 1/2)$ to $(1/2, -\sqrt{3}/2)$

9. a. Show $\alpha(-a) = \alpha(2\pi - a)$. (*Hint:* Use property 4.)
   b. Show $\alpha(a + \pi) = \alpha(a - \pi)$.
   c. Show $\alpha(a + (2n - 1)\pi) = \alpha(a + \pi)$ where $n \in N$.

10. For each domain element used in Exercise 2.5.1 find the value of the image under each function below.

   a. $p_1 \circ \alpha$ e. $\dfrac{p_1 \circ \alpha}{p_2 \circ \alpha}$

   b. $p_2 \circ \alpha$

   c. $\dfrac{1}{p_1 \circ \alpha}$ f. $\dfrac{p_2 \circ \alpha}{p_1 \circ \alpha}$

   d. $\dfrac{1}{p_2 \circ \alpha}$ g. $(p_1 \circ \alpha)^2 + (p_2 \circ \alpha)^2$

11. Repeat Exercise 2.5.10 where the domain elements are those used in

   a. Exercise 2.5.4   b. Exercise 2.5.6

12. Repeat Exercise 2.5.10 where

    a. $\alpha(a) = (3/5, 4/5)$ f. $\alpha(a) = (-8/17, 15/17)$
    b. $\alpha(a) = (-3/5, 4/5)$ g. $\alpha(a) = (-7/25, 24/25)$
    c. $\alpha(a) = (-5/13, -12/13)$ h. $\alpha(a) = (24/25, -7/25)$
    d. $\alpha(a) = (12/13, -5/13)$ i. $\alpha(a) = (-7/25, -24/25)$
    e. $\alpha(a) = (8/17, -15/17)$

13. Compute the values of the six functions in parts a.–f. of Exercise 2.5.10 at $-a$ where a is as given in each part of Exercise 2.5.12.

14. Find a period for each of the functions of Exercise 2.5.10. (It would ably be helpful to determine the natural domain in each case.)

15. Suppose $\alpha(a) = (x, y)$. Show (using property 6) that $\alpha(\pi \quad a) = (-x, y)$, $\alpha(\pi + a) = (-x, -y)$, and $\alpha(2\pi - a) = (x, -y)$.

16. Show that if $\alpha(a) = (x, y)$ and $b = \pi/2 - a$, then $\alpha(b) = (y, x)$.

17. Given numbers a and b, we can determine points on the unit circle given by
$$\left(\frac{a}{(a^2 + b^2)^{1/2}}, \frac{b}{(a^2 + b^2)^{1/2}}\right), \left(\frac{b}{(a^2 + b^2)^{1/2}}, \frac{a}{(a^2 + b^2)^{1/2}}\right)$$
Determine such a pair of points for the following pairs of numbers.

    a. 1, 1   d. 8, 15
    b. 1, $\sqrt{3}$ e. 7, 24
    c. 5, 12

18. Determine the following points.

    a. $\alpha(7\pi/12)$
    b. $\alpha(11\pi/12)$
    c. $\alpha(143\pi/12)$
    d. $\alpha(13\pi/12)$

19. In Exercise 2.5.10 you were asked to compute certain values given in terms of composition of functions. Attempt to sketch a graph of each composition in Exercise 2.5.7. We might help a bit by discussing, for example, the sketch of $p_1 \circ \alpha$.

    We may envision our starting the sketch by looking at $p_1 \circ \alpha(0)$ and then proceed through the positive domain elements. Since $\alpha(0) = (1, 0)$ and p_1 picks out the first coordinate, $p_1 \circ \alpha(0) = p_1 \circ \alpha = 1$. As a goes from 0 through $\pi/2$, $\alpha(a)$ goes from $(1, 0)$ to $(0, 1)$, whence $p_1 \circ \alpha(a)$ goes from 1 to 0, steadily decreasing. Similarly, as a goes from $\pi/2$ to π, $\alpha(a)$ traverses that part of the unit circle in the second quadrant and $p_1 \circ \alpha(a)$ decreases from 0 to -1. Continuing, we see that as a passes from π to $3\pi/2$, $p_1 \circ \alpha(a)$ increases from -1 to 0 and as a continues to 2π, $p_1 \circ \alpha(a)$ continues increasing to 1. If we retrace our steps from 2π back to zero, we see exactly what happens as a goes from 0 to -2π (why?). The graph we sketch appears in Figure 2.5.6. The sketch indicates the effect of the periodicity of α.

    Complete the other sketches.

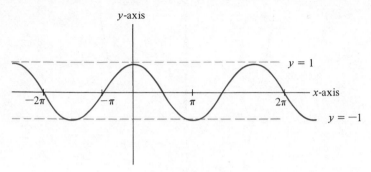

Figure 2.5.6

20. Compute the equation of the line through the origin and

    a. the points in Exercise 2.5.1.
    b. the points in Exercise 2.5.3.
    c. the points in Exercise 2.5.4.
    d. the points in Exercise 2.5.6.
    e. the points in Exercise 2.5.12.
    f. the points in Exercise 2.5.17.

21. From Exercise 2.5.17 we find that given a point $(a, b) \neq (0, 0)$, there is a unique point $(a/(a^2 + b^2)^{1/2}, b/(a^2 + b^2)^{1/2})$ on the unit circle. Show that this point is on the line passing through the origin and (a, b). Show further that the new point is on the ray through (a, b) emanating from $(0, 0)$. In this sense we are tying each point different from the origin to a single point on the unit circle. In fact, all points on a given ray emanating from the origin are related to the same point. From this idea, discover how each point (not the origin) is related to α. [*Hint:* Show that for each $(x, y) \neq (0, 0)$, $(x, y) = (rp, rq)$ where $(x^2 + y^2)^{1/2} = r$ and $(p, q) = \alpha(a)$ for some real number a.]

2.6 SOME REAL FUNCTIONS

We wish now to examine a few real functions that will be the basis for most of our studies of the elementary functions. Recall that certain exercises dealt with i_R, i_R^2, and various other powers and scalar products of powers of that *identity* function. We wish to further examine such functions.

Let $n \in N$ and let $i_R^n: R \to R$ be denoted by $(\ )^n: R \to R$. The reason for this notation is simple: $(\ )^n(x) = (x)^n = x^n$. The notation makes clear the substitution process used in evaluation of functions.

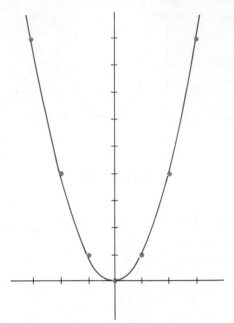

Figure 2.6.1 $f(x) = x^2$.

EXAMPLE 2.6.1.   Consider $(\ )^2 : R \to R$ and its graph.

Now, for each real number x, $(\ )^2$ has the value x^2. That is, $(\ )^2 = \{(x, x^2) : x \in R\}$. Thus $(1, 1)$, $(0, 0)$, $(2, 4)$, $(3, 9)$, and so forth belong to $(\ )^2 = i_R^2$. Furthermore, $x^2 = (-x)^2$, whence $(\ )^2$ is an even function and is "symmetric" about the y-axis. Consequently, $(-1, 1)$, $(-2, 4)$, and $(-3, 9)$ are points of the graph (function). Figure 2.6.1 shows the location of these sample points and the "rough in" of the rest of the function.

How accurate is the "rough in"? It tells in general what the function is doing. As $x > 0$ increases, x^2 increases. As $x < 0$ decreases, x^2 again increases. The figure points out these facts together with the idea of symmetry. Very little else is given (except for the location of the few points determined).

Figures 2.6.2 and 2.6.3 show sketches of $(\ )^4$ and $(\ )^6$. Note how they resemble the graph of $(\ )^2$. Both $(\ )^4$ and $(\ )^6$ are symmetric with respect to the y-axis, share the points $(1, 1)$, $(-1, 1)$, and $(0, 0)$ and have the same general shape (relative to the concept of "rising" and "falling").

If n is an even (positive) integer, $(\ )^n$ is an even function and has the same general shape as $(\ )^2$, sharing the points $(1, 1)$, $(-1, 1)$, and $(0, 0)$.

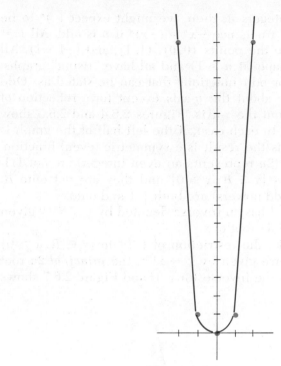

Figure 2.6.2 $f(x) = x^4$.

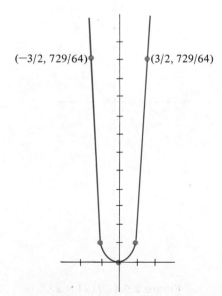

Figure 2.6.3 $f(x) = x^6$.

For odd positive integers n, then, we might expect $(\ )^n$ to be an odd function. This is true since $-x^n = (-x)^n$ if n is odd. All $(\ )^n$ for odd, positive n share the points $(0,0)$, $(1,1)$, and $(-1,-1)$. All have much the same shape (if $n > 1$) and all have "rising" graphs. There is a symmetry for odd functions that can be stated as: Odd functions are symmetric about the y-axis except for a reflection of the negative portion about the x-axis. Figures 2.6.4 and 2.6.5 show graphs of $(\ )^3$ and $(\ )^5$. In each case, if the left half of the graph is reflected about the x-axis the result is a symmetric (even) function.

The functions $(\ )^{2n}$ ($2n$ represents an even integer) are not $1\!:\!1$. Moreover, their range is $\{x \in R : x \geqslant 0\}$ and they are not onto R. The functions $(\ )^{2n+1}$ (odd powers) are both $1\!:\!1$ and onto.

Consequently, $(\ )^{2n+1}$ has an inverse denoted by $(\ )^{1/(2n+1)}$ given by $x \rightarrow x^{1/(2n+1)}$, the $(2n + 1)$ root of x.

Similarly, if we make the restriction of $(\ )^{2n}$ to $\{x \in R : x \geqslant 0\}$ $= P$, $(\ )^{2n}|P$ has an inverse given by $x \rightarrow x^{1/2n}$, the *principal* $2n$ root of x. Figure 2.6.6 shows the inverse for $(\ )^3$ and Figure 2.6.7 shows

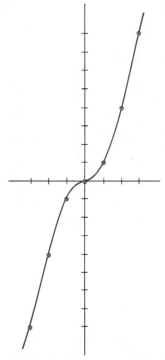

Figure 2.6.4 $f(x) = x^3$.

( )²|P and [( )²|P]⁻¹. Observe that the graphs of the inverse images can be formed from the graph of the functions by fixing the *y*-axis horizontally and the *x*-axis vertically.

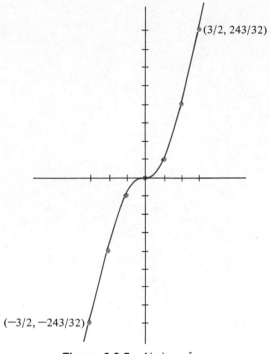

Figure 2.6.5 $f(x) = x^5$.

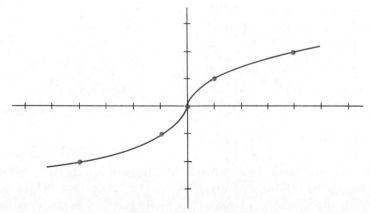

Figure 2.6.6 $f(x) = x^{1/3}$.

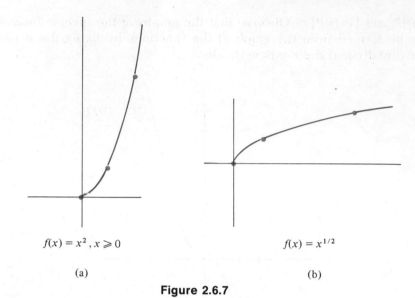

$$f(x) = x^2, x \geqslant 0 \qquad\qquad\qquad f(x) = x^{1/2}$$

(a)   (b)

Figure 2.6.7

EXERCISES 2.6

1. Sketch a graph for each of the functions f given below.

   a. $f(x) = 3x$ d. f^{-1} from part a
   b. $f(x) = x^2/2$ e. f^{-1} from part c
   c. $f(x) = -x^3/3$

2. Find each of the following.

   a. $f(7)$, f in Exercise 2.6.1a.
   b. $f^{-1}(4)$, f in Exercise 2.6.1a.
   c. $f(3)$, f in Exercise 2.6.1b.
   d. $f(-2)$, f in Exercise 2.6.1b.
   e. $f(2)$, f in Exercise 2.6.1c.
   f. $f(-2)$, f in Exercise 2.6.1c.
   g. $f^{-1}(81)$, f in Exercise 2.6.1c.
   h. $f^{-1}(-81)$, f in Exercise 2.6.1c.

3. Let $f = (\ )^{2n+1}$, $n \in N$. Show that f^{-1} is an odd function.

4. Let $f(x) = ax^n$, $g(x) = bx^k$; $n, k \in N$. When is $h = f + g$ odd? Even?

2.7 EXPONENTIAL FUNCTIONS

A very important class of real functions consists of those of the type suggested by the equality $f(x) = a^x$, $a > 0$, $a \neq 1$. The function f (having domain R and range $\{y \in R: y > 0\}$) is an *exponential* function having base a. It is assumed that we understand the import of the symbol a^x, even though our experiences are limited to the special case that x is rational.

The rules for exponential behavior dictate some behavior patterns for f (as above).

EXAMPLE 2.7.1. If $f(x) = a^x$, $a > 0$, $a \neq 1$, then $f(x + y) = f(x) f(y), f(x - y) = f(x)/f(y)$, and $f(nx) = [f(x)]^n$. These are simply seen to result from $a^x a^y = a^{x+y}$, $a^x/a^y = a^{x-y}$, and $(a^x)^n = a^{xn}$ respectively. A graph of f is given in Figure 2.7.1 for $a > 1$ and in Figure 2.7.2 for $0 < a < 1$.

The general behavior of the exponential functions is linked to that of two other classes of functions. A real function f is *increasing* if for each x and y in its domain, $x \leq y$ implies that $f(x) \leq f(y)$. Similarly, a real function g is *decreasing* if whenever domain elements x and y satisfy $x \leq y$, it follows that $g(x) \geq g(y)$.

It is (intuitively) clear that f is an increasing function if $a > 1$ [if $x < y$, $a^y = a^x(a^{y-x}) > a^x$ since $a^{y-x} > 1$] and decreasing if $0 < a < 1$ [if $x < y$, $a^y = a^x(a^{y-x}) < a^x$ since $0 < a^{y-x} < 1$]. Thus f is a 1:1 function and has an inverse. The inverse for f is from a class of functions as important as the class to which f itself belongs; we discuss such functions in the next section.

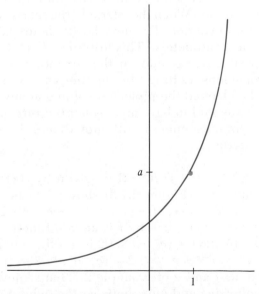

Figure 2.7.1 $f(x) = a^x$, $a > 1$.

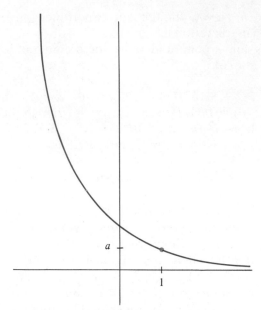

Figure 2.7.2 $f(x) = a^x$, $0 < a < 1$.

EXAMPLE 2.7.2. A student observes that his "live mass" is expanding in volume. When he started, the mass was about one cubic centimeter in volume. The next day (24 hours later) the volume was three cubic centimeters. This tripling effect lasted for seven days. His experiment was due on the seventh day along with his report and projections. Letting t be the time in days after the experiment started, he claimed the projected volume at any time t to be 3^t. Thus the mass was said to have an *exponential rate of growth*.

This is not an uncommon result; growth and decay are known to follow such patterns.

EXAMPLE 2.7.3. Let $f: R \to R$ be given by $f(x) = (e^x - e^{-x})/2$, $1 < e$. The graph of f (Figure 2.7.3) shows f to be increasing and 1:1, and $f(0) = 0$. Moreover, $f(-x) = (e^{(-x)} - e^{-(-x)})/2 = (e^{-x} - e^x)/2 = -(e^x - e^{-x})/2 = -f(x)$ whereby f is an *odd* function. The function $g: R \to R$ given by $g(x) = (e^x + e^{-x})/2$ is, on the other hand, an *even* function $(g(-x) = (e^{-x} + e^{-(-x)})/2 = (e^{-x} + e^x)/2 = g(x))$.

The functions f and g of Example 2.7.3 are known respectively as the *hyperbolic sine* and *hyperbolic cosine* (where e is a particular irrational number slightly greater than 2.7). The expressions $f(x)$ and $g(x)$ are written *sinh x* and *cosh x*, respectively. The graph of the

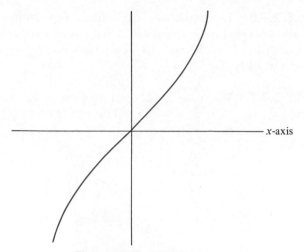

Figure 2.7.3 $f(x) = \sinh x$.

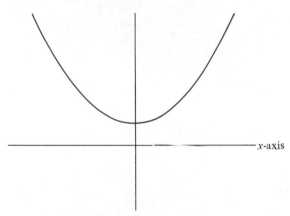

Figure 2.7.4 $f(x) = \cosh x$.

hyperbolic cosine as seen in Figure 2.7.4 illustrates the nonnegative-ness of that function ($\cosh x > 0$).

EXAMPLE 2.7.4. The rules of exponents yield some unexpected results: $\cosh^2 x - \sinh^2 x = [(e^x + e^{-x})/2]^2 - [(e^x - e^{-x})/2]^2 = [(e^x)^2 + 2e^x e^{-x} + (e^{-x})^2]/4 - [(e^x)^2 - 2e^x e^{-x} + (e^{-x})^2]/4$. However, $e^x e^{-x} = e^0 = 1$, whence $\cosh^2 x - \sinh^2 x = 1$.

EXAMPLE 2.7.5. Equations such as $3^{3x} = 9^{x+1}$ can be easily solved. We need observe two things. First, $9 = 3^2$ whereby $9^{x+1} = (3^2)^{x+1} = 3^{2(x+1)}$. Secondly, $a^x = a^y$ if and only if $x = y [a^x = f(x)$ is a 1:1 function]. Thus $3x = 2(x + 1) = 2x + 2$ or $x = 2$.

EXAMPLE 2.7.6. Inequalities are (like equations) solvable because of the increasing or decreasing nature of exponential functions. For example, $7^x < 49$ can be rewritten in the form $7^x < 7^2$. This holds if and only if $x < 2$.

EXAMPLE 2.7.7. We can graph solution sets for inequalities [such as $\{(x, y) : -1 < x < 2, y > 3^x)\}$]. The set $\{(x, y) : y > 3^x\}$ con-

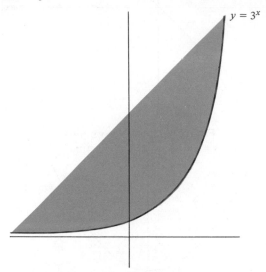

$y = 3^x$

Figure 2.7.5 $\{(x, y) : y > 3^x\}$.

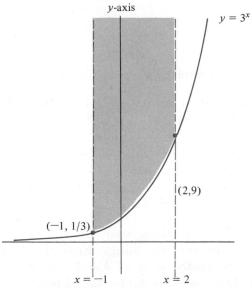

y-axis

$y = 3^x$

$(2,9)$

$(-1, 1/3)$

$x = -1$ $x = 2$

Figure 2.7.6 $\{(x, y) : -1 < x < 2, y > 3^x\}$.

sists of those points "above" the graph of $y = 3^x$. This is seen in Figure 2.7.5, while the more restricted set $\{(x, y) : -1 < x < 2,\ y > 3^x\}$ is shown in Figure 2.7.6.

EXERCISES 2.7

1. Sketch graphs of μ where $\mu(x)$ is given by

   a. 2^x c. 1^x e. 5^x
   b. $(1/2)^x$ d. 3^x f. $(-2)^x$

2. Evaluate each of the exponential functions from Exercise 2.7.1 where x has the value:

   a. 1   c. 2   e. 1/2
   b. 0   d. -2 f. $-1/2$

3. Superimpose graphs of the functions given by the equations: $f(x) = 2^x$, $g(x) = 2^{(x+1)}$, $h(x) = 2^{(x-1)}$, and $t(x) = 2^{-x}$.

4. Superimpose graphs of f and f^{-1} where $f(x) = 5^x$. Give the domain for f^{-1}.

5. From Exercise 2.7.4 compute the following values.

   a. $f(0)$ d. $f(-2)$ g. $f(3)$
   b. $f(-1)$ e. $f(2)$ h. $f(-4)$
   c. $f(1)$ f. $f(-3)$ i. $f(4)$

6. From Exercises 2.7.4 and 2.7.5 calculuate:

   a. $f^{-1}(1)$ d. $f^{-1}(1/25)$ g. $f^{-1}(125)$
   b. $f^{-1}(1/5)$ e. $f^{-1}(25)$ h. $f^{-1}(1/625)$
   c. $f^{-1}(5)$ f. $f^{-1}(1/125)$ i. $f^{-1}(625)$

7. Given values $f(x)$ and $f(x + 1)$ for $f(x) = a^x$, how would you determine a?

8. Using the fact that exponential functions are $1:1$, solve the following equations.

   a. $2^{3x} = 2^{x+1}$ d. $5^{7x} = 1$
   b. $10^{x+1} = 10^{2x-1}$ e. $27(3^x) = 81$
   c. $3^{3x} = 9^{x+1}$ f. $2(8^{x+1}) = 4^{x-1}$

9. Find the values of x satisfying each of the following inequalities.

   a. $2^x < 1$ c. $(1/2)^x > 1$
   b. $3^x > 1/8$ d. $(1/3)^x < 243$

10. Graph each of the following point sets.

   a. $\{(x, y) : y \leqslant 2^x\}$ c. $\{(x, y) : 0 \leqslant x \leqslant 3,\ y \leqslant 10^x\}$
   b. $\{(x, y) : y \geqslant 2^{x-1}\}$ d. $\{(x, y) : |x| > 1,\ y > 3^x\}$

11. Define the hyperbolic tangent, cotangent, secant, and cosecant by: $\tanh x = \sinh x/\cosh x$, $\coth x = 1/\tanh x$, $\operatorname{sech} x = 1/\cosh x$, and $\operatorname{csch} x = 1/\sinh x$. (See Examples 2.7.3 and 2.7.4.)

a. Graph each of the hyperbolic functions defined above. Show that the following hold:
b. $\tanh^2 x + \mathrm{sech}^2 x = 1$
c. $\coth^2 x - \mathrm{csch}^2 x = 1$
d. $\sinh(x + y) = \sinh x \cosh y + \cosh x \sinh y$
e. $\cosh(x + y) = \cosh x \cosh y + \sinh x \sinh y$
f. $\tanh(x + y) = (\tanh x + \tanh y)/(1 + \tanh x \tanh y)$
g. The hyperbolic secant, tangent, and cotangent are odd functions.
h. The hyperbolic secant is an even function.

2.8 LOGARITHMIC FUNCTIONS

In the last section we viewed the exponential functions, noting that they were 1:1 and had range $P = \{y \in R : y > 0\}$. In this section we examine the inverses for those functions.

Let $f(x) = a^x$, $a > 0$, $a \neq 1$. We denote f^{-1} by $\log_a(\ )$. (See Figure 2.8.1.) That is, $f^{-1}(x) = \log_a(x)$, $(\log_a : P \to R)$. The *logarithmic* functions — those of the form $\log_a(\ )$ — are necessarily 1:1 functions. The following examples illustrate some further properties. Figures 2.8.2 and 2.8.3 show graphs of $\log_a(\ )$ for $a < 1$ and $a > 1$.

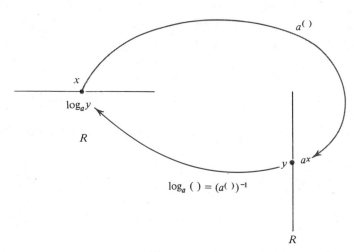

Figure 2.8.1

EXAMPLE 2.8.1. Let $f(x) = 3^x$. Then we compute $f^{-1}(y)$ in the following way. We write $y = 3^t$; that is, we write y as a power of the base 3. This illustrates the fact that $y = 3^t = f(t)$. Then $f^{-1}(y) = f^{-1}(3^t) = f^{-1}(f(t)) = t$. Thus, $f^{-1}(y)$ is the exponent for 3 yielding y.

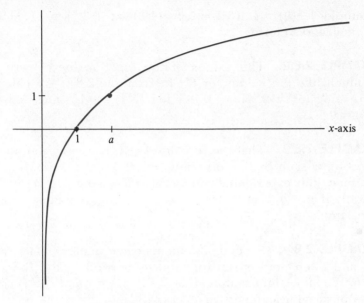

Figure 2.8.2 $f(x) = \log a^x$, $a > 1$.

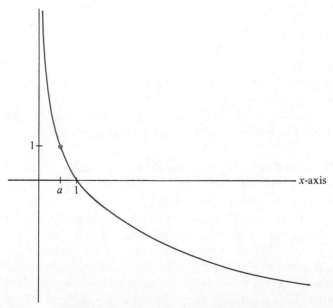

Figure 2.8.3 $f(x) = \log a^x$, $0 < a < 1$.

For example: $f^{-1}(9) = f^{-1}(3^2) = 2$, $f^{-1}(1/3) = f^{-1}(3^{-1}) = -1$, $f^{-1}(1) = f^{-1}(3^0) = 0$, and so forth.

EXAMPLE 2.8.2. The values of the function $\log_5(\ )$ are computed much like the values for f^{-1} in Example 2.8.1. We calculate: $\log_5(25) = \log_5(5^2) = 2$, $\log_5(1/5) = \log_5(5^{-1}) = -1$, and $\log_5(1) = \log_5(5^0) = 0$.

EXAMPLE 2.8.3. That the function $\log_a(\ )$, $a > 1$, is an increasing function is seen by the following. Let $0 < x < y$ with $x = a^p$ and $y = a^q$. Since this exponential function is increasing $(a > 1)$, $a^p < a^q$ if and only if $p < q$ (that is, $\log_a x = p < q = \log_a y$, whence, $\log_a(\ )$ is increasing).

EXAMPLE 2.8.4. Several statements occur in elementary mathematics and often appear startling unless viewed in the functional notation. We know, for example, that $f(f^{-1}(x)) = x$ and $f^{-1}(f(x)) = x$ for functions having an inverse. Necessarily, $\log_a a^x = x$ and $a^{\log_a x} = x$. [Write $f(x) = a^x$, $f^{-1}(x) = \log_a(x)$ and $\log_a a^x$ becomes $f^{-1}(f(x))$ while $a^{\log_a x}$ becomes $f(f^{-1}(x))$.]

EXAMPLE 2.8.5. We find that certain rules for the mechanical behavior of $\log_a(\ )$ exist just as was the case for the associated exponential function. Since $a^{x+y} = a^x a^y$, $\log(a^x a^y) = x + y = \log_a a^x + \log_a a^y$. Thus if $5 = a^{.7}$ and $2 = a^{.3}$, $\log_a 10 = \log_a(5 \cdot 2) = \log_a 5 + \log_a 2 = .7 + .3 = 1$. The statement $\log_a 10 = 1$ means that $a^1 = 10$, whence $a = 10$.

The *logarithm of a product* of two numbers *is the sum of the two individual logarithms*.

Similarly, we can show that $\log_a x/y = \log_a x - \log_a y$ and $\log_a x^p = p \log_a x$. (See the exercises at the end of this section.)

EXAMPLE 2.8.6. The hyperbolic sine function discussed in the last section has an inverse. If $f(x) = \sinh x$, $f^{-1}(y) = \text{argsinh } y$, the *inverse hyperbolic sine* evaluated at y.

Let $x = \text{argsinh } y$. Then $\sinh x = \sinh (\text{argsinh } y) = y$. [Remember that $\sinh (\text{argsinh } y) = f(f^{-1}(y)) = y$.] Thus $y = (e^x - e^{-x})/2$ or $2y = e^x - e^{-x}$. Multiplying both sides by e^x, we find $2ye^x = (e^x)^2 - 1$ or alternatively, $(e^x)^2 - 2y(e^x) - 1 = 0$. Substituting z for e^x, we see the equation $z^2 - (2y)z - 1 = 0$. As we will find in the next chapters, $z = e^x$ must be either $y + [y^2 + 1]^{1/2}$ or $y - [y^2 + 1]^{1/2}$. However, $e^x > 0$ dictates that $z = e^x = y + [y^2 + 1]^{1/2}$. Now, $x = \log_e e^x = \log_e(y + [y^2 + 1]^{1/2})$. That is, argsinh $y = \log_e(y + [y^2 + 1]^{1/2})$. That the inverse of this hyperbolic function is a logarithmic function was to have been anticipated since the hyperbolic function involves the exponential.

Equations and inequalities can be handled much as with the exponential function.

EXAMPLE 2.8.7. The equation $\log_2 x = \log_8 y$ is handled by looking at equivalent statements involving exponentials. Now $\log_2 x = z = \log_8 y$ implies $2^z = x$ and $y = 8^z = (2^3)^z = 2^{3z} = (2^z)^3 = x^3$. Recapping, we see that $x^3 = y$. Moreover, $8 = 2^3$ might suggest that there is a relationship existing that is dependent on the bases involved.

EXAMPLE 2.8.8. Find $\log_a x$ where $\log_b x$ and $\log_a b$ are known. Now, $b = a^{\log_a b}$ whereby $x = b^{\log_b x} = (a^{\log_a b})^{\log_b x}$ and $\log_a x = \log_a (a^{\log_a b \, \log_b x}) = \log_a b \, \log_b x$. We might say that the term $\log_a b$ is a "conversion" factor. We have shown the relationship

$$\log_a x = (\log_a b)(\log_b x)$$

Similarly, $a^x = (b^{\log_b a})^x = b^{x \, \log_b a}$, forming a means of converting from one exponential base to another.

EXAMPLE 2.8.9. The inequality $\log_a(x + 1) \geq \log_a(2x)$, $a > 1$ is solved by observing that $\log_a(\ )$ is an increasing function. Thus, $\log_a(x + 1) \geq \log_a(2x)$ if and only if $x + 1 \geq 2x$ or $1 \geq x$. Figure 2.8.4 illustrates this fact by showing the superimposed curves.

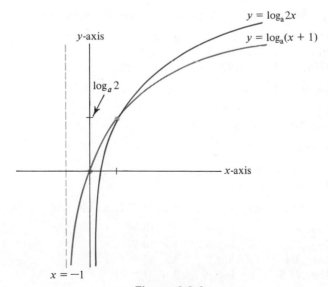

Figure 2.8.4

EXAMPLE 2.8.10. A graph of $\{(x, y): \log_2(x + 1) \leqslant y \leqslant \log_2(2x - 1)\}$ is seen to be a graph of $\{(x, y): \log_2(x + 1) \leqslant y\} \cap \{(x, y): y \leqslant \log_2(2x - 1)\}$. The former of these sets is that part of $R \times R$ "above" the curve $y = \log_2(x + 1)$ while the latter is that portion "below" $y = \log_2(2x - 1)$. Figure 2.8.5 illustrates.

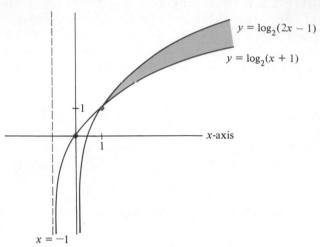

Figure 2.8.5 $\{(x, y): \log_2(x + 1) \leqslant y \leqslant \log_2(2x - 1)\}$

EXERCISE 2.8

1. Let $f = \log_2(\ )$ and $g = \log_7(\ )$. Compute the following:

    a. $f(1)$ d. $g(343)$ g. $f^{-1}(5)$
    b. $g(1)$ e. $f(1/1024)$ h. $g^{-1}(2)$
    c. $f(16)$ f. $g(1/49)$ i. $gf(128)$

2. Evaluate:

    a. $3^{\log_3 5}$ c. $9^{\log_3 5}$
    b. $\log_3 3^{.79}$ d. $\log_3 27^{.62}$

3. Show that:

    a. $\log_a(x/y) = \log_a x - \log_a y$
    b. $\log_a x^p = p \log_a x$

4. Knowing that $\log_{10} 2 = .301$, $\log_{10} 3 = .477$, and $\log_3 10 = 2.10$, compute:

    a. $\log_{10} 4$ e. $\log_3 2$ i. $10^{.301}$ m. $27^{2.10}$
    b. $\log_{10} 6$ f. $\log_3 6$ j. $10^{.477}$
    c. $\log_{10} 3/4$ g. $\log_9 2$ k. $3^{2.10}$
    d. $\log_{10} 27/8$ h. $\log_{27} 6$ l. $9^{4.20}$

5. Use $\log_{10} 7 = .845$ to calculate:

   a. $\log_{10} 7000$ c. $\log_{10} .07$ e. $\log_{10}(7 \cdot 10^n)$
   b. $\log_{10} 70$ d. $\log_{10} .0007$

6. Solve the following for the appropriate values of x.

   a. $\log_2 x = 6$ d. $\log_5 x + \log_5 x^2 = 3$
   b. $\log_2(x - 1) + \log_2 5 = 0$ e. $\log_5(2x - 1) - \log_5(x + 1) = 0$
   c. $\log_2 x + 3 \log_4 x = 10$

7. Find the solution set for each of the following.

   a. $\log_2 x \leqslant \log_2(3x - 1)$
   b. $\log_3(2x - 1) \geqslant \log_3(x + 1)$
   c. $\log_{1/5}(x + 1) \leqslant \log_{1/5}(2x - 2)$
   d. $\log_3(4x - 1) > \log_9 x^2$

8. Sketch a graph of the following sets.

   a. $\{(x, y): 2 < y < \log_3 x\}$
   b. $\{(x, y): 3 < x < 5, \log_3 x < y < \log_2 x\}$

2.9 REVIEW EXERCISES

1. What is the completeness property for the real numbers?

2. Show that if $A \subset R$ is bounded, A is contained in a closed interval. What is the "smallest" closed interval containing A?

3. Under what conditions is

   a. $[a, b] \cup [c, d]$ an interval?
   b. $[a, b] \cap [c, d]$ an interval?
   c. $[a, b] - [c, d]$ an interval?

4. What is (a, a)? $[a, a]$?

5. Using the information from Appendix B if necessary, sketch solution sets for each inequality:

   a. $\{x \in R: 2x < 4\}$
   b. $\{x \in R: x + 7 < 2\}$
   c. $\{x \in R: 3x - 2 < 4\}$

6. Which of the solution sets in Exercise 2.9.5 are bounded above? bounded below? bounded? List all l.u.b.'s and g.l.b.'s for these sets.

7. Find the distance between the following pairs of points.

   a. $(0, 0), (5, 12)$ c. $(2, 1), (-5, -23)$
   b. $(1, 2), (9, -3)$ d. $(3, 5)(2, 6)$

8. Sketch graphs of $\{(x, y): x \text{ and } y \text{ satisfy the conditions specified below}\}$.

   a. $0 < x < 1, 0 < y < x$ c. $-2 \leqslant x \leqslant 1, 0 \leqslant y \leqslant [x]$
   b. $-1 < x < 0, y \leqslant x$

9. Sketch a representation of the following binary operations on R.

    a. $f: R \times R \to R$ given by $f(a, b) = 2a + b$
    b. $f: R \times R \to R$ given by $f(a, b) = a^2 - b^2$
    c. $f: R \times R \to R$ given by $f(a, b) = a^3 + b^3$

10. Are any of the functions in Exercise 2.9.9 1:1? onto?

11. Write f from b. and c. of Exercise 2.9.9 as a product of two functions $h: R \times R \to R$ and $k: R \times R \to R$.

12. Sketch a graph of the following functions.

    a. $f: R \to R$ given by $f|[0, 1] = i_{[0, 1]}$ and f is periodic of period 1.
    b. $f: R \to R$ given by $f|[0, 2] = [\ ]|_{[0, 2]}$ and f is periodic of period 2.
    c. $f: R \to R$ given by $f|[-1, 1] = |\ ||_{[-1, 1]}$ and f is periodic of period 2.
    d. $f: R \to R$ given by $f(x) = 1/x$ if $-1 < x < 0$ or $0 < x < 1$, $f(0) = 0$, and f is periodic of period 2.

13. Are the functions from Exercise 2.9.12 bounded? even? odd? onto? 1:1?

14. Give the ranges of the functions from Exercise 2.9.12.

15. Write the equations of the following loci.

    a. Circle of center $(2, -3)$ and radius 2.
    b. Circle of center $(1, 1)$ and passing through $(6, 13)$.
    c. $\{(x, y) = P: d(P, P_0) + d(P, P_1) = 5, P_0 = (-2, 0), P_1 = (2, 0)\}$
    d. $\{(x, y) = P: d(P, P_0) = x + 2, P_0 = (2, 0)\}$
    e. $\{(x, y) = P: d(P, P_0) = d(P, P_1) = 2, P_0 = (-1, 0), P_1 = (1, 0)\}$

16. Find the length of the arc given by

    a. 1/2 the circle of Exercise 2.9.15a.
    b. 1/8 the circle of Exercise 2.9.15b.

17. Find the following points.

    a. $\alpha(-3\pi/4)$ d. $\alpha(-79\pi/6)$
    b. $\alpha(491\pi/6)$ e. $\alpha(-31\pi/4)$
    c. $\alpha(231\pi/3)$ f. $\alpha(596\pi)$

18. Find $\alpha^{-1}[A]$ where A is the set:

    a. $A = \{\alpha(\pi/4)\}$
    b. $A = \{\alpha(3\pi/4)\}$
    c. $A = \{\alpha(5\pi/4)\}$
    d. $A = \{\alpha(-\pi/4)\}$
    e. $\{(1/\sqrt{2}, 1/\sqrt{2}), (-1/\sqrt{2}, -1/\sqrt{2}), (-1/\sqrt{2}, 1/\sqrt{2}), (1/\sqrt{2}, -1/\sqrt{2})\}$
    f. $\{(0, 1), (0, -1)\}$
    g. $\{(1, 0), (-1, 0)\}$

19. Find $f^{-1}[\{0\}]$ where f is as in each case of Exercise 2.9.12.

20. Find $f^{-1}[\{0\}]$ where f is:

    a. $[\ ]$ b. $|\ |$ c. i_R

21. Let $f(x) = x^2$ [that is, $f = (\ )^2$], $g(x) = x$, and $h(x) = 2$ be functions from R into R. Calculate

   a. $f(0)$ g. $(f+g)(0)$ m. $f \circ g(x)$
   b. $g(0)$ h. $(f+g)(2)$ n. $g \circ f(x)$
   c. $h(0)$ i. $(fg)(2)$ o. $g \circ h(x)$
   d. $f(2)$ j. $(fh)(2)$ p. $h \circ g(x)$
   e. $g(2)$ k. $(f/g)(2)$ q. $f \circ h \circ g(x)$
   f. $h(2)$ l. $(f/g)(0)$ r. $g \circ h \circ f(x)$

22. Find f, g, and h from Exercise 2.9.21.

   a. $(g+h)^{-1}[\{0\}]$ d. $(f-h)^{-1}[\{0\}]$
   b. $(fg)^{-1}[\{0\}]$ e. $(f^2)^{-1}[\{x\}] = (ff)^{-1}[\{x\}]$
   c. $(f+g)^{-1}[\{0\}]$ f. $[(g+h)/h]^{-1}[\{0\}]$

23. Calculate

   a. $\log_2 64$ e. $\log_3 \sqrt{27}$
   b. $\log_2 1/32$ f. $\log_3 1$
   c. $\log_{10} .0001$ g. $\log_3 3^a$
   d. $\log_{10} \sqrt{10}$ h. $3^{\log_3 a}$ where $a > 0$

24. Sketch

   a. $\{(x, y): 1 \leqslant x \leqslant 2, 0 < y < \log_2 x\}$
   b. $\{(x, y): 0 \leqslant x \leqslant 1, x^2 \leqslant y \leqslant x\}$
   c. $\{(x, y): 0 \leqslant x \leqslant 1, x \leqslant y \leqslant 2^x\}$

25. Show that the following hold.

   a. $\sinh(x - y) = \sinh x \cosh y - \cosh x \sinh y$
   b. $\cosh(x - y) = \cosh x \cosh y - \sinh x \sinh y$
   c. $\tanh(x - y) = (\tanh x - \tanh y)/(1 - \tanh x \tanh y)$

2.10 QUIZ

Work as much of the quiz as you can in one hour and then check your solutions.

1. Is a constant function even? Odd? Periodic? Bounded? Does a constant function have an inverse?

2. Give the definitions for:

   a. The open interval from a to b $(a < b)$
   b. $(-\infty, b)$
   c. $|x|$
   d. $[x]$
   e. $f: R \to R$ is periodic of period p
   f. $f: A \to R$ is bounded
   g. A binary operation on R

3. Let $f = (\ )$, $g = (\ )^2$, $t = k(k \in R)$, and let α, p_1, and p_2 be as in the text. Calculate:

   a. $f(x), f(x + h)$, and $[f(x + h) - f(x)]/h$ for $h \in R$
   b. $g(x), g(x + h)$, and $[g(x + h) - g(x)]/h$ for $h \in R$
   c. $t(x), t(x + h)$, and $[t(x + h) - t(x)]/h$ for $h \in R$
   d. $p_2 \circ \alpha \circ f(\pi/3)$
   e. $p_1 \circ \alpha \circ g(\sqrt{\pi})$
   f. $f \circ g(2)$
   g. $(f + g)(3)$
   h. $(g - t)(0)$
   i. $(gt)(1)$
   j. $(g/t)(5)$

4. Sketch $\{(x, y) : 0 \leqslant x \leqslant 2, 0 \leqslant y \leqslant f(x)\}$ where $f(x) = [x]$ if $0 \leqslant x \leqslant 1/2$, $f(x) = 1 - x$ if $1/2 < x \leqslant 1$, and f has period 1.

5. Solve the equations:

   a. $\log_2 x + \log_2 2^{.7} = 1.7$ b. $e^x = e^{2x-1}$

2.11 ADVANCED EXERCISES

1. If $f: R \to Y$ has periods p and q, then $p + q$, np, and $np + mq$ are also periods for f with $n, m \in Z$. $p + q$ was shown to be a period. If we can show np to be a period also, we know that mq is a period as well, and $np + mq$ is necessarily a period. Show, then, that np is a period for f. [*Hint:* The theorem is obvious for $n = 0$. Use induction to show that np is a period for f where $n \in N$. It follows that $-np$ is a period since $f(x - np) = f((x - np + np)) = f(x)$. In particular, since 2π is a period, the exercise shows that $\alpha(x + 2n\pi) = \alpha(x)$, that is, $2n\pi$ is a period for α if $n \in Z$.

2. Consider the point $(1, 3)$. How can we describe by equation the set of all circles of radius 5 that pass through $(1, 3)$? [*Hint:* each circle has equation $(x - h)^2 + (y - k)^2 = 25$. What are the conditions describing h and k?]

3. In Figure 2.11.1 we see that a finite number of points of the circle $x^2 + y^2 = r^2$ are chosen and that the line segment connecting "successive" points forms an inscribed polygon. The perimeter of the polygon is the sum of the lengths of its sides. Define:

   $S = \{p \in R : p$ is the perimeter of a polygon inscribed in the circle $x^2 + y^2 = r^2\}$

   Now S is the set formed of the perimeters of all polygons inscribed in our circle. Show that S is bounded above by

   (a) showing that each polygon as in Figure 2.11.1 has perimeter not greater than the associated polygon in Figure 2.11.2, and

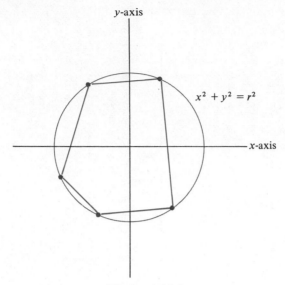

Figure 2.11.1

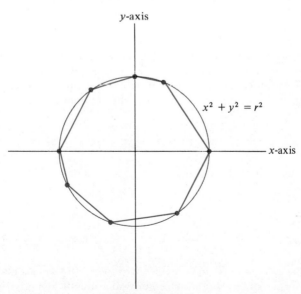

Figure 2.11.2

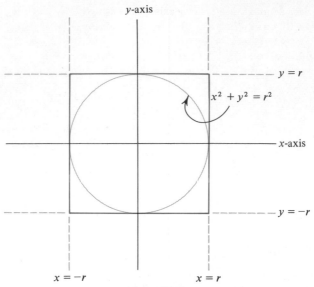

Figure 2.11.3

(b)   showing that each polygon in Figure 2.11.2 has perimeter less than the square in Figure 2.11.3.

Let C be the l.u.b. of S and define $C/2r = \pi$, whence we have the equation $C = 2\pi r$.

4.   The following is a symbolism widely used in mathematics.

$$\sum_{i=1}^{n}(a_i) = a_1 + a_2 + a_3 + \cdots + a_n$$

This is read, "The sum of the a_i as i goes from 1 to n." In other words, the *index* i is allowed to take on all integral values from the lower value 1 to the upper value n. The *sum* of all expressions obtained in this process is the value indicated by the Σ notation on the left. As an example,

$$\sum_{i=1}^{5}(1/i) = 1/1 + 1/2 + 1/3 + 1/4 + 1/5 = 137/60$$

Evaluate each of the following.

a. $\displaystyle\sum_{i=0}^{10}(i)$ c. $\displaystyle\sum_{i=1}^{4}(i^3)$ e. $\displaystyle\sum_{j=1}^{3}(j^2 - j)$

b. $\displaystyle\sum_{i=1}^{7}(i^2)$ d. $\displaystyle\sum_{i=1}^{5}\left(\frac{i^3 - 1}{i^2}\right)$ f. $\displaystyle\sum_{k=1}^{5}[(k) - (k+1)]$

5.   The spaces R^1 and R^2 have been introduced in the text material. These are two special cases of a more general concept, namely the *Euclidean n-space*, R^n. The generalization is completely analogous to what has

been done. A point in R^1, for example, was a single real number while a point in R^2 was an ordered pair of real numbers. What, then, is a point in R^3? It must be an ordered triple of real numbers; one such being $(3, -1, 0)$. The number 3 is the first coordinate value, -1 the second, and 0 the third. The concept of order in the triple is the same as that in the ordered pair of the plane. R^3, then, is the set of all ordered triples of real numbers.

For the completely general case, let n be any positive integer. The following symbol is known as an *n-tuple* of real numbers:

$$(x_1, x_2, \ldots, x_n)$$

The designation n-tuple comes from the fact that there are n of the subscripted x's appearing. The term "real numbers" in the name is a result of each subscripted x being a real number. For clarification:

1. $(1, -2, 3)$ is an ordered triple of real numbers.
2. $(-1/2, 14)$ is an ordered pair.
3. $(-1, 6, 15, -3, 1/2, 56, -90, 12)$ is an ordered 8-tuple.

Each position in the symbol (position being given by the subscript or by counting from left to right) is called a *coordinate* of the n-tuple. The quadruple (x_1, x_2, x_3, x_4) has first coordinate value x_1, second coordinate value x_2, and so on. The values of the numbers in the coordinate positions are also called *projections* (onto the respective axis). In the quadruple just above, x_1 is the projection onto the first axis, x_2 the projection onto the second axis, and so on. The significance (geometric and intuitive) of the axis will be seen shortly.

Two n-tuples are said to be *equal* if and only if they are equal coordinate by coordinate. This means that $(1, 2, 3)$ and $(3, 2, 1)$ are not the same (equal) pairs since they differ in the first and third coordinates. In this sense the order of the numbers (order of appearance) in the n-tuple is important. For this reason such a definition makes the n-tuples into *ordered* n-tuples of real numbers.

Now, R^n denotes: $\{(x_1, x_2, \ldots, x_n) : (x_1, x_2, \ldots, x_n)$ is an ordered n-tuple of real numbers$\}$. The ordered n-tuples will be, as one might suppose, the points in the space. The n in R^n denotes the dimension of the space. The mathematician thinks nothing of working a space of dimension 3,000,000,000,000,000. He may have no suggestive model but the consequences of such a deficiency are not of major proportions.

A distance concept can be extended in a manner completely analogus to that for R^1 and R^2. Using the notation of Exercise 2.11.4, we may define *distance* relative to R^n.

Let $P_1 = (x_1, x_2, x_3, \ldots, x_n)$ and $P_2 = (y_1, y_2, \ldots, y_n)$. Then:

$$d(P_1, P_2) = \left[\sum_{i=1}^{n} (x_i - y_i)^2 \right]^{1/2}$$

This distance is calculated in the same manner as done in R^1 and R^2. We simply:

1. compute the difference between the points coordinate by coordinate,
2. square each difference,
3. sum the squares, and finally,
4. compute the principal square root of this sum.

For illustration, find the distance between P_1 and P_2 where $P_1 = (1, -1, 3, 0, 1)$ and $P_2 = (2, 0, 0, 1, 3)$. (*Note:* P_1 and P_2 are in R^5.)

$$d(P_1, P_2) = [(1-2)^2 + (-1-0)^2 + (3-0)^2 + (0-1)^2 + (1-3)^2]^{1/2}$$
$$= [1 + 1 + 9 + 1 + 4]^{1/2} = \sqrt{16} = 4$$

Several fundamentals should be observed:

1. $d(P_1, P_2) \geqslant 0$ since the principal square root is indicated.
2. $d(P_1, P_2) = 0$ if and only if $P_1 = P_2$.
3. $d(P_1, P_2) = d(P_2, P_1)$.
4. $d(P_1, P_2) \leqslant d(P_1, P_3) + d(P_3, P_2)$, P_3 any point in R^n.

The proof of the validity of 1 is self-evident from the definition of principal square root. A distance satisfying 1–4 is called a *metric*.

a. Prove 2.
b. Prove 3.

A proof of 4 is quite complicated in nature and will not be asked of you. A proof can be found in many texts covering more advanced analysis. Property 4 is referred to quite often as the *triangle inequality*. The geometric reason for this is clear.

Find the distances between the following pairs of points (making sure both points in each pair are from the same dimension space).

c. $(1, 0, 3)$, $(2, -3, -1)$
d. $(1/2, 1/3, 1/4, -2)$, $(-1/2, -2/3, -3/4, -1)$
e. $(5, 0, 6, -2, 3, 7)$, $(1, 1, 1, 1, 1, 1)$
f. $(3, 0, 5, 6)$, $(4, 0, 2, 7, 1)$
g. For each point in c.–f., give the value of the projection onto the first axis projection onto the third axis and projection onto the fifth axis where appropriate.

Figure 2.11.4 shows a two-dimensional picture of a model of R^3. There are three axes (the same as the number of dimensions). A point $(1, 2, -2)$ is located. The procedure is analogous to that for locating points in a model of the plane, R^2.

Sketch a picture of a model of R^3 and locate the following points.

h. $(0, 0, 0)$ j. $(0, 1, 0)$ l. $(-1, -1, -1)$
i. $(1, 0, 0)$ k. $(0, 0, 1)$ m. $(4, 2, 3)$

6. In the plane define a new *distance*. Let $P_1 = (x, y)$ and $P_2 = (a, b)$. Define the distance between P_1 and P_2 by

$$t(P_1, P_2) = |x - a| + |y - b|$$

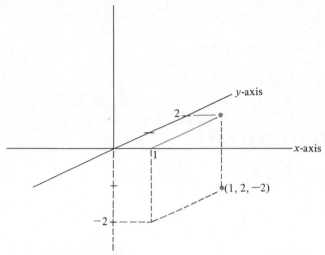

Figure 2.11.4 $R^3 = R \times R \times R$.

This is sometimes called the "taxicab metric." As an example, the distance, in this context, between $(1, 5)$ and $(2, 6)$ is given by $|1 - 2| + |5 - 6| = 1 + 1 = 2$.

Find the distances (in the taxicab metric) between the following pairs of points.

a. $(1, 6)$, $(2, 8)$
b. $(0, 0)$, $(3, 5)$
c. $(3, 5)$, $(2, -3)$
d. $(2, 1)$, $(-3, -6)$
e. Show that the taxicab metric has all the properties of a metric.

CHAPTER 3

Polynomial Functions

INTRODUCTION

In this section we propose to examine a certain collection of real functions known as *polynomials*. In particular, we will be interested in *zeros* of the polynomials (that is, numbers for which the polynomial is zero in value) and graphs. Inequalities involving polynomials will also come under investigation. In the process of studying these concepts, we will have occasion to look at the ideas of factoring and division of polynomials.

Since we normally graph functional values as the second coordinate values, a zero for a function is a point of the form $(x, 0)$ and therefore is a point of the x-axis (or horizontal axis). Thus when we find zeros for polynomials (or any other function) we are in reality finding points where the graph crosses the horizontal axis. We might expect that, conversely, graphs will prove useful in (approximately) locating zeros for polynomials. With these things in mind, let us introduce the concept of the polynomial.

3.1 POLYNOMIALS

In this section we wish to discuss a combination of items previously given. In particular, we examine the results of combining functions of the form ( )^n by the processes of addition, multiplication, and scalar multiplication.

First, let us define $(\ )^0 : R \to R$ by the correspondence $x \to 1$. That is, $(\ )^0$ is the constant function which mates every real number to the number 1.

A *monomial* function (or simply monomial) is a function of the form $k(\ )^n$ where $k \in R$ and $n \in N$ or $n = 0$. The real number k is the *coefficient* of the monomial.

EXAMPLE 3.1.1. Examples of monomials are: $2(\ )^3, 6(\ )^4, 2(\ )^0$, $0(\ )^1$, and $-(\ )^5$. That is, $M(x) = 2x^3$, $N(x) = 6x^4$, $Q(x) = 2$, $S(x) = 0$, and $T(x) = -x^5$ describe monomials.

If $f = a(\ )^n$, $f(x) = ax^n$. The expression ax^n is also called a monomial (in x).

EXAMPLE 3.1.2. If $f = 3(\ )^2$, $g = -2(\ )^1$, and $h = 3(\ )^0$, then:

$$f(1) = 3(1)^2 = 3$$
$$g(1) = -2(1)^1 = -2$$
$$h(1) = 3 \cdot 1 = 3$$
$$(f + g)(2) = 3(2)^2 - 2(2)^1 = 8$$
$$(fg)(2) = 3(2)^2 [-2(2)^1] = -48$$
$$(6f)(2) = 6 \cdot 3(2)^2 = 72$$
$$f(x) = 3x^2$$
$$g(x) = -2x$$
$$h(x) = 3$$
$$(f + g + h)(x) = 3x^2 - 2x + 3$$
$$f(x + a) = 3(x + a)^2$$
$$[f(x + a) - f(x)]/a = 6x + 3a \quad \text{if} \quad a \neq 0$$

The coefficients of f, g, and h respectively are $3, -2$, and 3.

If $f = a(\ )^n$ is a monomial with $a \neq 0$, the *degree* of f is n. If $a = 0$, the degree of f is zero. In Example 3.1.2, the respective degrees of f, g, and h are 2, 1, and 0.

Before continuing the discussion of these functions, we need to introduce some symbolism. The expression $\Sigma_{i=1}^{n} a_i$ will mean $a_1 + a_2 + a_3 + \cdots + a_n$. In other words, we let the *index i* take on all integral values from the *lower index* (1 in this case) through the *upper index* (n for our example). These values for i are substituted into the expression a_i to the right of the Σ. Thus as i takes on successively the values $1, 2, 3, \ldots$, and n, the terms a_i become $a_1, a_2, a_3, \ldots$, and a_n respectively. We then add all such generated terms, giving $a_1 + a_2 + \cdots + a_n$.

EXAMPLE 3.1.3.

$$\sum_{i=0}^{5} (i) = 0 + 1 + 2 + 3 + 4 + 5 = 15$$

$$\sum_{i=2}^{3} (i^2) = 2^2 + 3^2 = 13$$

$$\sum_{i=4}^{4} (i^3 + 1) = 4^3 + 1 = 65$$

Using this notation, we formulate the following. A function $f: R \rightarrow R$ is a *polynomial* function (or polynomial) if f is of the form $\sum_{i=0}^{n} a_i(\ )^i$ where the $a_i \in R$ and $n \in N$ or $n = 0$. That is, $f = a_0(\ )^0 + a_1(\ )^1 + a_2(\ )^2 + \cdots + a_n(\ )^n$.

A polynomial is nothing more than the sum of a (finite) number of monomials.

To be more nearly correct, our polynomials should be called polynomials in one real variable. However, until necessary to do so, we will simply say polynomial.

The polynomial form above may also be expressed by the notation

$$f(x) = \sum_{i=0}^{n} a_i x^i$$

or

$$f(x) = a_n x^n + a_{n-1} x^{n-1} + \cdots + a_1 x + a_0$$

It will prove helpful for us to retain the notation $a_i(\ )^i$ for a few brief examples. Gradually we will switch to the latter symbolism.

We shall always assume that in $\sum_{i=0}^{n} a_i(\ )^i$, the coefficient a_n is not zero unless $n = 0$. Then, we define the degree of $\sum_{i=0}^{n} a_i(\ )^i$ to be n. The *degree of a polynomial* is therefore seen to be the largest among the degrees of its monomials. A polynomial is *linear* if it is of degree 1, *quadratic* if of degree 2, *cubic* if of degree 3, and so on. In Example 3.1.2, the degree of $f + g + h$ is 2.

The degree of a polynomial is of interest because, as we will find out later, a polynomial of degree $n > 0$ has at most n zeros. Thus a linear polynomial crosses the x-axis at most once, a quadratic function at most twice, and so forth.

EXAMPLE 3.1.4.   Let $f = \sum_{i=0}^{3} (i + 1)(\ )^i$, $g = \sum_{i=2}^{5} (i - 1)(\ )^i$, and $h = 3(\ )^1 + (\ )^0$.

Then $f = 1(\ )^0 + 2(\ )^1 + 3(\ )^2 + 4(\ )^3$, $f(x) = 1 + 2x + 3x^2 + 4x^3$, and $f(2) = 1 + 2 \cdot 2 + 3 \cdot 2^2 + 4 \cdot 2^3 = 49$. The degree of f is 3 and the coefficients of $(\ )^0$, $(\ )^1$, $(\ )^2$, and $(\ )^3$ are respectively 1, 2, 3, and 4.

Even though the lower limit in the Σ expression for g is 2, g is a polynomial [consider the first two powers $(\ )^0$ and $(\ )^1$ to have coefficients zero]. Thus $g(x) = 4x^5 + 3x^4 + 2x^3 + x^2 = 4x^5 + 3x^4 + 2x^3 + x^2 + 0x + 0$ and $g(0) = 4 \cdot 0^5 + 3 \cdot 0^4 + 2 \cdot 0^3 + 0^2 = 0$.

The function h is a linear function and $h(x) = 3x + 1$. Furthermore, $h(-1/3) = 0$.

Since polynomials are functions, we can expect to be able to add, multiply, and compose polynomials as well as form scalar multiples. If, indeed, f and g are polynomials, $f + g$, $f - g$, fg, kf, $f \circ g$, and $g \circ f$ are also polynomials. The expression f/g, however, may fail to be a polynomial. Functions like f/g are said to be *rational functions*; the name is indicated by the (quotient) form.

EXAMPLE 3.1.5. Suppose $f = 2(\ )^3 + 3(\ )^2 + (\ )^1 + 2(\ )^0$ while $g = 3(\ )^1 + 2(\ )^0$. That is, $f(x) = 2x^3 + 3x^2 + x + 2$ and $g(x) = 3x + 2$.

Then $f(x) + g(x) = 2x^3 + 3x^2 + x + 2 + 3x + 2 = 2x^3 + 3x^2 + 4x + 4$ and $f(x) - g(x) = 2x^3 + 3x^2 + x + 2 - (3x + 2) = 2x^3 + 3x^2 + x + 2 - 3x - 2 = 2x^3 + 3x^2 - 2x$. The addition and subtraction are performed by adding similar powers of x according to the distributive property. Hence, $f + g = 2(\ )^3 + 3(\ )^2 + 4(\ )^1 + 4(\ )^0$ and $f - g = 2(\ )^3 + 3(\ )^2 - 2(\ )^1$. The degrees of $f + g$ and $f - g$ are *not greater than* the larger of the individual degrees of f and g.

EXAMPLE 3.1.6. With f and g as in Example 3.1.5, $(fg)(x)$ and $(kf)(x)$ are given by $(2x^3 + 3x^2 + x + 2)(3x + 2) = 6x^4 + 9x^3 + 3x^2 + 6x + 4x^3 + 6x^2 + 2x + 2 = 6x^4 + 13x^3 + 9x^2 + 8x + 2$ and $k(2x^3 + 3x^2 + x + 2) = 2kx^3 + 3kx^2 + kx + 2k$ respectively. Thus $fg = 6(\ )^4 + 13(\ )^3 + 9(\ )^2 + 8(\ )^1 + 2(\ )^0$ and $kf = 2k(\ )^3 + 3k(\ )^2 + k(\ )^1 + 2k(\ )^0$. The degree of fg is the *sum* of the degrees of f and g while the degree of kf is the *same* as the degree of f if $k \neq 0$.

EXAMPLE 3.1.7. Let f and g be as in Example 3.1.5. The composition $(f \circ g)$ is given by $(f \circ g)(x) = f(g(x)) = f(3x + 2) = 2(3x + 2)^3 + 3(3x + 2)^2 + (3x + 2) + 2 (f = 2(\ )^3 + 3(\ )^2 + (\ ) + 2$ whence $f(g(x))^3 + 2(g(x))^3 + 3(g(x))^2 + (g(x)) + 2)$. Thus $(f \circ g)(x) = 2(27x^3 + 54x^2 + 36x + 8) + 3(9x^2 + 12x + 4) + (3x + 2) + 2 = 54x^3 + 135x^2 + 111x + 32$, whereby $f \circ g = 54(\ )^3 + 135(\ )^2 + 111(\ )^1 + 32(\ )^0$. The degree of $f \circ g$ and $g \circ f$ is the *product* of the degrees of f and g if neither degree is zero.

EXAMPLE 3.1.8. With f and g as in Example 3.1.5, $(f/g)(x) = (2x^3 + 3x^2 + x + 2)/(3x + 2)$ and $f/g = [2(\ )^3 + 3(\ )^2 + (\ )^1 + 2(\ )^0]/[3(\ )^1 + 2(\ )^0]$. The function f/g is not a polynomial. It is, however, a rational function.

It is very easy to determine whether certain numbers are zeros of a polynomial. Three such numbers are $1, -1$, and zero.

Let $f(x) = a_n x^n + a_{n-1} x^{n-1} + \cdots + a_1 x + a_0$. Since $f(0) = a_0$, we find that $f(0) = 0$ if and only if the *constant term* $a_0 = 0$. That is, 0 is a zero for f if and only if the constant term is zero.

The expression $f(1)$ is seen to be $a_n + a_{n-1} + a_{n-2} + \cdots + a_1 + a_0$. Thus $f(1) = 0$ if and only if the coefficients add to zero.

Now $(-1)^n$ is 1 if n is even and (-1) if n is odd. Consequently, $f(-1)$ is zero if and only if the odd coefficients (coefficients with odd subscripts) add to the sum of the even coefficients (coefficients with even subscripts). This follows since

1. if n is odd, $f(-1) = -a_n + a_{n-1} - a_{n-2} + \cdots + a_2 - a_1 + a_0$
   and

2. if n is even, $f(-1) = a_n - a_{n-1} + a_{n-2} - \cdots - a_1 + a_0$.

EXAMPLE 3.1.9. Are any among 0, 1, and -1 zeros for $f(x) = 6x^5 - 3x^4 + x^3 - 3x^2 + x - 2$?

The constant term is -2 whence $f(0) = -2 \neq 0$. The coefficients add to $6 - 3 + 1 - 3 + 1 - 2 = 0$ so that $f(1) = 0$. However, the sum of the odd coefficients is $6 + 1 + 1 = 8$ while the sum of the even coefficients is $-3 - 3 - 2 = -8$, rendering $f(-1) \neq 0$.

EXERCISE 3.1

1. Which of the following indicate monomial functions?

   a. $f = 6(\ )^5$
   b. $f(x) = 2x^3$
   c. $f = 3(\ )^{1/2}$
   d. $f = (\ )^0$
   e. $f(x) = 0$
   f. $f(x) = 3x^2 + 2x$

   g. $f = \sum_{i=0}^{3} (i)(i-1)(i-2)(\ )^i$
   h. $f(x) = (2x + 1)/3$
   i. $f(x) = (x^2 + 1)/(\dot{x} - 2)$

2. Which of those in Exercise 3.1.1 are polynomials? rational functions?

3. Find each of the following for each function f in Exercise 3.1.1.

   a. $f(0)$
   b. $f(1)$
   c. $f(-1)$
   d. $f(4)$
   e. $f(2)$
   f. $f(-2)$

4. Graph each of the functions in Exercise 3.1.1.

5. Find the degree of each polynomial in Exercise 3.1.1.

6. Find the coefficient of each monomial involved in Exercise 3.1.1.

7. Let $f = 2(\ )^2 + 3(\ )^1 + 2$ and $g = -(\ )^1 + 7$ and compute:

   a. $f + g$ e. $g \circ f$ i. $g(1)$ m. $g(1/2)$
   b. $f - g$ f. $f(0)$ j. $f(-1)$ n. $f(-1/2)$
   c. fg g. $g(0)$ k. $g(-1)$ o. $g(-1/2)$
   d. $f \circ g$ h. $f(1)$ l. $f(1/2)$

8. Sketch a graph of f, g, and fg in Exercise 3.1.7.

9. Sketch the set $\{(x, y) : y \leqslant f(x)\}$ for f as in each case in Exercise 3.1.1.

10. Sketch the set $\{(x, y) : -5 < x < 7, \ g(x) - 2 < y < g(x)\}$ for g as in Exercise 3.1.7.

11. Is any element from $\{0, 1, -1\}$ a zero for

    a. $15x^5 + 2x^2 + 7x$ c. $x^5 + x^4 + x^3 + x^2 + x + 1$
    b. $3x^3 - 3x^2 + 2x + 2 - 4$ d. $x^7 + 3x^6 - 4x^5 - 2x^3 + 2$

12. Let f and g be polynomials of degree $m > 0$ and $n > 0$, respectively. How do the degrees of the following compare with m and n?

    a. $f + g$ d. $f \circ g$
    b. $f - g$ e. $g \circ f$
    c. fg

    Do you think that $g \circ f$ and $f \circ g$ are the same? Why or why not? Do you believe that $f \circ (g \circ h) = (f \circ g) \circ h$ for h a polynomial? Why or why not?

13. Find a zero for each polynomial expression.

    a. $x + 2$ e. $9x^2 - 1$
    b. $3x - 1$ f. $x(x^2 - 1)$
    c. x^2 g. $(x - 3)(x - 4)(x - 1)$
    d. $x^2 - 1$

3.2 FACTORING AND DIVISION OF POLYNOMIALS

We know that if a and b are real numbers, $a \cdot b = 0$ if and only if $a = 0$ or $b = 0$. Similarly, if $f = gh$ then $f(x) = 0$ if and only if $g(x)h(x) = 0$, which is to say that $g(x) = 0$ or $h(x) = 0$. The idea of being able to write a polynomial (or any other function for that matter) as a product of two polynomials (functions) can prove useful in determining zeros. Such a process is called *factoring*.

EXAMPLE 3.2.1. Find the three zeros for f where $f(x) = x^3 - 8x^2 + 19x - 12$.

The zeros of f are not immediately apparent. We will develop techniques for handling this situation. However, our concern here is with the value of factoring. Since $f(x) = (x-3)(x-4)(x-1)$, the zeros are easily seen to be the zeros of the individual factors. Thus, $f(3) = f(4) = f(1) = 0$. The factored form [factored into a product of *linear* factors $(x-a)$] displays the zeros readily.

Most of our work in this section deals with determining linear factors for a polynomial. Let us remark here that *every polynomial in* x *can be written as the product of linear and quadratic factors. Consequently, if we can find zeros for linear and quadratic polynomials, we have a strong tool to use in the search for zeros of general polynomials.*

The following (unproved) well-known theorem sets the stage for our study.

> *Theorem 3.2.1.*   *If* f *is a polynomial and* a *is a real number, there is a unique polynomial* g *and a real number* r *whereby*

$$f(x) = (x-a)g(x) + r$$

We recognize the r as being the "remainder upon division of $f(x)$ by $(x-a)$." No division is involved in the statement of Theorem 3.2.1 but if $x \neq a$, the conclusion may be restated as

$$f(x)/(x-a) = g(x) + r/(x-a)$$

Thus $x-a$ is a factor of $f(x)$ [divides $f(x)$ evenly] if and only if $r = 0$ and we have

> *Theorem 3.2.2.*   *The linear polynomial* $(x-a)$ *is a factor of* f(x) *if and only if* f(a) = 0.

> *Proof.*   Since $f(x) = (x-a)g(x) + r$, $f(a) = r$.   Necessarily, $(x-a)$ divides $f(x)$ evenly if and only if $0 = r = f(a)$.

Theorem 3.2.2 is sufficient to justify an earlier remark. The value $x = a$ is a zero for $f(x) = a_nx^n + a_{n-1}x^{n-1} + \cdots + a_1x + a_0$ if and only if $(x-a)$ divides $f(x)$, whence f has zeros $z_1, z_2, \ldots, z_k$ means that $(x-z_1)(x-z_2)(x-z_3) \cdots (x-z_k)q(x) = f(x)$ for some q (use induction). However, the degree of $f(x)$ is the sum of the degrees of $(x-z_1)$, $(x-z_2)$, $\ldots$, $(x-z_k)$, and q. That the degree of q is nonnegative means that $k \leqslant n$. Conclusion? f has no more than n zeros, n being the degree of f.

EXAMPLE 3.2.2.   Let $f(x) = 9x^2 + 6x + 1$ and determine whether any numbers in $\{0, 1, -1, -1/3\}$ are zeros for f.

Because the constant term is 1, $f(0) = 1 \neq 0$. Furthermore, $9 + 6 + 1 = 16$ and $9 + 1 \neq 6$, whence neither 1 nor -1 is a zero for f.

However, $f(-1/3) = 9(-1/3)^2 + 6(-1/3) + 1 = 0$ indicates that $-1/3$ is a zero for f.

From this example we can further deduce that $(x - (-1/3)) = (x + 1/3)$ is a factor of $f(x)$. However, writing $f(x) = (x + 1/3)g(x)$, we find that $g(x)$ is not known to us.

We may set about finding $g(x)$ by long division. Alternatively, we may simplify the form by synthetic division, a process we will introduce shortly.

EXAMPLE 3.2.3. Write $9x^2 + 6x + 1$ in the form $(x + 1/3)g(x)$, $g(x)$ a polynomial.

The following indicates the step-by-step process of long division.

$$
\begin{array}{r}
9x \phantom{{}+6x+1} \\
x + 1/3 \overline{\smash{\big)}\ 9x^2 + 6x + 1} \\
-(9x^2 + 3x) \\
\hline
3x
\end{array}
$$

$$
\begin{array}{r}
9x \phantom{{}+6x+1} \\
x + 1/3 \overline{\smash{\big)}\ 9x^2 + 6x + 1} \\
-(9x^2 + 3x) \\
\hline
3x + 1
\end{array}
$$

$$
\begin{array}{r}
9x + 3 \phantom{{}x+1} \\
x + 1/3 \overline{\smash{\big)}\ 9x^2 + 6x + 1} \\
-(9x^2 + 3x) \\
\hline
3x + 1 \\
-(3x + 1) \\
\hline
0
\end{array}
$$

At each stage, a subtraction took place. In the following, we write $-1/3$ rather than $+1/3$ and replace subtraction by addition. In fact, we do not even need the powers of x. We can proceed using only the coefficients. The expression $9x^2 + 6x + 1$ is replaced by 9   6   1.

$$
\begin{array}{r|ccc}
-1/3 & 9 & 6 & 1 \\
\hline
 & 9 & &
\end{array}
$$
Bring down the leftmost term.

Multiply the lowered term by $(-1/3)$.

$$
\begin{array}{r|ccc}
-1/3 & 9 & 6 & 1 \\
 & & -3 & \\
\hline
 & 9 & 3 &
\end{array}
$$
Add the product to the second coefficient.

$$
\begin{array}{r|ccc}
-1/3 & 9 & 6 & 1 \\
 & & -3 & -1 \\
\hline
 & 9 & 3 & \boxed{0}
\end{array}
$$
Multiply the sum by $(-1/3)$ and add to the next coefficient.

The last term on the right (0 in this case) is the remainder while the 9   3 is interpreted as $9x + 3$. Thus $9x^2 + 3x + 1 = (x + 1/3)(9x + 3) + 0$. This is the process of synthetic division.

EXAMPLE 3.2.4. Use synthetic division to divide $x^3 + 3x^2 - 2x + 7$ by $x + 2 = x - (-2)$.

$$
\begin{array}{r|rrrr}
-2 & 1 & 3 & -2 & 7 \\
\hline
   & 1 & & &
\end{array}
$$

$$
\begin{array}{r|rrrr}
-2 & 1 & 3 & -2 & 7 \\
   &   & -2 & & \\
\hline
   & 1 & 1 & &
\end{array}
$$

$$
\begin{array}{r|rrrr}
-2 & 1 & 3 & -2 & 7 \\
   &   & -2 & -2 & \\
\hline
   & 1 & 1 & -4 &
\end{array}
$$

$$
\begin{array}{r|rrr|r}
-2 & 1 & 3 & -2 & 7 \\
   &   & -2 & -2 & +8 \\
\hline
   & 1 & 1 & -4 & 15
\end{array}
$$

We write $x^3 + 3x^2 - 2x + 7 = (x + 2)(x^2 + x - 4) + 15$

EXAMPLE 3.2.5. Let $f(x) = 3x^4 + 2x^2 + 1$ and write $f(x)$ in the form $f(x) = (x - 2)g(x) + r$, using synthetic division.

Now, writing $x - 2$ in the form $x - r$, $r = 2$. Moreover, filling in the "missing" powers of x, $f(x) = 3x^4 + 0x^3 + 2x^2 + 0x + 1$. Our synthetic division calculation has the appearance:

$$
\begin{array}{r|rrrr|r}
2 & 3 & 0 & 2 & 0 & 1 \\
  &   & 6 & 12 & 28 & 56 \\
\hline
  & 3 & 6 & 14 & 28 & 57
\end{array}
$$

Therefore $f(x) = (x - 2)(3x^3 + 6x^2 + 14x + 28) + 57$.

EXAMPLE 3.2.6. Show that $x + 2$ is a factor of $x^3 + 3x^2 + 3x + 2$.

If $x + 2$ is a factor of $x^3 + 3x^2 + 3x + 2$, then there are two ways of determining this. First we need only calculate $f(-2)$. Alternatively, the remainder upon division by $x + 2$ must be zero.

$$f(-2) = (-2)^3 + 3(-2)^2 + 3(-2) + 2 = -8 + 12 - 6 + 2 = 0.$$

Thus, $x + 2$ is a factor of $f(x)$. By synthetic division we may check this conclusion and find the other "factor."

$$
\begin{array}{r|rrr|r}
-2 & 1 & 3 & 3 & 2 \\
   &   & -2 & -2 & -2 \\
\hline
   & 1 & 1 & 1 & 0
\end{array}
$$

Consequently $f(x) = (x + 2)(x^2 + x + 1)$.

Earlier we mentioned that a polynomial could be factored into a product of linear and quadratic polynomials. The process of synthetic division can be used.

EXAMPLE 3.2.7. Write f as a product of linear and quadratic polynomials where $f(x) = x^5 - x$.

Clearly $f(x) = x(x^4 - 1)$. Examine the *depressed* polynomial $f_1(x) = x^4 - 1$. Now, $f_1(1) = f_1(-1) = 0$ so that $x + 1$ and $x - 1$ are factors of $f_1(x)$. By synthetic division,

$$
\begin{array}{r|rrrrr}
1 & 1 & 0 & 0 & 0 & -1 \\
  &   & 1 & 1 & 1 & 1 \\
\hline
  & 1 & 1 & 1 & 1 & \underline{0} \\
\end{array}
$$

so that $f_1(x) = (x^3 + x^2 + x + 1)(x - 1)$. Let $f_2(x) = x^3 + x^2 + x + 1$ be the depressed polynomial for f_1.

$$
\begin{array}{r|rrrr}
-1 & 1 & 1 & 1 & 1 \\
   &   & -1 & 0 & -1 \\
\hline
   & 1 & 0 & 1 & \underline{0} \\
\end{array}
$$

whereby the depressed polynomial for f_2 is f_3 given by $f_3(x) = x^2 + 1$. Since $x^2 \neq -1$ for any real number x, we cannot factor $x^2 + 1$. As a result we write

$$f(x) = x(x + 1)(x - 1)(x^2 + 1)$$

a product of linear and quadratic polynomials. (In a later chapter we will remove the deficiency inherent in not being able to factor $x^2 + 1$.)

EXERCISE 3.2

1. Write each in the form $(x - a)g(x) + r$ where a is as given.

    a. $x^3 - 4x^2 - 4x - 5$; $a = 5$
    b. $x^2 + x - 1$; $a = \sqrt{2}$
    c. $2x^3 - 3x^2 - 11x + 6$; $a = 1/2, -2, 3$
    d. $x^4 + (1/2)x^3 - x - 1/2$; $a = -1/2$
    e. $x^4 - 16$; $a = 1, 2, -2$
    f. $x^3 + 8$; $a = -1, 2, -2$
    g. $x^4 - 2x^2 + 1$; $a = -1, 1$
    h. $\alpha x^2 + \beta x + \gamma$; $a = \dfrac{-\beta + \sqrt{\beta^2 - 4\alpha\gamma}}{2\alpha}, \dfrac{-\beta - \sqrt{\beta^2 - 4\alpha\gamma}}{2\alpha}$

        (What must be true about $\beta^2 - 4\alpha\gamma$?)

2. Write each expression in Exercise 3.2.1 as a product.

3. Let $f(x)$ be given as in each case of Exercise 3.2.1. Compute

    a. $f(1/3)$ d. $f(8)$
    b. $f(6)$ e. $f(-19)$
    c. $f(-4)$ f. $f(-13)$

4. From your experience in Exercise 3.2.3, when is it easier to use substitution to find $f(x)$ and when is it easier to use synthetic division?

5. In each case below, find a value for k rendering the given linear polynomial a factor of t.

    a. $t(x) = x^5 + 3x^4 + 2x^3 + 3x^2 + x + k;\ x - 1$
    b. $t(x) = x^5 + 3x^4 + 2x^3 + 3x^2 + x + k;\ x - 2$
    c. $t(x) = x^6 - x^3 + x^2 + 3x + k;\ x + 3$
    d. $t(y) = y^2 + 2by + k;\ y + b$
    e. $t(z) = z^3 + k;\ z - 2$

3.3 LINEAR POLYNOMIALS

The previous section indicated how the study of zeros of polynomials could be reduced to finding zeros of linear and quadratic polynomials. Accordingly, we will study the algebraic and geometric properties of linear functions f given by $f(x) = ax + b$, $a \neq 0$, and then proceed to a similar study for quadratic polynomials.

Finding the zero for a linear function is clear. If $f(x) = ax + b = 0$, $x = -b/a$. It is just as easily seen that $f(-b/a) = 0$. Thus, the unique zero for a linear function is found. A graph of the function f given by $f(x) = ax + b$ is seen in Figure 3.3.1. The figure appears to be a straight line. To insure that our intuition and knowledge of geometry coincide, we define: l is a *straight line* means that l is a vertical line, a horizontal line, or (the graph of) a linear function f [$f(x) = ax + b$, $a \neq 0$].

Since we have examined the horizontal and vertical lines, let us examine the general line given by $f(x) = ax + b$, $a \neq 0$.

The functional value $f(0)$ is seen to be b, whence b is called the *y-intercept* [$(0, b)$ is a point of the function]. The coefficient a is called the *slope* of the line. The following analysis indicates why this is the case.

If $(x, y) = (x, f(x))$ and $(u, v) = (u, f(u))$ belong to f, we see that $(y - v)/(x - u) = (ax - au)/(x - u) = a(x - u)/(x - u) = a$. The slope, then, is the ratio of the differences of the respective coordinates of any pair of its points (see Figure 3.3.2). If we please, the slope indicates the inclination of the line with respect to the first axis (x-axis).

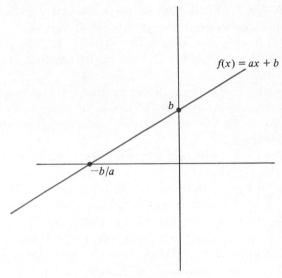

Figure 3.3.1

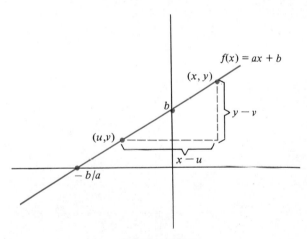

Figure 3.3.2

EXAMPLE 3.3.1. Suppose we are to find the correspondence f representing a straight line passing through the point $(1, 2)$ and having slope 3.

Now $f(x) = ax + b$ and $a = 3$, whence $f(x) = 3x + b$. Furthermore, $2 = f(1) = 3 \cdot 1 + b = 3 + b$ and $b = -1$. We therefore see that f is given by $f(x) = 3x - 1$.

EXAMPLE 3.3.2. Find the correspondence f whereby f is a straight line passing through the points $(1, 3)$ and $(2, 5)$.

Since $f(x) = ax + b$ and $a = (5 - 3)/(2 - 1) = 2$, $f(x) = 2x + b$. Further, $3 = f(1) = 2 \cdot 1 + b = 2 + b$, implying that $b = 1$. Consequently $f(x) = 2x + 1$.

Examples 3.3.1 and 3.3.2 illustrate that we can graph a straight line if we know one point and the slope, or two points on the line. Quite often we wish to know where a straight line given by $f(x) = ax + b$ crosses the x-axis.

Such values for x are merely zeros for f, and we are clearly in command of the situation.

We could just as easily have asked the question: For what values x is $f(x) = k$, $k \in R$? Equivalently, what are the zeros for $f(x) - k$ (that is, when is $f(x) - k = 0$)? Graphically, we are interested in knowing where f and the horizontal line $y = k$ intersect. (See Figure 3.3.3.) Solutions are found exactly as the zeros were found above. [$f(x)$ and $f(x) - k$ are compared in Figure 3.3.4.]

We can extend this last thought somewhat by asking the following. If $f(x) = ax + b$ and $g(x) = cx + d$, is there a value x so that $f(x) = g(x)$ (that is, $ax + b = cx + d$)?

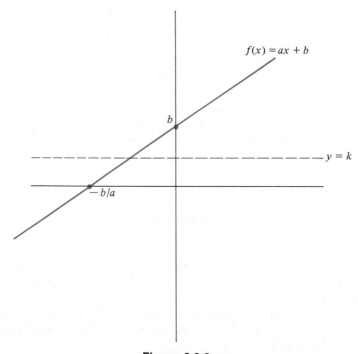

Figure 3.3.3

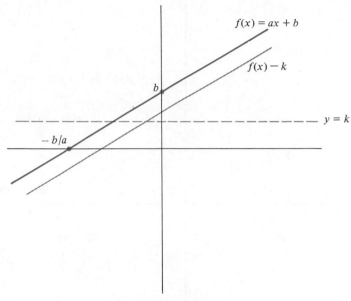

Figure 3.3.4

Solving the equation, we see that if such a simultaneous solution exists, x must be $(d - b)/(a - c)$. This is a real number if and only if $a \neq c$ (that is, if and only if f and g have different slopes). Furthermore, if there is such a solution, there is exactly one. Alternatively, if f and g have different slopes, they intersect exactly once and conversely, if f and g intersect not at all or more than once, they have the same slope. We conclude, then, that *through any pair of points, there can be drawn exactly one straight line* (any two such lines have the same slope and two points in common). Thus two straight lines are seen to intersect once (they are simultaneous), not at all (parallel), or they are the same line. See Figure 3.3.5.

EXAMPLE 3.3.3. Find the points of intersection of f and g where $f(x) = 3x + 2$ and $g(x) = -x + 1$. The equation $3x + 2 = -x + 1$ gives $4x = -1$ or $x = -1/4$. $f(-1/4) = 3(-1/4) + 2 = 5/4$ and $g(-1/4) = -(-1/4) + 1 = 5/4$ so that $(-1/4, 5/4)$ is the intersection point. See Figure 3.3.6.

EXAMPLE 3.3.4. Find the points of intersection for f and g where $f(x) = 3x + 1$ and $g(x) = 3x + 2$. Since f and g both have slope 3, they are the same function or they represent parallel lines. The functions are obviously different $[f(0) = 1 \neq 2 = g(0)]$, whence they

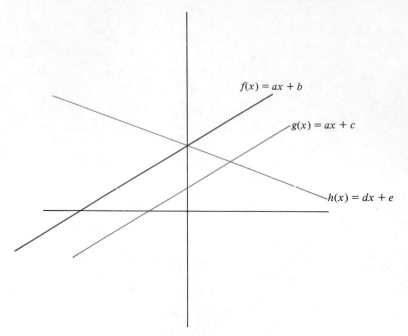

Figure 3.3.5

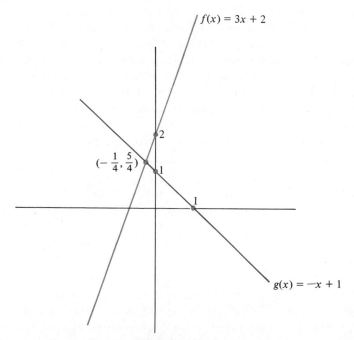

Figure 3.3.6

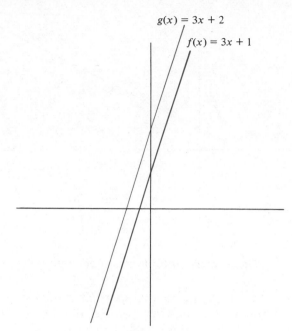

$g(x) = 3x + 2$

$f(x) = 3x + 1$

Figure 3.3.7

represent parallel lines. The two functions have no intersection point. [If we were to set $f(x) = g(x)$, we would observe: $3x + 1 = 3x + 2$ or $1 = 2$.] (See Figure 3.3.7.)

We can do away with the functional notation in examining straight lines in the plane. Since we usually denote points in the plane by (x, y) we alternatively define the following.

If l is a straight line, there are constants a, b, and c (not all zero) such that $l = \{(x, y) \in R^2: ax + by + c = 0\}$.

If $a = 0$, the line is horizontal while if $b = 0$, the line is vertical.

With these two special cases having been examined, let us turn our attention to the general case $ax + by + c = 0$ where $a \neq 0$ and $b \neq 0$. The equation may be rewritten in the form

$$y = (-a/b)x - c/b$$

The quantity $-a/b$ is the *slope* of the line. Furthermore, setting $x = 0$, we see that $y = -c/b$ results. Hence $(0, -c/b)$ is one point of the line, namely, the *y-intercept*. For this reason, the form $y = mx + p$ is known as the *slope-intercept* form of the equation of a straight line. If we write $y = f(x)$, we see that our original form for the line results. The term *slope* is highly suggestive.

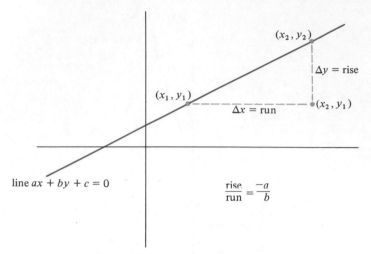

Figure 3.3.8 Interpretation of slope

Let (x_1, y_1) and (x_2, y_2) be two distinct points of the line in question. Since the line is neither vertical nor horizontal, $x_1 \neq x_2$ and $y_1 \neq y_2$. Figure 3.3.8 portrays such a situation.

Define Δy to be $y_2 - y_1$ and Δx to be $x_2 - x_1$. Notice that in some sense, Δy represents the *directed y-distance from* (x_1, y_1) *to* (x_2, y_2) while Δx represents the corresponding *directed x-distance*. Examine the ratio of these two quantities.

$$\Delta y / \Delta x = (y_2 - y_1)/(x_2 - x_1) = \text{rise/run}$$

Since (x_1, y_1) and (x_2, y_2) are on the line, they must satisfy the equation of the line, whence

$$y_1 = -ax_1/b - c/b$$
$$y_2 = -ax_2/b - c/b$$

Subtracting y_1 from y_2, we note that

$$y_2 - y_1 = (-a/b)(x_2 - x_1)$$

or

$$\frac{y_2 - y_1}{x_2 - x_1} = -a/b$$

Substituting into the above ratio $\Delta y / \Delta x$, we conclude:

$$\Delta y / \Delta x = -a/b$$

Consequently, the slope as defined earlier has this geometric significance: the slope is a measure of the inclination of the line relative to the x-axis (axis 1). Further, the slope is defined independently of any given point on the line. Necessarily $\Delta y/\Delta x$ is the same regardless of the two points chosen. This fact then allows us to create the equation of a line passing through

1. two known points or

2. find the equation of a line having a given slope and passing through a particular point.

We might consider worrying about whether or not the slope can be zero. In our construction above, zero slope was not the case ($y_1 \neq y_2$). However, in the form $y = mx + b$, $m = 0$ makes sense, implying that the line is horizontal. Conversely, any horizontal line is of the form $y = b$, which is to say $m = 0$. Thus we may (and we will) *associate horizontal lines with a slope of zero*.

We cannot define a slope for vertical lines, hence we say that *a vertical line has no associated slope*. If we say we want the equation of the line of slope m passing through the point P, we automatically imply that the line is not vertical.

EXAMPLE 3.3.5. As an example of one of the situations above, find the equation of the line passing through (x_0, y_0) and having slope m.

Let (x, y) be any point (of the desired line) different from the fixed point (x_0, y_0). Then $\Delta y = y - y_0$ and $\Delta x = x - x_0$, whereby $m = \Delta y/\Delta x = (y - y_0)/(x - x_0)$. Since $x - x_0 \neq 0$, the equation may be altered to appear in the form

$$y - y_0 = m(x - x_0)$$

The equation developed in Example 3.3.5 is known as the *point-slope* form of the equation of a line. It is well to observe that (x_0, y_0) satisfies the equation even though (x_0, y_0) cannot be substituted into the expression $m = (y - y_0)/(x - x_0)$.

EXAMPLE 3.3.6. Find an equation depicting a line through $(6, 4)$ with slope 2.

Letting (x, y) denote a general point on the line, we find that

$$\frac{y - 4}{x - 6} = 2$$

or

$$y - 4 = 2(x - 6)$$

whence

$$y = 2x - 8$$

The next to last form of the equation may be had directly from the result of Example 3.3.5 by setting $(x_0, y_0) = (6, 4)$ and $m = 2$.

EXAMPLE 3.3.7. Should we be given two distinct points (x_0, y_0) and (x_1, y_1) (instead of a single point and a slope) we ought not despair, since

$$m = (y_0 - y_1)/(x_0 - x_1)$$

It has been tacitly assumed that $x_0 \neq x_1$. If this were not the case, the equation would be immediate, as a vertical line results. Hence in the nonvertical case the point-slope form becomes the *two point form*

$$y - y_0 = (y_0 - y_1)(x - x_0)/(x_0 - x_1)$$

EXAMPLE 3.3.8. Find an equation of the line through the points $(1, 5)$ and $(2, 8)$.

The slope m is determined from the two points by

$$\frac{8 - 5}{2 - 1} = 3$$

Thus if (x, y) is an arbitrary point on the line,

$$\frac{y - 5}{x - 1} = 3$$

or

$$y - 5 = 3(x - 1)$$

whence

$$y = 3x + 2$$

We might also note that the equation can be derived by using the slope and the point $(2, 8)$ rather than the point $(1, 5)$.

$$\frac{y - 8}{x - 2} = 3$$

$$y - 8 = 3(x - 2)$$

$$y = 3x + 2$$

EXERCISE 3.3

1. Find the zeros for each of the following.

a. $f(x) = 3x + 7$ c. $f(x) = (-2/3)x + 1$ e. $f(x) = (x + 1)/3$
b. $f(x) = 2(\ )^1$ d. $f = (-1/2)(\ )^1 - 2$ f. $f(x) = -2x + 1$

2. Graph each of the linear functions in Exercise 3.3.1.

3. Find the y-intercepts in each of Exercise 3.3.1.

4. Find the slopes of the functions in Exercise 3.3.1.

5. Find all appropriate values for x where f is from each of a.–f. of Exercise 3.3.1 and:

   a. $f(x) = 2$ c. $f(x) = 1$
   b. $f(x) = -1$ d. $f(x) = 1/3$

6. Write the equation of each indicated straight line.

   a. Passing through $(-3, 4)$ with slope $-1/2$.
   b. Passing through $(1, 0)$ with slope 6.
   c. Passing through $(1, 1)$ and $(2, 1)$.
   d. Slope 3 and y-intercept -2.
   e. Passing through $(1, 2)$ with slope 4.
   f. Passing through $(1/2, 1/2)$ with slope 1.
   g. Passing through $(-1, -4)$ with slope 2.
   h. Passing through $(1/3, -1)$ with slope -1.
   i. Passing through $(1, 1)$ and $(3, 5)$.
   j. Passing through $(-1, 1)$ and $(0, 0)$.
   k. Passing through $(1, -2)$ and $(2, -2)$.
   l. Slope -1 and y-intercept 1.
   m. y-intercept 1 and x-intercept 1.

7. Find the points of intersection of the following pairs. Tell when the pairs are parallel.

   a. $f(x) = 3x + 1$, $g(x) = 2x - 7$
   b. $f(x) = (1/2)x + 2$, $g(x) = 7x - 1$
   c. $f(x) = -2x + 1$, $g(x) = (1/2)x + 1$
   d. $f(x) = x - 1$, $g(x) = x + 1$

8. Show that if $f(x) = ax + b$, $a \neq 0$, then $f: R \to R$ is $1:1$ and onto. (That is, show that if $x_1 \neq x_2$, $f(x_1) \neq f(x_2)$, and if $y \in R$ there is an x such that $f(x) = y$.)

9. The *midpoint* between (a, b) and (c, d) is defined to be the point (x, y) where $x = (a + c)/2$ and $y = (b + d)/2$. Show that (a) the midpoint is on the line determined by (a, b) and (c, d) and (b) the midpoint is equidistant from (a, b) and (c, d).

10. Given the points $(5, 6)$ and $(11, -2)$ find the equation of the line passing through the midpoint between the two (see Exercise 3.3.9) and having slope the negative reciprocal of the slope of the line containing the original two points. Show that any point (x, y) on this new line is equidistant from the two original points $(5, 6)$ and $(11, -2)$. Can you think of another name for this line?

11. Graph the solution set for each inequality.

   a. $\{(x, y): y \leqslant f(x)\}$, $f(x)$ as in Exercise 3.3.1, parts a.–f.

b. $\{(x, y) : g(x) \geqslant y, f(x) \geqslant y\}$ f, g as in Exercise 3.3.7, parts a.–d.
c. $\{(x, y) : 0 \leqslant x, \ 0 \leqslant y, \ y \leqslant g(x), \ y \leqslant f(x)\}$ f, g as in Exercise 3.3.7, parts a.–d.

12. We know that inequalities can be handled in some respects much the way we manipulate equations, that is, if $3x + 2 < 2x + 1$; $3x + 2 - 2x < 2x + 1 - 2x$; $x + 2 < 1$; $x + 2 - 2 < 1 - 2$; whence $x < -1$. Conversely, if $x < -1$; $x + 2x + 2 < -1 + 2x + 2$; whence $3x + 2 < 2x + 1$. Using such techniques, find the solution set (set of real numbers satisfying the inequality) for each case. Graph the solution set.

a. $3x - 1 < 2$ d. $2x < 7 - 3x$
b. $2x + 1 < 7 - x$ e. $ax + b > 0, \ a > 0$
c. $-3 \ \ 2x \geqslant -5 - x$ f. $ax + b > 0, \ a < 0$

13. Let $f(x) = mx + b$. What is $[f(x + h) - f(x)]/h$? $[f(x) - f(a)]/(x - a)$? $\Delta f / \Delta x$, where Δf is the change in $f(x)$ caused by a given change in x. (What do these words mean intuitively?)

3.4 QUADRATIC POLYNOMIALS

In this section we wish to continue the pursuit of zeros of functions. In particular we want to explore means of finding zeros for quadratic polynomials (that is, functions of the form $f = a(\ )^2 + b(\ )^1 + c(\ )^0$ or $f(x) = ax^2 + bx + c, \ a \neq 0$). Two general means of attacking the problem are studied.

Suppose that $f(x) = x^2 + 3x + 2$. We can factor f, since $f(x) = x^2 + 3x + 2 = (x + 2)(x + 1)$. Thus if $g(x) = x + 2$ and $h(x) = x + 1$, we see that f is the product gh of the linear functions g and h. Recall that the product of two numbers, however, is zero if and only if one of the two factors is zero. Hence, $f(x) = 0$ if and only if $g(x) = 0$ or $h(x) = 0$. That is, the zeros of f are the zeros of g together with the zeros of h. In such a case the problem of finding the zeros of the quadratic function is reduced to finding the zeros of two linear functions.

EXAMPLE 3.4.1. Find the zeros of f where $f(x) = x^2 + 3x + 2$. We know that $x^2 + 3x + 2 = (x + 2)(x + 1)$, whereby we need only examine $x + 2 = 0$ and $x + 1 = 0$. These two equations have solutions $x = -2$ and $x = -1$ respectively; our zeros for f lie in $\{-2, -1\}$. Examination shows that $f(-2) = f(-1) = 0$. See Figure 3.4.1 for a graphic interpretation.

It is not always possible to factor a quadratic expression. In any case, the trouble and difficulty involved may not make the effort worthwhile. The following mechanical gymnastics reveal a general means for finding existing zeros of any quadratic function.

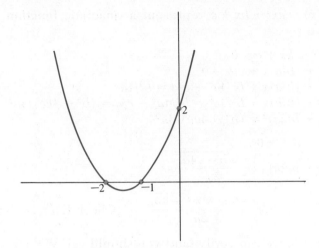

Figure 3.4.1 $f(x) = x^2 + 3x + 2$.

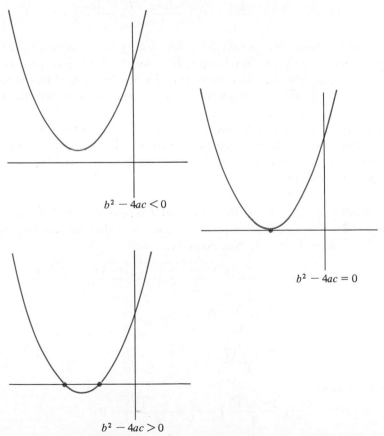

$b^2 - 4ac < 0$

$b^2 - 4ac = 0$

$b^2 - 4ac > 0$

Figure 3.4.2 $f(x) = ax^2 + bx + c;\ a > 0$.

Let $f(x) = ax^2 + bx + c$ represent a quadratic function. Setting $f(x) = 0$, we see:

$$ax^2 + bx + c = 0$$
$$x^2 + (b/a)x + c/a = 0$$
$$x^2 + (b/a)x + b^2/4a^2 + c/a = b^2/4a^2$$
$$x^2 + (b/a)x + b^2/4a^2 = b^2/4a^2 - c/a = (b^2 - 4ac)/4a^2$$
$$(x + b/2a)^2 = (b^2 - 4ac)/4a^2$$

Thus if $b^2 - 4ac \geqslant 0$,

$$(x + b/2a) = \frac{\pm\sqrt{b^2 - 4ac}}{2a}$$

$$x = \frac{-b \pm \sqrt{b^2 - 4ac}}{2a} \qquad \text{(Check Exercise 3.2.1h.)}$$

If $b^2 - 4ac \geqslant 0$, we can verify (and you should verify) that

$$f\left(\frac{-b + \sqrt{b^2 - 4ac}}{2a}\right) = f\left(\frac{-b - \sqrt{b^2 - 4ac}}{2a}\right) = 0$$

We therefore know that when $b^2 - 4ac > 0$, f has *two* zeros. If $b^2 - 4ac = 0$, f has exactly *one* zero and if $b^2 - 4ac < 0$, f has *no* real zeros. The expression $b^2 - 4ac$ is called the *discriminant* and the general curve of the quadratic polynomial is known as a *parabola*. (See Figures 3.4.2 and 3.4.3.)

The resultant form of the zeros, $x = (-b \pm \sqrt{b^2 - 4ac})/2a$, is often called the *quadratic formula*. Observe that a is the coefficient of the second-degree monomial, b is the coefficient of the linear term, and c is the constant.

EXAMPLE 3.4.2. Find the zeros of f where $f(x) = 2x^2 - 3x - 1$. Since $f = 2(\ )^2 + (-3)(\ )^1 + (-1)(\ )^0$, we see that for our example, $a = 2$, $b = -3$, and $c = -1$. Necessarily,

$$x = \frac{-(-3) \pm \sqrt{(-3)^2 - 4(2)(-1)}}{2 \cdot 2}$$

$$= \frac{3 \pm \sqrt{9 + 8}}{4}$$

$$= \frac{3 \pm \sqrt{17}}{4}$$

The expression

$$f\left(\frac{3 + \sqrt{17}}{4}\right) = f\left(\frac{3 - \sqrt{17}}{4}\right) = 0$$

holds and we know that the zeros for f are found. (See Figure 3.4.4.)

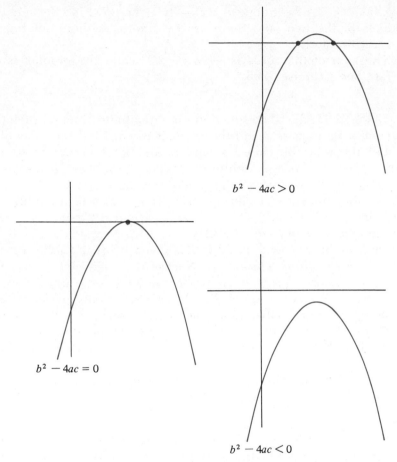

$b^2 - 4ac > 0$

$b^2 - 4ac = 0$

$b^2 - 4ac < 0$

Figure 3.4.3 $f(x) = ax^2 + bx + c;\ a < 0.$

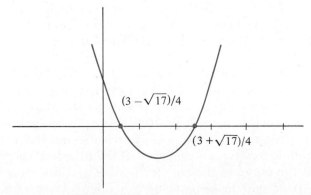

$(3 - \sqrt{17})/4$

$(3 + \sqrt{17})/4$

Figure 3.4.4 $f(x) = 2x^2 - 3x - 1.$

EXAMPLE 3.4.3. Suppose $f(x) = 3x^2 + 2x - 7$. Find the sum and products of the two (and there are two) zeros without solving for the zeros themselves.

The sum of the zeros is $-b/a = -2/3$ while the product is $c/a = -7/3$. (See Exercise 3.4.5.)

EXAMPLE 3.4.4. The function f is a quadratic function such that the sum of its roots is 2 and the product is -3. Find the roots of f.

Let the roots be called r_1 and r_2 and let $f(x) = ax^2 + bx + c$. Then $r_1 + r_2 = -b/a = 2$, while $r_1 r_2 = c/a = -3$. This leads us to $-3 = r_1 r_2 - r_1(2 - r_1) = 2r_1 - r_1^2$ which may be given by $r_1^2 - 2r_1 - 3 = 0$. By the process of factoring, $(r_1 - 3)(r_1 + 1) = 0$, implying that $r_1 = 3$ or $r_1 = -1$. Clearly, if $r_1 = 3$, $r_2 = -1$ and if $r_1 = -1$, $r_2 = 3$, so that the two zeros of f are -1 and 3.

One might ask, what is f? If f is given by $f(x) = ax^2 + bx + c$, $-b/a = r_1 + r_2$ while $c/a = r_1 r_2$. Necessarily, $b = -ar_1 - ar_2$ and $c = ar_1 r_2$. Thus $f(x) = ax^2 + bx + c = ax^2 + (-ar_1 - ar_2)x + ar_1 ar_2 = a(x^2 + (-r_1 - r_2)x + r_1 r_2) = a(x - r_1)(x - r_2)$. This demonstrates the fact that every polynomial having real zeros can be factored. Consequently, f has roots r_1 and r_2 if and only if there is a real number $a \neq 0$ such that $f(x) = a(x - r_1)(x - r_2)$. We can extend this idea further.

Two quadratic polynomials f and g have the same zeros if and only if each is a scalar multiple of the other.

EXAMPLE 3.4.5. The quadratic function f has zeros 2 and 1/2. Find the form of the function f. Well, $f = a(x - 2)(x - 1/2) = a[x^2 - (5/2)x + 1]$ for some $a \in R$, $a \neq 0$.

EXAMPLE 3.4.6. Find the expression for f in Example 3.4.5 where $f(1) = 1/2$. We know that $f(1) = a[1^2 - (5/2)(1) + 1] = (-1/2)a$. Thus $(-1/2)a = 1/2$ or $a = -1$. We now know that $f(x) = -x^2 + (5/2)x - 1$.

The graphs of quadratic functions have the general forms as shown in Figures 3.4.2 and 3.3.3. In each instance, there is a *maximum* (high) point for the function or a *minimum* (low) point. If the coefficient of the $(\ )^2$ term is positive, the extreme point is a minimum point while the extreme point is a maximum if the coefficient of the second degree term is negative. If the function has a minimum we say that the curve is *concave* (or opens) upward while if it has a maximum, it is said to be concave (or open) downward. Observe the following argument that is similar in nature to the derivation of the quadratic formula.

$$f(x) = ax^2 + bx + c$$
$$= ax^2 + bx + b^2/4a + c - b^2/4a$$
$$= a(x^2 + (b/a)x + b^2/4a^2) + \left(\frac{4ac - b^2}{4a}\right)$$
$$= a(x + b/2a)^2 + \frac{4ac - b^2}{4a}$$

Now $(x + b/2a)^2$ has a minimum value of zero when $x = -b/2a$ [for any other x, $(x + b/2a)^2 > 0$]. Then $f(-b/2a) = (4ac - b^2)/4a$. If $a > 0$, $a(x + b/2a)^2$ has a minimum value when $x = -b/2a$ so that f has a minimum value at $x = -b/2a$. This minimum value $f(-b/2a) = (4ac - b^2)/4a$.

If, however, $a < 0$, $a(x + b/2a)^2$ has its maximum value at $x = -b/2a$ and that maximum value is once again $(4ac - b^2)/4a$. In conclusion, f has a minimum if $a > 0$, a maximum if $a < 0$, the extreme value occurs when $x = -b/2a$, and the extreme value is $(4ac - b^2)/4a$. The point of extreme excursion is called the *vertex* of the parabola. It is worth noting that $-b/2a$ is one-half the sum of the zeros of f (if f has zeros). This indicates that the vertex occurs midway between the zeros.

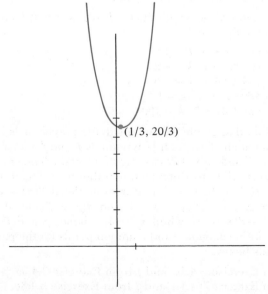

Figure 3.4.5 $f(x) = 3x^2 - 2x + 7$.

EXAMPLE 3.4.7. Find the extreme (minimum or maximum) value of $f(x) = 3x^2 - 2x + 7$.

The function has a minimum value, since $a = 3 > 0$. Furthermore, the minimum occurs at $x = -b/2a = 2/6 = 1/3$. The minimum value for $f(x)$ is

$$\frac{4ac - b^2}{4a} = \frac{4 \cdot 3 \cdot 7 - (-2)^2}{4 \cdot 3} = \frac{84 - 4}{12} = 80/12 = 20/3$$

Since the minimum of $f(x)$ is positive, we may also conclude that f has no zeros. (See Figure 3.4.5.)

EXERCISE 3.4

1. Find the zeros for each polynomial expression. Sketch each polynomial.

   a. $P(x) = 5x^2$
   b. $P(x) = x^2 + x$
   c. $P(x) = x^2 - 36$
   d. $P(x) = x^2 + 5x + 6$
   e. $P(x) = -x^2 + 7x - 12$
   f. $P(x) = +x^2 - x - 6$
   g. $P(x) = x^2 + 2x + 1$
   h. $P(x) = x^2 - 10x + 25$

   i. $P(x) = x^2 + 4x - 1$
   j. $P(x) = 3x^2 - x - 2$
   k. $P(x) = x^2 - x - 7$
   l. $P(x) = x^2 - x + 3$
   m. $P(x) = x^2 + 1$
   n. $P(x) = x^2 + x + 1$
   o. $P(x) = x^2 - x + 1$

2. Where the polynomials from Exercise 3.4.1 have zeros, write them as the product of two linear polynomials.

3. Find the values x for which $f(x) = k$ where f and k are given below. Sketch each polynomial.

   a. $f(x) = x^2 + 5x + 6$, $k = 12$
   b. $f(x) = x^2 + 2x - 1$, $k = -2$
   c. $f(x) = x^2 + 2x + 1$, $k = -3$
   d. $f(x) = 3x^2 + 2x - 1$, $k = 1$
   e. $f(x) = -x^2 + 4x - 5$, $k = 100$

4. It is possible that graphs of two quadratic polynomials intersect twice, once, or not at all. Two such polynomials P and f intersect (have points in common) if and only if $P(x) = f(x)$ for some x. The problem of finding simultaneous solutions (intersections) then is reduced to finding zeros for $P(x) - f(x)$ or $f(x) - P(x)$. For example, if $P(x) = x^2 + x$ and $f(x) = x^2 + 2x + 1$, $f(x) - P(x) = x + 1$ and $f(x) - P(x) = 0$ when $x = -1$. Thus, $f(x) = P(x)$ only when $x = -1$, whence f and P have only the point $(-1, 0)$ in common. Find common points for the pairs of quadratic polynomials below.

   a. P from Exercise 3.4.1c and f from Exercise 3.4.3a.
   b. P from Exercise 3.4.1n and f from Exercise 3.4.3c.
   c. P from Exercise 3.4.1j and f from Exercise 3.4.3a.

    d. P from Exercise 3.4.1j and f from Exercise 3.4.3d.
    e. P from Exercise 3.4.1n and f from Exercise 3.4.3d.
    f. P from Exercise 3.4.1n and f from Exercise 3.4.3e.
    g. P from Exercise 3.4.1h and f from Exercise 3.4.3e.

5.   Show that the sum and product of the zeros of $f(x) = ax^2 + bx + c$ are $-b/a$ and c/a respectively [where $f(x) = 0$ has roots (solutions)].

6.   Find the maximum or minimum point for each function from Exercises 3.4.1 and 3.4.3.

7. a. If the minimum value for the quadratic polynomial $f(x)$ is positive, what can you say about the set $\{x \in R : f(x) \leqslant 0\}$? $\{x \in R : f(x) \geqslant 0\}$?
    b. If the maximum value of the quadratic $f(x)$ is negative, what is the set $\{x \in R : f(x) \leqslant 0\}$? $\{x \in R : f(x) \geqslant 0\}$? Find the sets $\{x \in R : P(x) < 0\}$ and $\{x \in R : P(x) > 0\}$ where P is from:

    c. 3.4.1d   f. 3.4.1m
    d. 3.4.1i   g. 3.4.1n
    e. 3.4.1l   h. 3.4.1o

8.   Tell how many zeros each of the following functions has. Do not solve for the zeros. Tell what the sum and product of the zeros are in each case where zeros exist.

    a. $3x^2 + 7x + 1$ c. $-x^2 + 5x + 8$ e. $-x^2 + 5x - 3$
    b. $3x^2 + 7x - 5$ d. $-x^2 - 5x + 3$ f. $5x^2 - 43x + 2$

9.   It is often of value to know the set of values x for which $f(x) < k$ or $f(x) > k$. If f is a linear polynomial the process is easy (see the problem set for the last section). If f is a quadratic polynomial the problem is somewhat more complicated but not too difficult. The inequalities $f(x) < k$ and $f(x) > k$ can be changed to $f(x) - k < 0$ and $f(x) - k > 0$. Now, $f(x) - k$ is a polynomial so that if we know how to solve the latter set of inequalities, we can handle all related inequalities. From Figures 3.4.2 and 3.4.3 discuss the solution sets for $f(x) < 0$ and $f(x) > 0$ in terms of a, b, and c where $f(x) = ax^2 + bx + c$, and in terms of the zeros of f.

10. If $f(x) = ax^2 + bx + c$ has zeros it can be factored into $(\alpha x + \beta)(\gamma x + \delta)$. Then $f(x) < 0$ can be studied in terms of $(\alpha x + \beta)(\gamma x + \delta) < 0$. The product of two real numbers is negative if and only if one is negative and the other is positive. Thus we are faced with two choices:

        1. $\alpha x + \beta < 0$ and $\gamma x + \delta > 0$, or
        2. $\alpha x + \beta > 0$ and $\gamma x + \delta < 0$

Our conclusions concerning the x satisfying $f(x) < 0$ come from our analysis of 1 and 2. For example, find the values x for which $x^2 + 2x - 3 < 0$.

    Since $x^2 + 2x - 3 = (x - 1)(x + 3)$ we have two possibilities:

        1. $x - 1 < 0$ and $x + 3 > 0$, or
        2. $x - 1 > 0$ and $x + 3 < 0$

From 1 we see that $x < 1$ and $x > -3$ (that is, the solution set for 1 is $(-\infty, 1) \cap (-3, \infty) = (-3, 1)$) while from 2, $x > 1$ and $x < -3$. However, the intersection of $(1, \infty)$ and $(-\infty, -3)$ is empty. Thus, the solution set for the inequality is a subset of $(-3, 1) \cup \phi = (-3, 1)$. Consequently, the only x satisfying $x^2 + 2x - 3 < 0$ come from $(-3, 1)$. Moreover, all such x satisfy the inequality.

a. Describe conditions analogous to 1 and 2 for the inequality $f(x) = (\alpha x + \beta)(\gamma x + \delta) > 0$.

Find the solution set for the following:

b. $P(x) < 0$, P from Exercise 3.4.1a.
c. $P(x) < 0$, P from Exercise 3.4.1b.
d. $P(x) < -11$, P from Exercise 3.4.1c.
e. $P(x) < 12$, P from Exercise 3.4.1d.
f. $P(x) < 25/4$, P from Exercise 3.4.1f.
g. $P(x) > 0$, P from Exercise 3.4.1a.
h. $P(x) > 0$, P from Exercise 3.4.1b.
i. $P(x) > -11$, P from Exercise 3.4.1c.
j. $P(x) > -12$, P from Exercise 3.4.1d.
k. $P(x) > -25/4$, P from Exercise 3.4.1f.

11. Graph the solution sets for the following conditions.

a. $P(x) < y < 125$, P from Exercise 3.4.1a.
b. $P(x) \leq y \leq 13$, P from Exercise 3.4.1c.
c. $0 \leq y \leq P(x)$, P from Exercise 3.4.1d.
d. $0 \leq y \leq P(x)$, P from Exercise 3.4.1e.
e. $-x^2 \leq y \leq x^2$
f. $x^2 - 7x - 12 \leq y \leq -x^2 + 7x + 12$
g. $x^2 + 5x + 6 \leq y \leq -x^2 + 7x + 12$

12. Find the values x for which $f(x) = P(x)$ where $f(x)$ is a linear polynomial and $P(x)$ is a quadratic polynomial as given below.

a. $P(x) = x^2 + 6x + 6$, $f(x) = x$
b. $P(x) = x^2 + x + 1$, $f(x) = 2x$
c. $P(x) = x^2$, $f(x) = x - 1$
d. $P(x) = x^2 - x - 7$, $f(x) = 2x + 1$

13. Sketch each pair of functions from Exercise 3.4.12. In each case where the line intersects the parabola, an area is enclosed. Sketch the set of all points $A = \{(x, y) : (x, y) \text{ is enclosed as described.}\}$ Then determine $p_1[A]$ and $p_2[A]$.

14. The following is another example of a locus problem. Write an equation to denote the set

$$\{(x, y) : d((x, y), (1, 0)) + d((x, y), (-1, 0)) = 5\}$$

Sketch a graph of this set.

15. The hyperbolic functions (Example 2.7.3 and Exercise 2.7.11) have inverses (see Example 2.8.6). Determine [in terms of $\log_e (\ )$] the following inverse functions.

a. argcosh (domain $= (1, \infty)$)
b. argtanh (domain $= (-1, 1)$)
c. argcoth (domain $= (-\infty, 1) \cup (1, \infty)$)

16. A ball is thrown in the air and its height at any time t is given by $h(t) = 6 + 8t - t$. What is its maximum height?

17. A rocket is launched at time $t = 0$ and its velocity $v(t)$ at any time t is given by $v(t) = 4000t - 5t$. What is its maximum velocity? Minimum velocity?

3.5 ZEROS FOR GENERAL POLYNOMIALS

We wish now to examine general polynomials and their zeros. As we might suspect, certain types of polynomials will be easier to handle than others and certain zeros will be easier to find than others.

Most of our polynomials have been such that their coefficients have been integers. This type is the easiest to examine. The *rational zeros* (rational numbers p/q such that p/q is a zero) for such polynomials are quickly determined.

Let $f(x) = a_n x^n + a_{n-1} x^{n-1} + \cdots + a_1 x + a_0$ be a polynomial of degree n such that each coefficient is an integer. Furthermore, suppose that p/q is a rational zero with p/q in lowest terms. That is,

$$0 = f(p/q) = a_n p^n/q^n + a_{n-1} p^{n-1}/q^{n-1} + \cdots + a_1 p/q + a_0$$

or, multiplying by q^n, we have

$$0 = a_n p^n + a_{n-1} p^{n-1} q + \cdots + a_1 p q^{n-1} + a_0 q^n.$$

But this yields

$$-a_n p^n = a_{n-1} p^{n-1} q + \cdots + a_1 p q^{n-1} + a_0 q^n$$
$$= q(a_n p^{n-1} + \cdots + a_1 p q^{n-2} + a_0 q^{n-1})$$

Since p/q was in lowest terms, no factor of q will divide p, meaning that q divides into (is a factor of) a_n. This follows from the fact that

$$-a_n p^n/q = a_n p^{n-1} + \cdots + a_1 q^{n-1} + a_0 q^{n-1}$$

is an integer. In addition, the form

$$0 = a_n p^n + a_{n-1} p^{n-1} q + \cdots + a_1 q^{n-1} + a_0 q^n$$

can be written

$$-a_0 q^n = p(a_n p^{n-1} + a_{n-1} p^{n-2} q + \cdots + a_1 q^{n-1})$$

whereby p is a factor of a_0. Our conclusion:

Theorem 3.5.1. If $f(x) = a^n x^n + a_{n-1} x^{n-1} + \cdots + a_1 x + a_0$ *has only integral coefficients, each rational zero* p/q *(in lowest terms) is such that* p *is a factor of the constant term while* q *is a factor of the leading coefficient (coefficient of the term of highest degree).*

EXAMPLE 3.5.1. Let $f(x) = 2x^3 + 3x^2 - x - 1$. Find the zeros of f.

If f has any rational zeros, they must be among ± 1 and $\pm 1/2$, since the factors of the constant term are 1 and -1 while the factors of the leading coefficient are ± 1 and ± 2. A quick check (by substitution or synthetic division) shows that ± 1 and $1/2$ are not zeros. However,

$$
\begin{array}{r|rrrr}
-1/2 & 2 & 3 & -1 & -1 \\
     &   & -1 & -1 & 1 \\
\hline
     & 2 & 2 & -2 & \underline{0} \\
\end{array}
$$

shows that $f(x) = 2x^3 + 3x^2 - x - 1 = (x + 1/2)(2x^2 + 2x - 2) = 2(x + 1/2)(x^2 + x - 1) = (2x + 1)(x^2 + x - 1)$.

The polynomial $f(x) = 0$ if and only if $2x + 1 = 0$ or $f_1(x) = x^2 + x - 1 = 0$. The value $x = -1/2$ is a known zero and remaining zeros for f may be found by examining the depressed function $f_1(x)$. The quadratic formula shows the zeros of f_1 (the remaining zeros of f) to be

$$
\frac{-1 \pm \sqrt{1+4}}{2} = \frac{-1 \pm \sqrt{5}}{2}.
$$

EXAMPLE 3.5.2. Find all the zeros of f where $f(x) = 2x^3 + (17/3)x^2 + (8/3)x + 1/3$. We, of course, would like to be able to use the simple criterion for finding all possible rational zeros for f. However, the coefficients of f are not all integral. We are not hampered in the least, though, for $f(x) = (1/3)(6x^3 + 17x^2 + 8x + 1)$ and the zeros for $f(x)$ and $6x^3 + 17x^2 + 8x + 1$ are identical. Thus the possible rational roots for f are -1, $-1/2$, $-1/3$, and $-1/6$. (There can be no positive zeros since the coefficients are all positive.)

$$
\begin{array}{r|rrrr}
-1/3 & 6 & 17 & 8 & 1 \\
     &   & -2 & -5 & -1 \\
\hline
     & 6 & 15 & 3 & \underline{0} \\
\end{array}
$$

shows that $-1/3$ is a zero for f and $f(x) = (1/3)(x + 1/3)(6x^2 + 15x + 3) = (1/3)(3x + 1)(2x^2 + 5x + 1)$. A check of the depressed function given by $2x^2 + 5x + 1$ shows the remaining zeros of f to be $\dfrac{-5 \pm \sqrt{17}}{4}$.

All polynomials with rational coefficients can be investigated in an analogous manner.

EXAMPLE 3.5.3. Find the zeros for $f(x) = x^4 - 2x^3 + 2x^2 - 2x + 1$. The only possible rational roots for f are ± 1. Since the coefficients of f add to zero, $z = 1$ must be a zero for f.

$$
\begin{array}{r|rrrrr}
1 & 1 & -2 & 2 & -2 & 1 \\
  &   & 1 & -1 & 1 & -1 \\
\hline
  & 1 & -1 & 1 & -1 & \underline{} 0 \\
\end{array}
$$

The coefficients of the depressed polynomial $x^3 - x^2 + x - 1$ also add to zero so that we can divide it by $x - 1$.

$$
\begin{array}{r|rrrr}
1 & 1 & -1 & 1 & -1 \\
  &   & 1 & 0 & 1 \\
\hline
  & 1 & 0 & 1 & \underline{} 0 \\
\end{array}
$$

We may write $f(x) = (x - 1)(x - 1)(x^2 + 1)$, and say that $x = 1$ is a *double zero*, for $f[(x-1)^2$ divides $f(x)]$. The second depressed polynomial $x^2 + 0x + 1 = x^2 + 1$ has no real zeros, meaning that $x = 1$ is the only zero for f.

In our examples above we found the zeros of each function f by finding the rational zeros and computing depressed functions until arriving at a quadratic depressed polynomial. However if we cannot find enough rational zeros (that is, if f does not have enough rational zeros) to form a quadratic depressed function, the problem becomes a little more difficult.

Our next alternative is to look for certain characteristics of the function that will give us clues as to how we may go about searching out the approximate values of zeros. We begin by stating a number of theorems (we do not offer proofs) that prove useful in our endeavor.

Theorem 3.5.2. **(Intermediate Value Theorem)** *If f is a polynomial, x_1 and x_2 are real numbers, and if k is any real number between $f(x_1)$ and $f(x_2)$ then there is a number ξ between x_1 and x_2 with $f(\xi) = k$.*

Figure 3.5.1 illustrates this concept. We observe that k is an arbitrary real number between $f(x_1)$ and $f(x_2)$. By the theorem, the horizontal line through $(0, k)$ intersects the graph of f in such a way that the projection onto the x-axis is between x_1 and x_2. Intuitively, there are no "jumps" in the value of $f(x)$. The graph moves "smoothly" from $(x_1, f(x_1))$ to $(x_2, f(x_2))$.

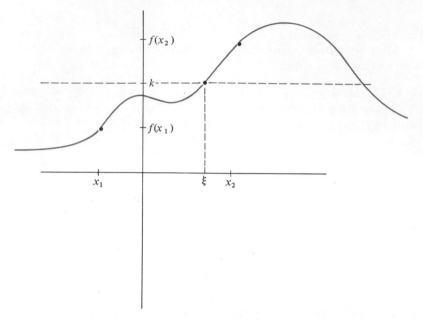

Figure 3.5.1 Intermediate value property.

As an application of Theorem 3.5.2 we observe that if $f(x_1) > 0$ and $f(x_2) < 0$, 0 lies between $f(x_1)$ and $f(x_2)$ so that $f(\xi) = 0$ for some ξ between x_1 and x_2. We might say that we are "bracketing" zeros of f. We are simply noticing that if f is positive at x_1 and negative at x_2, then $f(x)$ is zero somewhere between x and y. Exactly where the zero occurs is not determined directly by this process.

EXAMPLE 3.5.4. Locate a zero for f between two numbers differing by no more than 1 where $f(x) = x^4 + x^3 - x^2 - 2x - 2$.
Observe:

$$f(0) = -2$$
$$f(1) = -3$$
$$f(2) = 14$$

There is then, a zero (for f) between 1 and 2. What number is the zero for f? It is not clear. However, $f(3/2) = 19/16 > 0$ whence the zero in question lies between 1 and 3/2. This bracketing process can be used repeatedly to approximate the zero to any degree of accuracy desired [for example, next evaluate $f(5/4)$].

The function f above can be expressed in the form $f(x) = (x^2 - 2)(x^2 + x + 1) = (x - \sqrt{2})(x + \sqrt{2})(x^2 + x + 1)$ so that $x = \sqrt{2}$ is the zero we were attempting to approximate.

Another aid in determining where to look for zeros is given in the following theorem.

Theorem 3.5.3. (*Descartes' Rule of Signs*). *Let* $f(x) = a_n x^n + \cdots + a_1 x + a_0$. *The number of positive zeros for* f *is equal to (1) the number of sign changes occurring between successive coefficients, or (2) that number diminished by an even positive integer. Similarly, the number of negative zeros for* f *is equal to (1) the number of sign changes in* $f(-x)$, *or (2) that number reduced by an even natural number.*

EXAMPLE 3.5.5. Find the number of positive zeros and negative zeros for f where $f(x) = x^5 + x^4 + x^3 + x^2 - x - 1$.

There is only one sign change in the coefficients for $f(x)$, namely the change between the second-degree term and the first-degree term. Thus there is precisely one positive zero for f (one cannot be meaningfully diminished by any even positive integer).

Now, $f(-x) = (-x)^5 + (-x)^4 + (-x)^3 + (-x)^2 - (-x) - 1 = -x^5 + x^4 - x^3 + x^2 + x - 1$, whence the number of sign changes is 4. Consequently, f has either 4, 2, or no negative zeros.

We may proceed by further observing that the only possible rational roots are ± 1. The number 1 is not a zero but -1 is.

$$
\begin{array}{r|rrrrrr}
-1 & 1 & 1 & 1 & 1 & -1 & -1 \\
   &   & -1 & 0 & -1 & 0 & 1 \\
\hline
   & 1 & 0 & 1 & 0 & -1 & 0
\end{array}
$$

Thus $f(x) = (x + 1)(x^4 + x^2 - 1)$. If in $x^4 + x^2 - 1$ we replace x^2 by y we have $y^2 + y - 1$ which has zeros $\dfrac{-1 \pm \sqrt{5}}{2} = y$. Consequently, zeros for f occur where $x^2 = \dfrac{-1 \pm \sqrt{5}}{2}$ or $x = \pm \left[\dfrac{-1 + \sqrt{5}}{2}\right]^{1/2}$ or $x = \pm \left[\dfrac{-1 - \sqrt{5}}{2}\right]^{1/2}$. However, since $-1 - \sqrt{5} < 0$, the only possible zeros for f are $x = \pm \left[\dfrac{-1 + \sqrt{5}}{2}\right]^{1/2}$ (together with $x = -1$). We see that f has two negative and one positive zero.

In the example given here, we were able to find the zeros for the quartic polynomial, since no odd powers of x occurred. This is a useful technique since it does away with the need for bracketing irrational zeros.

EXERCISE 3.5

1. Find the possible number of positive zeros for f where:

a. $f(x) = 6x^3 + 3x^2 + 2x - 1$

    b. $f(x) = x^2 + 2x + 2$
    c. $f(x) = x^3 + 2x$
    d. $f(x) = x^5 + 2x^2 - 3x + 7$
    e. $f(x) = x^3 - x^2 - 3x + 1$
    f. $f(x) = x^3 + 2x^2 - 5x - 6$
    g. $f(x) = x^4 - 5x^3 + 9x^2 - 8x - 4$

2.   Find the possible number of negative zeros for f in each case in Exercise 3.5.1.

3.   Find all rational zeros for each function f in Exercise 3.5.1.

4.   Let $u > 0$ and suppose further that f is a polynomial with $f(x) = (x - u) g(x) + r_u$ where all the coefficients of $g(x)$ and the number r_u are of the same sign. Then u is an upper bound of the zeros of f. [*Note:* To say that the coefficients of $g(x)$ and the term r_u have the same sign is to say that the bottom row in the synthetic division of $f(x)$ by $x - u$ consists in only nonnegative or nonpositive numbers.] Find upper bounds for the zeros for each f in Exercise 3.5.1.

5.   If f is a polynomial and $l < 0$ with $f(x) = (x - l)g(x) + r_l$, l is a lower bound of the zeros for f if the coefficients of $g(x)$ and r_l alternate (that is, if the bottom row in the synthetic division of $f(x)$ by $x - l$ consists of numbers alternating in sign: 0 may be considered as +0 or −0 as needed). Find a lower bound for the zeros of each polynomial f in Exercise 3.5.1.

6.   Find all zeros for each f from Exercise 3.5.1 where possible and locate the others between two successive integers.

7.   Let n be any positive integer and $r \in R$. Show that $x - r$ is a factor of $x^n - r^n$ and $x + r$ is a factor of $x^n + r^n$ if, in the latter case, n is odd. (*Hint:* Use mathematical induction.)

8.   Show that if all the a_i are positive and n is odd, $f(x) = a_n x^n - a_{n-1} x^{n-1} + a_{n-2} x^{n-2} - \cdots + a_1 x - a_0$ has no negative roots.

9.   Let the a_i be as in Exercise 3.5.8. Show that $f(x) = a_n x^n + a_{n-1} x^{n-1} + \cdots + a_1 x + a_0$ has no positive roots.

10.   Why must a polynomial of odd degree have at least one zero?

Find the roots in Exercises 3.5.11–3.5.15.

11. $2x^3 - 3x^2 - 11x + 6 = f(x)$

12. $x^4 - 16x^3 + 86x^2 - 176x + 105 = f(x)$

13. $x^4 - 4x^3 + 4x - 1 = f(x)$

14. $8x^5 - 12x^4 + 14x^3 - 13x^2 + 6x - 1 = f(x)$

15. $x^6 - 64 = f(x)$

Find all the zeros for $f(x)$ to within a tolerance of $1/100$ in Exercises 3.5.16–3.5.18.

16. $x^4 - x^3 + 2x^2 - 3x - 2 = f(x)$

17. $x^4 - 2x^3 - x^2 - 1 = f(x)$
18. $x^4 - 4x^3 - 4x + 12 = f(x)$
19. Using the methods of this chapter, find each of the following to within a tolerance of $1/1000$.

    a. $\sqrt[3]{6}$ b. $\sqrt[4]{2}$ c. $\sqrt[5]{-9}$

3.6 REVIEW EXERCISES

1. Locate all zeros of the following polynomials and sketch a graph of each.

    a. $f(x) = 2x - 3$
    b. $g(x) = x^2 + 20x + 51$
    c. $\beta(x) = x^2 + 3x + 6$
    d. $\eta(x) = x^3 - 1$
    e. $\lambda(x) = 6x^3 - 29x^2 - 6x + 5$
    f. $\mu(x) = x^4 + 3x^3 + 3x^2 - 3x - 4$
    g. $\delta(x) = x^4 + 6x^2 + 9$

2. Solve for all x where:

    a. $f(x) = 7$, f from Exercise 3.6.1a.
    b. $g(x) = h(x)$, where $h(x) = 20x + 100$, g is from Exercise 3.6.1b.
    c. $\beta(x) = \beta(x) - x^2$, β from Exercise 3.6.1c.

3. Locate two zeros for each polynomial below to within a tolerance of $1/10$.

    a. $x^4 + x^3 - x^2 - 2x - 2$
    b. $x^3 + 8x^2 - 5x - 40$
    c. $x^4 - 10x^2 + 21$
    d. $x^4 - 4x^2 - x + 2$
    e. $5x^5 + 3x^4 - 34x^3 + 38x^2 - 39x + 35$
    f. $x^6 + 2x^5 - 10x^4 - 22x^3 - 12x^2 + 11$

4. How many positive zeros can each polynomial in Exercise 3.6.3 have? Negative zeros?

5. Find the equations of the following straight lines.

    a. Passing through $(6, 1)$ and $(6, 2)$
    b. Passing through $(6, 1)$ and $(5, 1)$
    c. Passing through $(2, 3)$ and $(3, 5)$
    d. Passing through $(1, 7)$ and $(3, 1)$
    e. Passing through $(3, -2)$ with slope $-2/3$
    f. Passing through $(-1, 2)$ with slope $4/3$
    g. Slope $1/2$ and y-intercept 2
    h. Slope $2/3$ and x-intercept -4

6. Let P_1 be (x_1, x_2) and P_2 be (z_1, z_2). Find the midpoint between P_1 and P_2 and call it P_3. Find the midpoints P_4 and P_5 between P_1 and P_3 and P_3 and P_2 respectively. Write the equation of l_1 passing through P_1 and P_2. Then write the equations of the following lines with slope m, $m/2$,

and $m/4$ respectively: the line l_2 passing through P_3, l_3 passing through P_4 and l_4 passing through P_5. Under what conditions will l_2, l_3, and l_4 be parallel? Can two of these lines be parallel without all three satisfying the parallel condition?

7. Let l and l' be lines of slopes m and m' respectively (it is implied that m and m' are real numbers).

   a. Show that if $mm' = -1$, l and l' intersect.

   b. Let the situation in a. be the case and let $P_1 = (a, b)$ be the point of intersection. Let $P_2 = (x, y)$ be on l but different from P_1 with $P_3 = (p, q)$ on l' different from P_1. Show that the triangle formed by these three points is a right triangle in the sense of the converse of the Pythagorean theorem.

   c. To examine a converse of b, let $P_1 = (a, b)$ be the intersection of l and l' with slopes m and m' respectively. Suppose that whenever $P_2 = (x, y) \in l$ and $P_3 = (p, q) \in l'$ with neither the same as P_1, it follows that the triangle formed is a right triangle in the sense of the Pythagorean theorem. Prove then that $mm' = -1$. [*Note:* The lines in b. and c. are usually said to be *perpendicular.*]

   d. Find an equation of the line perpendicular to $y = 3x + 2$ at the point $(0, 2)$.

   e. Find an equation of the perpendicular to the line passing through $(1, 5)$ and $(2, 7)$ such that the perpendicular passes through the origin.

   f. A *tangent* to a circle at a point (a, b) is the line through (a, b) perpendicular to the radius line drawn through the center of the circle and (a, b). Find the line tangent to the circle of center $(1, 3)$ and radius 17 at the point $(0, 7)$.

   g. Find the equation of a tangent to the circle $(x - 2)^2 + (y + 3)^2 = 25$ so that the tangent is horizontal (has slope zero). There are two such.

   h. Find the equation of two tangents to the circle $(x - h)^2 + (y - k)^2 = r^2$ where the tangents do not have a defined slope.

8. a. Given the line $x = -5$ and the point $(5, 0) = P_0$, find the equation representing the points $(x, y) = P$ where $d(P, P_0)$ is equal to the distance from (x, y) to $(-5, y)$. The distance from (x, y) to $(-5, y)$ is seen to be the distance along the line perpendicular to $x = -5$ and passing through (x, y). By "distance along the perpendicular" is meant the distance from the given point to the intersection of the given line with the constructed perpendicular. [*Note:* The curve generated in this problem is called a *parabola.* That curve in Exercise 3.3.10 is the *perpendicular bisector* of the line segment from $(1, 3)$ to $(-1, 2)$.]

   The curve of Exercise 3.4.14 is an *ellipse.* The curve in the following problem is a *hyperbola.*

   b. Consider the two points $P_1 = (3, 0)$ and $P_2 = (-3, 0)$. Write an equation in x and y depicting the set of points

   $$\{P = (x, y) : |d(P, P_1) - d(P, P_2)| = 2\}$$

9. Find in each case the point(s) of intersection of the given line and circle.

    a. line $y = 3x - 2$, circle $x^2 + y^2 = 4$
    b. line $y + 2 = x$, circle $x^2 + (y - 1)^2 = 25$
    c. line through $(5, 10)$ with slope 1 and circle of center $(5, 10)$ and radius 1.

10. Let (a, b) and (c, d) be two points. The point (x, y) is *between* (a, b) and (c, d) if there is a $\lambda \in R$ with $0 < \lambda < 1$ and $(x, y) = (a + \lambda(c - a)$, $b + \lambda(d - b)) = (\lambda c + (1 - \lambda)a, \lambda d + (1 - \lambda)b)$. Show that each point between (a, b) and (c, d) lies on the line containing the two points. Furthermore, if (x, y) is between (a, b) and (c, d), $d((a, b), (x, y)) + d((x, y), (c, d)) = d((a, b), (c, d))$.

11. Refer to Exercise 3.6.10 for definitions. The *ray emanating from* (a, b) *and passing through* (c, d) is the set

    $$\{(x, y) : (x, y) = (a, b), (x, y) \text{ is between } (a, b) \text{ and } (c, d)$$

    or

    $$(c, d) \text{ is between } (a, b) \text{ and } (x, y)\}$$

    Show that the ray is a subset of the line through (a, b) and (c, d).

12. Use Exercise 3.6.11 to show that if a ray emanates from the origin and passes through (c, d), and if (x, y) is on the ray, x and c are both positive or both negative. Also, d and y are both positive or both negative (that is, x and c have the same sign while d and y have the same sign).

13. Suppose (a, b) and (c, d) lie on the same ray (Exercise 3.6.11). Then, (a, b) belongs to the circle given by $x^2 + y^2 = a^2 + b^2$ and (c, d) lies on the circle $x^2 + y^2 = c^2 + d^2$. How does the arc length from $((a^2 + b^2)^{1/2}, 0)$ to (a, b) compare with the arc length from $((c^2 + d^2)^{1/2}, 0)$ to (c, d)? Your answer may be formed by your answer to the last question of Exercise 3.6.3. Observe that each of the arc lengths above can be compared to that from $(1, 0)$ to $(a/(a^2 + b^2)^{1/2}, b/(a^2 + b^2)^{1/2})$ along the unit circle. Show that $(a/(a^2 + b^2)^{1/2}, b/(a^2 + b^2)^{1/2})$ is on the unit circle and on the ray through (a, b) and (c, d). In Exercise 3.6.7, show that if $a^2 + b^2 < 1$ and $c^2 + d^2 > 1$, then $(a/(a^2 + b^2)^{1/2}, b/(a^2 + b^2)^{1/2})$ lies between (a, b) and (c, d).

14. The *line segment* between (a, b) and (c, d) is the set

    $$\{(x, y) : (x, y) = (a, b), (x, y) = (c, d)$$

    or

    $$(x, y) \text{ is between } (a, b) \text{ and } (c, d)\}.$$

    The line segment between (a, b) and (c, d) is "very nearly" the intersection of two rays. Describe this line segment in terms of the intersection of two rays. (There are two ways of doing it.)

15. Refer to Exercise 3.6.14. Describe the line segment between each pair of points in terms of two rays, describing the rays completely.

a. $(0, 1)$ and $(0, 5)$ d. $(3, 5)$ and $(9, 11)$
b. $(1, 1)$ and $(2, 2)$ e. $(2, 4)$ and $(1, 6)$
c. $(-5, 4)$ and $(-5, 12)$ f. $(-2, -3)$ and $(-4, 1)$

3.7 QUIZ

Give yourself one hour to work this quiz. Then check your solutions.

1. Write $f(x)$ in the form $(x - 2)g(x) + r$ where $f(x) = x^5 + 2x^3 + x^2 + x - 3$.

2. How many positive zeros can f have? negative zeros?

3. What are the choices for rational zeros of f? Does f have any rational zeros?

4. Does f have any zeros greater than 3?

5. Does f have any zeros less than -3?

6. Locate a zero of f to within a tolerance of $1/10$.

7. Find the points of intersection of f and q where $f(x) = 3x - 7$ and $q(x) = 5x^2 + 5x - 8$.

8. Graph the function in problem 7 and shade in the set $\{(x, y) : q(x) \leqslant y \leqslant f(x)\}$.

9. For what x is the following inequality satisfied?

$$|x - 5| < 3$$

[*Hint:* $|a| = \sqrt{a^2}$ and $0 \leqslant \sqrt{p} < \sqrt{q}$ implies $p < q$.]

3.8 ADVANCED EXERCISES

1. Prove the assertions about upper and lower bounds for zeros of polynomials made in Exercises 3.5.4 and 3.5.5.

2. Use a linear function to show that a $1:1$ onto function exists between any two open intervals (a, b) and (c, d) $(a < b$ and $c < d)$.

3. Use Exercise 3.8.2 to show that $[a, b) \sim [c, d)$ (that is, there is a $1:1$ onto function between $[a, b)$ and $[c, d)$.

4. Use Exercise 3.8.2 to show that $[a, b] \sim [c, d]$.

5. Show that $[a, b) \sim (a, b]$.

6. Find a $1:1$ onto function from (a, b) to $[c, d)$.

7. Show that $[a, b) \sim [a, b]$.

8. From Exercises 3.8.2–3.8.7 show that $(a, b) \sim [a, b]$.

9. Use a technique analogous to that in Exercises 3.8.2–3.8.7 to show that $(a, \infty) \sim [a, \infty) \sim (-\infty, b) \sim (-\infty, b]$.

10. The removal of a straight line l from the plane results in two disjoint collections of points (see Figure 3.8.1). We define: P_1 and P_2 (neither on l) lie on the *same side* of l if the line segment between P_1 and P_2 does not contain a point of l. Show that if P_1 and P_2 lie on the same side of l and if P_2 and P_3 lie on the same side of l, then P_1 and P_3 lie on the same side of l. (This shows that the relation "lies on the same side of l as" is an equivalence relation on $R^2 - l$.) If P_1 and P_2 (neither on l) do not lie on the same side of l, they lie on *opposite sides* of l. Show that l has exactly two sides and that these sides may be determined by (1) picking P_1 and P_2 (neither on l) on opposite sides, and (2) showing that:

$A = \{P_3 : P_3 \text{ and } P_1 \text{ lie on the same side of } l\}$
$B = \{P_3 : P_3 \text{ and } P_2 \text{ lie on the same side of } l\}$

satisfy $A \cap B = \phi$ and $A \cup B = R^2 - l$.

11. Show that if l is given by the equation $y = mx + b$, then
$$A = \{(x, y) : y < mx + b\}$$
and
$$B = \{(x, y) : y > mx + b\}$$
form the two sides of l.

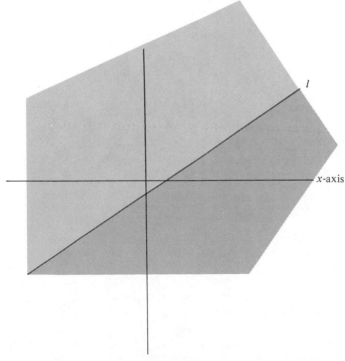

Figure 3.8.1

12.   Let $f(x) = ax^2 + bx + c$ and denote by A and B the sets

$$A = \{(x, y) : y < f(x)\}$$
$$B = \{(x, y) : y > f(x)\}$$

Show that if $P_1 = (\alpha, \beta) \in A$ and $P_2 = (\lambda, \delta) \in B$, the line segment joining P_1 and P_2 intersects f. This may be extended to arbitrary polynomials f.

CHAPTER 4

Other Real Functions

4.1 RATIONAL FUNCTIONS

We have already introduced as *rational* those functions expressible as quotients of polynomials. As has been our custom, we are to be interested in finding zeros for such functions and we are to be curious about their graphs.

The rational functions in general will not have R as a domain. If $f = g/h$, g and h polynomials, $f(x)$ fails to be defined whenever $h(x) = 0$. Consequently the domain of f is $R - H$ where H is the set of zeros for h.

The quotient a/b of real numbers is zero if and only if $a = 0$ and $b \neq 0$. We easily see, therefore, that the zeros of $f = g/h$ are those zeros of g that are not zeros of h (that is, those x for which $g(x) = 0$ but $h(x) \neq 0$). It is clear that we must know the zeros of both g and h. Our knowledge of these zeros must come directly from our study of the preceding chapter.

The zeros of h (in $f = g/h$) are known as *poles* of f and if the multiplicity of a zero a of h is greater than the multiplicity of a as a zero of g, then $x = a$ is a *vertical asymptote* for f. In general, a graph of f will "approach" the vertical asymptote in some sense without intersecting it. Diagrammatically, if $x = a$ is a vertical asymptote, the values of $f(x)$ become unboundedly large (in absolute value) as x nears a. Example 4.1.1 illustrates the concept of the zeros and vertical asymptotes for a rational function.

EXAMPLE 4.1.1. Let $f(x) = (x-3)(x+2)/(x+2)(x-1)$. Find the zeros, poles, and vertical asymptotes for f.

The zeros for $(x-3)(x+2)$ are 3 and -2 while the zero set for $(x+2)(x-1)$ is $\{1,-2\}$. Thus, the only zero for f is $x=3$ ($x=-2$ is not a domain value for f; its domain is $R - \{1,-2\}$).

The poles for f are -2 and 1. The vertical asymptotes are found by examining the poles. Since -2 is a zero of order one for both the numerator and denominator, that value must be discarded as a possible vertical asymptote. The value 1, however, is not a zero of the numerator, whence $x=1$ is a vertical asymptote.

Figure 4.1.1 shows a graph of f with the vertical asymptote shown in all its intuitive significance. A "hole" appears in the graph of f where $x=-2$ occurs. This happens because $f(-2)$ is not defined. The graph behaves nicely near $x=-2$ since if $x \neq -2$, $f(x) = (x-3)/(x-1)$. The expression $(x-3)/(x-1)$ has no pole at -2 indicating that the behavior there is not as wild as it is at a vertical asymptote.

Unlike polynomials, as x grows large in absolute value, rational functions can find themselves bounded. The criterion for determining

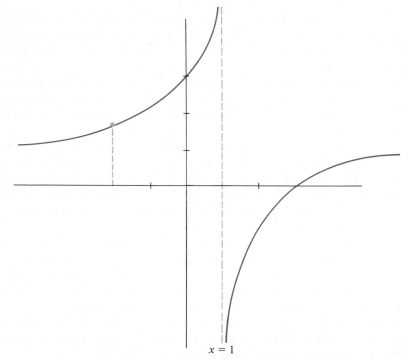

$x = 1$

Figure 4.1.1 $f(x) = (x-3)(x+2)/(x+2)(x-1)$.

the behavior of rational functions for large values of x is tied directly to the relative sizes of the degrees of the numerator and denominator polynomials.

If $f = g/h$ with g and h having the same degree, f has a *horizontal asymptote*. To see this concept, follow the analysis below.

Write

$$g(x) = a_n x^n + a_{n-1} x^{n-1} + \cdots + a_1 x + a_0$$

and

$$h(x) = b_n x^n + b_{n-1} x^{n-1} + \cdots + b_1 x + b_0$$

Then f is given by

$$f(x) = \frac{a_n x^n + a_{n-1} x^{n-1} + \cdots + a_1 x + a_0}{b_n x^n + b_{n-1} x^{n-1} + \cdots + b_1 x + b_0}$$

$$= \frac{a_n + a_{n-1}/x + \cdots + a_1/x^{n-1} + a_0/x^n}{b_n + b_{n-1}/x + \cdots + b_1/x^{n-1} + b_0/x^n} \qquad \text{if } x \neq 0.$$

The last form comes by dividing the numerator and denominator by x^n. Intuitively, we see that as x grows large, each term with an x in its denominator approaches zero leaving us with the fact that $f(x)$ approaches a_n/b_n. The horizontal line $y = a_n/b_n$ is called a *horizontal asymptote* for f. The expression a_n/b_n is merely the quotient of the two leading coefficients (of g and h). Unlike vertical asymptotes, horizontal asymptotes may intersect the graph.

EXAMPLE 4.1.2.   In Example 4.1.1,

$$f(x) = (x-3)(x+2)/(x+2)(x-1) = (x^2 - x - 6)/(x^2 - x - 2)$$

has a horizontal asymptote since the degree of both the numerator and denominator is 2. Find the horizontal asymptote.

The leading coefficient of the numerator is one, as is the leading coefficient of the denominator. Thus $y = 1/1 = 1$ is the horizontal asymptote. Figure 4.1.2 is the same as 4.1.1 but with the horizontal asymptote added. Its significance relative to the behavior of $f(x)$ as x grows large is clear.

The graph in Figure 4.1.2 is sketched easily using the information derived. First, the horizontal and vertical asymptotes are sketched in. Remember that the function cannot intersect its vertical asymptote. The graph of f cannot in this case, cross its horizontal asymptote since $(x-3)(x-2)/(x+2)(x-1) = 1$ implies that $-1 = -3$, an absurdity. Since the function has a zero at $x = 3$ and $f(0) = 3$, the locations of the two portions of the curve are known. The actual shape

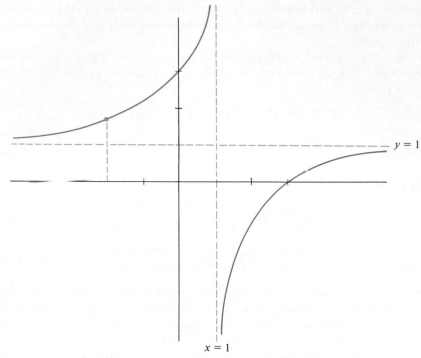

Figure 4.1.2 $f(x) = (x-3)(x+2)/(x+2)(x-1)$.

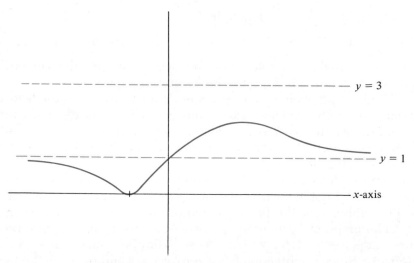

Figure 4.1.3 $f(x) = (x^2 + 2x + 1)/(x^2 + 1)$.

is a "rough" estimate of what the curve must look like. The "hole" comes from knowing that $x = -2$ is not a domain element for f.

EXAMPLE 4.1.3. Let $f(x) = g(x)/h(x) = (x^2 + 2x + 1)/(x^2 + 1)$. Find the zeros, poles, vertical asymptotes and horizontal asymptote for f.

Now h has no real zeros, whence there are no poles or vertical asymptotes. However, $x^2 + 2x + 1$ has a double zero at $x = -1$.

The horizontal asymptote is $y = 1$ as it was in Example 4.1.2 but there is one difference: $f(0) = 1$ (that is, the graph of f crosses the horizontal asymptote). Furthermore, f is a bounded function as is seen by the existence of the bounding lines in Figure 4.1.3.

If the degree of g is less than the degree of h and if $f = g/h$, we say that f has the horizontal asymptote $y = 0$. The analysis below shows why this ought to be.

Let $g(x) = a_n x^n + a_{n-1} x^{n-1} + \cdots + a_1 x + a_0$ with $h(x) = b_m x^m + b_{m-1} x^{m-1} + \cdots + b_1 x + b_0$. Then

$$f(x) = \frac{a_n x^n + a_{n-1} x^{n-1} + \cdots + a_1 x + a_0}{b_m x^m + b_{m-1} x^{m-1} + \cdots + b_1 x + b_0}$$

$$= \frac{a_n/x^{m-n} + a_{n-1}/x^{m-n+1} + \cdots + a_1/x^{m-1} + a_0/x^m}{b_m + b_{m-1}/x + \cdots + b_1/x^{m-1} + b_0/x^m}$$

The numerator of $f(x)$ in the last form tends to zero as x grows large $[(m - n) > 0]$ while the denominator tends to b_m. The quotient $f(x)$ then must tend to zero.

EXAMPLE 4.1.4. Let $f(x) = g(x)/h(x) = (x + 1)/(x + 2)(x + 3) = (x + 1)/(x^2 + 5x + 6)$.

The only zero for f is $x = -1$ while $x = -2$ and $x = -3$ are vertical asymptotes determined by the pole values -2 and -3. The domain for f is $R - \{-2, -3\}$.

Since the degree of g is one and the degree of h is two, f has $y = 0$ as a horizontal asymptote. Note that the graph of f (Figure 4.1.4) does intersect the horizontal asymptote. Horizontal asymptotes describe only the behavior of rational functions as x grows large.

The graph in Figure 4.1.4 was sketched by drawing in the axes, the vertical asymptotes, and the zero, observing that the x-axis is a horizontal asymptote, and noticing:

1. $f(x) < 0$ for $x < -3$
2. $f(x) > 0$ for $-3 < x < -2$, and
3. $f(x) < 0$ for $-2 < x < 1$

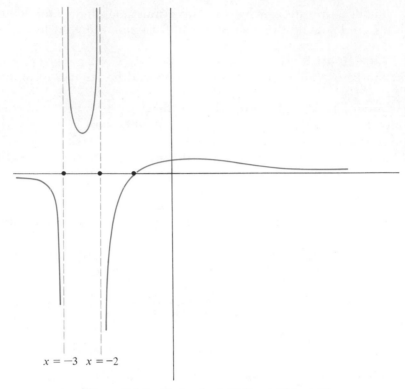

$x = -3$ $x = -2$

Figure 4.1.4 $f(x) = (x + 1)/(x + 2)(x + 3)$.

The shape of the right-hand portion follows from 1 and the knowledge of the horizontal asymptote. The center segment of the graph follows from 2 and the knowledge about the vertical asymptotes. The right-hand section is a result of 3, the zero at $x = -1$, and the fact concerning the horizontal asymptote.

Suppose now that the degree of g is $n + 1$, one more than the degree n of h. The rational function $f = g/h$ has an *oblique asymptote*.

Let $g(x) = a_{n+1}x^{n+1} + a_n x^n + \cdots + a_1 x + a_0$ and $h(x) = b_n x^n + b_{n-1}x^{n-1} + \cdots + b_1 x + b_0$. Then

$$f(x) = \frac{a_{n+1}x^{n+1} + a_n x^n + a_{n-1}x^{n-1} + \cdots + a_1 x + a_0}{b_n x^n + b_{n-1}x^{n-1} + \cdots + b_1 x + b_0}$$

$$= \frac{a_{n+1}x + a_n + a_{n-1}/x + \cdots + a_1/x^{n-1} + a_0/x^n}{b_n + b_{n-1}/x + \cdots + b_1/x^{n-1} + b_0/x^n}$$

As x grows large, $f(x)$ begins to approximate $(a_{n+1}x + a_n)/b_n$, a linear polynomial. The straight line $y = (a_{n+1}x + a_n)/b_n$ is called the *oblique asymptote* for f. Like horizontal asymptotes, oblique asymptotes may intersect the functions. Like horizontal asymptotes, oblique asymptotes tell the approximate behavior of rational functions for large domain values.

EXAMPLE 4.1.5. Let

$$f(x) = (x + 3)(x + 2)(x - 1)(x + 1)/(x + 2)^2(x - 3).$$

Find the zeros, poles, vertical asymptotes, and oblique asymptote for f.

The zeros for f are -3, $+1$, and -1. (Why is -2 not a zero?) The poles for f are -2 and 3. Since -2 is a double zero of the denominator and only a zero of order one of the numerator, $x = -2$ is a vertical asymptote as is $x = 3$.

Since $f(x) = (x^4 + 5x^3 + 5x^2 - 5x - 6)/(x^3 + x^2 - 8x - 12)$, $y = (x + 5)/1 = x + 5$ is the oblique asymptote.

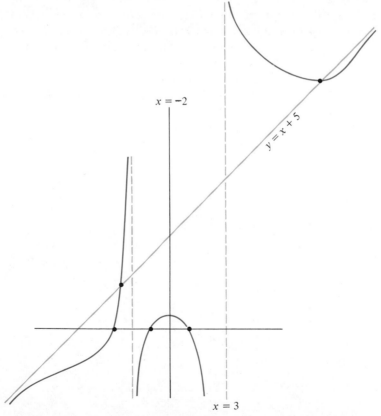

Figure 4.1.5 $f(x) = (x + 3)(x + 2)(x - 1)(x + 1)/(x + 2)^2(x - 3)$.

Does f intersect the oblique asymptote? This must be known in order to sketch accurately a graph of f. That is, does $f(x) = x + 5$ have a solution?

$$(x + 3)(x + 2)(x - 1)(x + 1)/(x + 2)^2(x - 3) = x + 5$$
$$(x + 3)(x + 2)(x - 1)(x + 1) = (x + 5)(x - 3)(x + 2)^2$$

Since $x = -2$ is not in the domain under consideration,

$$(x + 3)(x - 1)(x + 1) = (x + 5)(x - 3)(x + 2)$$
$$x^2 - 10x - 27 = 0$$
$$x = 5 \pm 2\sqrt{13}$$

We conclude that f does intersect its oblique asymptote twice. See Figure 4.1.5 for a sketch of f drawn from the knowledge of zeros and asymptotes.

EXAMPLE 4.1.6. Find the pertinent information about f where $f(x) = (x^3 + x - 1)/(3x^2 + 2)$.

The function f has no vertical asymptotes or poles. (Why?) It does have one irrational zero in $(0, 1)$. (Why?) Writing $f(x) = (x^3 + 0x^2 + x - 1)/(3x^2 + 0x + 2)$, we see that the oblique asymptote is given by $y = (x + 0)/3 = x/3$.

If we set $(x^3 + x - 1)/(3x^2 + 2) = x/3$, we get a solution $x = 3$. That is, $(3, 1)$ is a point of both the asymptote and the function. The

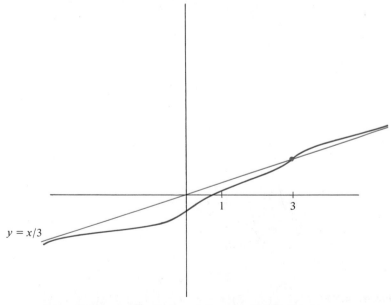

Figure 4.1.6 $f(x) = (x^3 + x - 1)/(3x^2 + 2)$.

function is sketched in Figure 4.1.6. Only the information given here is used in the sketch. Point by point plotting has given way to a knowledge of the behavior of rational functions.

EXAMPLE 4.1.7. Find points in common for the functions $f(x) = (x + 1)/(x - 1)$ and $g(x) = x + 2$.

Setting $f(x) = g(x)$, we have

$$(x + 1)/(x - 1) = x + 2$$
$$x + 1 = (x - 1)(x + 2) \qquad (x = -2 \text{ is not considered})$$
$$x^2 = 3$$
$$x = \pm\sqrt{3}$$

Now $g(\sqrt{3}) = 2 + \sqrt{3}$ while

$$f(\sqrt{3}) = (\sqrt{3} + 1)/(\sqrt{3} - 1)$$
$$= (\sqrt{3} + 1)(\sqrt{3} + 1)/(\sqrt{3} - 1)(\sqrt{3} + 1)$$
$$= (4 + 2\sqrt{3})/2 = 2 + \sqrt{3}$$

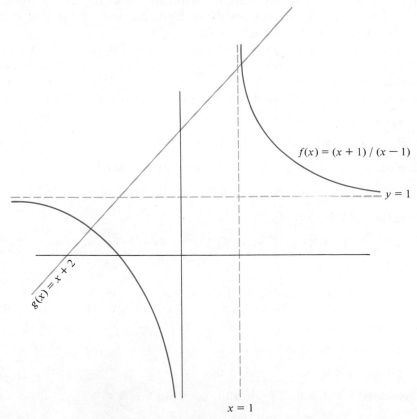

$$f(x) = (x + 1)/(x - 1)$$

$$y = 1$$

$$g(x) = x + 2$$

$$x = 1$$

Figure 4.1.7

We see that $g(\sqrt{3}) = f(\sqrt{3})$. Now $g(-\sqrt{3}) = 2 - \sqrt{3}$ while

$$f(-\sqrt{3}) = (1 - \sqrt{3})/(-1 - \sqrt{3})$$
$$= -(1 - \sqrt{3})(1 - \sqrt{3})/(1 + \sqrt{3})(1 - \sqrt{3})$$
$$= -(4 - 2\sqrt{3})/(-2) = 2 - \sqrt{3}$$

Thus $g(-\sqrt{3}) = f(-\sqrt{3})$. A sketch of the two rational functions (g is a rational function) is seen in Figure 4.1.7.

EXAMPLE 4.1.8.   Find all values x satisfying $(2x + 18)/(x + 4) <$ $x + 3$. This is equivalent to asking: for what domain values x is $f(x) < g(x)$ where $f(x) = (2x + 18)/(x + 4)$ and $g(x) = x + 3$. For $f(x)$ to be less than $g(x)$ implies that the particular portion of the graph of f is "below" g. (See Figure 4.1.8.)

Now if $(2x + 18)/(x + 4) < x + 3$, we may multiply by $x + 4$ to obtain an inequality involving only a quadratic polynomial, a type of inequality we have previously experienced. However, in order to multiply by $(x + 4)$ we must know (rather observe) whether $x + 4 < 0$ or $x + 4 > 0$. We will let A be the solution set obtained. If $x + 4 < 0$ and B that set resulting from the case $x + 4 > 0$, our solution set then comes from $A \cup B$.

Set A can be found by observing that if $x + 4 < 0$, $(2x + 18)/(x + 4)$ $< x + 3$ becomes $2x + 18 > (x + 4)(x + 3)$. Alternatively, the inequality can be written in the form $(x + 6)(x - 1) < 0$. There are, of course, two ways in which the inequality can result:

1. $x + 6 < 0$ and $x - 1 > 0$,   or
2. $x + 6 > 0$ and $x - 1 < 0$.

Keeping in mind that $x + 4 < 0$, the solution set for 1 is

$$(-\infty, -6) \cap (1, \infty) \cap (-\infty, -4) = \phi$$

while the solution set of 2 is

$$(-6, \infty) \cap (-\infty, 1) \cap (-\infty, -4) = (-6, -4)$$

Thus $A = (-6, -4)$.

We determine B similarly. If $x + 4 > 0$, $(2x + 18)/(x + 4) < x + 3$ becomes, after some calculation, $(x + 6)(x - 1) > 0$. This must result from either:

3. $x + 6 > 0$ and $x - 1 > 0$,   or
4. $x + 6 < 0$ and $x - 1 < 0$.

Thus, since $x + 4 > 0$, the solution set from 3 is

$$(-6, \infty) \cap (1, \infty) \cap (-4, \infty) = (1, \infty)$$

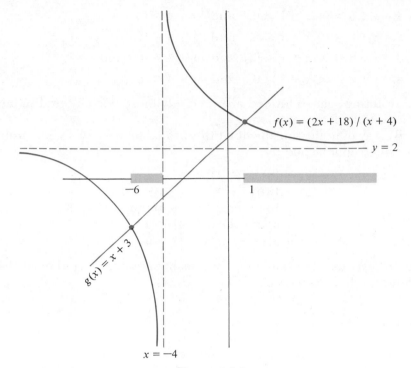

Figure 4.1.8

and the solution set for 4 is

$$(-\infty, -6) \cap (-\infty, 1) \cap (-4, \infty) = \phi.$$

Necessarily $B = (1, \infty)$.

Our solution set $A \cup B = (1, \infty) \cup (-6, -4)$.

The solution set seems to be correct in the light of Figure 4.1.8. We ought to be aware that we have really only shown that the solution set is a *subset* of $(-6, -4) \cup (1, \infty)$. You ought to verify that if $x \in (-6, -4)$ or $x \in (1, \infty)$ it does follow that $(2x + 18)/(x + 4) < x + 3$.

The problem of choosing $x + 4 > 0$ or $x + 4 < 0$ can be seemingly done away with by multiplying both sides of the inequality by $(x + 4)^2[(x + 4)^2 > 0]$. The following outlines this approach:

$$(2x + 18)/(x + 4) < x + 3$$
$$(2x + 18)(x + 4) < (x + 3)(x + 4)^2$$
$$0 < x^3 + 9x^2 + 4x - 24$$
$$0 < (x + 6)(x - 1)(x + 4)$$

The product of three numbers is positive if all three are positive or two are negative while the third is positive. Thus the possibilities below result.

1. $x + 6 > 0$, $x - 1 > 0$, and $x + 4 > 0$,
2. $x + 6 > 0$, $x + 1 < 0$, and $x + 4 < 0$,
3. $x + 6 < 0$, $x + 1 < 0$, and $x + 4 > 0$, and
4. $x + 6 < 0$, $x + 1 > 0$, and $x + 4 < 0$.

The resulting sets so formed are precisely those encountered in our earlier attack on the problem.

By way of indicating the flexibility available to us, this same problem can be approached by:

$$(2x + 18)/(x + 4) < x + 3$$
$$(2x + 18)/(x + 4) - (x + 3) < 0$$
$$[(2x + 18) - (x + 3)(x + 4)]/(x + 4) < 0$$
$$-(x + 6)(x - 1)/(x + 4) < 0$$
$$(x + 6)(x - 1)/(x + 4) > 0$$

The analysis can be completed in a manner similar to either of the techniques discussed earlier.

EXERCISES 4.1

In each part 1–10 determine all domains, zeros, poles, vertical asymptotes, horizontal asymptotes, and oblique asymptotes. Graph each function on the basis of this information. Determine whether each graph crosses its horizontal or oblique asymptote.

1. $f(x) = 1/(x - 3)$
2. $f(x) = 4/(x + 2)(x - 3)$
3. $f(x) = (2x - 4)/(x^2 - 9)$
4. $f(x) = (3x + 6)/(x^2 + 3x + 2)$
5. $f(x) = (x^2 - 4x + 4)/(x - 1)$
6. $f(x) = (x^3 - 27)/(x^2 - 1)$
7. $f(x) = (6x^2 + 1)/(2x^2 + 5x - 3)$
8. $f(x) = (x^2 - 4)/(x^2 - 9)$
9. $f(x) = (x^2 - 3x + 2)/(x^2 - 3x - 4)$
10. $f(x) = (x^2 + x - 2)/x(x^2 - 9)$
11. Are any of the functions in Exercises 4.1.1–4.1.10 bounded? What can you say about poles, asymptotes, and the like for a bounded rational function?
12. Find the points of intersection of the following pairs of functions. Graph each pair.

    a. $f(x) = 1/x$, $g(x) = 1/x^2$

b. $f(x) = (x + 3)/(x + 3)$, $g(x) = 1/x$
c. $f(x) = x^2/x$, $g(x) = 0$
d. $f(x) = (x - 1)/(x^2 + 4x - 5)$, $g(x) = (x + 2)/(x + 5)$
e. $f(x) = (x^2 + 1)/x$, $g(x) = x + 2$

13. Find the solution set for each inequality: $g(x) \leq y \leq f(x)$, f and g from Exercise 4.1.12.

14. Calculate $[f(x + h) - f(x)]/h$ for each function f in Exercises 4.1.1–4.1.10.

4.2 PARTIAL FRACTIONS

We remarked earlier that every polynomial can be written as a product of linear and quadratic factors. We also introduced the factor theorem: If f is a polynomial and a a real number, there exists a unique polynomial g and a real number r so that $f(x) = (x - a)g(x) + r$. The following *division algorithm* extends this idea.

Theorem 4.2.1. *If f and p are polynomials with the degree of p less than or equal to the degree of f, there are polynomials g and q with the degree of q less than the degree of p and*

$$f(x) = g(x)p(x) + q(x)$$

If $p(x) \neq 0$, the division algorithm may be interpreted as

$$f(x)/p(x) = g(x) + q(x)/p(x)$$

with $q(x)/p(x)$ a *proper rational function* in the sense that the degree of q is less than the degree of p.

EXAMPLE 4.2.1. Write $(3x^5 + 7x^4 + 2x^3 - x^2 - x - 3)/(x^2 + 1)$ as the sum of a polynomial and a proper rational function (that is, with $f(x) = 3x^5 + 7x^4 + 2x^3 - x^2 - x - 3$ and $p(x) = x^2 + 1$, find $g(x)$ and $q(x)$ in the division algorithm).
The following long-division process yields the desired form.

$$
\begin{array}{r}
3x^3 + 7x^2 - x - 8 \\
x^2 + 1 \,\overline{\big)\, 3x^5 + 7x^4 + 2x^3 - x^2 - x - 3} \\
-(3x^5 \qquad\quad + 3x^3) \\
\hline
7x^4 - x^3 - x^2 - x - 3 \\
-(7x^4 \qquad + 7x^2) \\
\hline
-x^3 - 8x^2 - x - 3 \\
-(-x^3 \qquad - x) \\
\hline
-8x^2 \qquad - 3 \\
-(-8x^2 \qquad -8) \\
\hline
5
\end{array}
$$

Thus

$$(3x^5 + 7x^4 + 2x^3 - x^2 - x - 3)/(x^2 + 1)$$
$$= 3x^3 + 7x^2 - x - 8 + 5/(x^2 + 1).$$

EXAMPLE 4.2.2. Consider the following problem. Write $(x^2 + 1)/(x + 1)(x - 1)(x + 2)$ in the form $a/(x + 1) + b/(x - 1) + c/(x + 2)$. First, if $x \notin \{1, -1, -2\}$,

$$a/(x + 1) + b/(x - 1) + c/(x + 2)$$

$$= \frac{a(x - 1)(x + 2) + b(x + 1)(x + 2) + c(x + 1)(x - 1)}{(x + 1)(x - 1)(x + 2)}$$

$$= \frac{(a + b + c)x^2 + (a + 3b)x + (-2a + 2b - c)}{(x + 1)(x - 1)(x + 2)}$$

If this expression is to be $(x^2 + 1)/(x + 1)(x - 1)(x + 2)$, then the two numerators must be identical. That is,

$$(a + b + c)x^2 + (a + 3b)x + (-2a + 2b - c) \equiv x^2 + 1$$

For this to occur, however, the coefficients of terms of like degree must be equal. In other words,

$$a + b + c = 1$$
$$a + 3b = 0$$
$$-2a + 2b - c = 1$$

We must now solve for a simultaneous solution to the three equations:

$$a + b + c = 1$$
$$\underline{-2a + 2b - c = 1}$$
$$-a + 3b = 2$$
$$\underline{a + 3b = 0}$$
$$6b = 2$$
$$b = 1/3$$
$$a + 3(1/3) = 0$$
$$a = -1$$
$$(-1) + 1/3 + c = 1$$
$$c = 5/3$$

Our conclusion appears to be

$$(x^2 + 1)/(x + 1)(x - 1)(x + 2) = -1/(x + 1) + 1/3(x - 1) + 5/3(x + 2)$$

A quick check shows the result to be correct.

This procedure can be extended to all rational functions: if f is a rational function, $f(x)$ can be written uniquely as the sum of a polynomial and terms of the form $a/(x-a)^k$ and $(ax+b)/(x^2+cx+d)^k$. This form for writing a rational function is known as its *partial-fraction expansion*.

That a partial-fraction expansion exists is hinted at by the previous statements concerning the factoring of polynomials and the division algorithm. One's ability to write rational functions in this form depends directly upon being able to factor the denominator into a product of linear factors and quadratic factors *having no real zero*.

There are several procedures that must be followed in order to obtain a partial-fraction expansion for f:

1. Write $f = p + g/h$ where p is a polynomial and g/h is a proper rational function.

2. Write h as a product of linear factors and quadratic factors having no real zero.

3. For each factor $(x-a)^k$ of h form the k terms $a_1/(x-a) + a_2/(x-a)^2 + a_3/(x-a)^3 + \cdots + a_k/(x-a)^k$.

4. For each factor $(ax^2+bx+c)^j$ form the j factors $(a_1x+b_1)/(ax^2+bx+c) + (a_2x+b_2)/(ax^2+bx+c)^2 + \cdots + (a_jx+b_j)/(ax^2+bx+c)^j$.

5. Solve for all the unknown terms so formulated.

6. Form the sum of all such terms.

EXAMPLE 4.2.3. Find the partial-fraction expansion for $f(x) = (x^8 + 2x^7 + 5x^6 - x^5 - 2x^4 - 3x^3 + 4x^2 + 7x + 5)/(x-1)^2(x^2+x+1)^2$.

The denominator of f is of degree 6 while the numerator is of degree 8, rendering f improper. Long division is in order.

$$
\begin{array}{r}
x^2 + 2x + 5 \\
\hline
x^6 - 2x^3 + 1 \, \big| \quad x^8 + 2x^7 + 5x^6 - x^5 - 2x^4 - 3x^3 + 4x^2 + 7x + 5 \\
-(x^8 \qquad\qquad - 2x^5 \qquad\qquad + x^2) \\
\hline
2x^7 + 5x^6 + x^5 - 2x^4 - 3x^3 + 3x^2 + 7x + 5 \\
-(2x^7 \qquad\qquad - 4x^4 \qquad\qquad + 2x) \\
\hline
5x^6 + x^5 + 2x^4 - 3x^3 + 3x^2 + 5x + 5 \\
-(5x^6 \qquad\qquad - 10x^3 \qquad\qquad + 5) \\
\hline
x^5 + 2x^4 + 7x^3 + 3x^2 + 5x
\end{array}
$$

Thus, $f(x)$

$$= (x^2 + 2x + 5) + (x^5 + 2x^4 + 7x^3 + 3x^2 + 5x)/(x-1)^2(x^2+x+1)^2.$$

The partial-fraction expansion for the rational expression on the right will be found.

Since $(x-1)$ appears as a double factor and (x^2+x+1) appears as a double factor of the denominator, we must write

$$a/(x-1) + b/(x-1)^2 + (cx+d)/(x^2+x+1) + (ex+g)/(x^2+x+1)^2$$
$$= (x^5 + 2x^4 + 7x^3 + 3x^2 + 5x)/(x-1)^2(x^2+x+1)^2$$

Adding the terms on the left, we have

$$[a(x-1)(x^2+x+1)^2 + b(x^2+x+1)^2 + (cx+d)(x-1)^2(x^2+x+1)$$
$$+ (ex+g)(x-1)^2]/(x-1)^2(x^2+x+1)^2$$

Thus

$$a(x-1)(x^2+x+1)^2 + b(x^2+x+1)^2 + (cx+d)(x-1)^2(x^2+x+1)$$
$$+ (ex+g)(x-1)^2$$
$$= x^5 + 2x^4 + 7x^3 + 3x^2 + 5x$$

If we substitute $x=1$, the expression reduces to

$$9b = 18$$
$$b = 2$$

Replacing b by 2,

$$a(x-1)(x^2+x+1)^2 + 2(x^2+x+1)^2 + (cx+d)(x-1)^2(x^2+x+1)$$
$$+ (ex+g)(x-1)^2$$
$$= x^5 + 2x^4 + 7x^3 + 3x^2 + 5x$$

Alternatively,

$$(a+c)x^5 + (2+a-c+d)x^4 + (4+a-d+e)x^3$$
$$+ (6-a-c-2e+g)x^2 + (4-a+c-d+e-2g)x$$
$$+ (2-a+d+g) = x^5 + 2x^4 + 7x^3 + 3x^2 + 5x$$

whereby we have the equations

$$a + c = 1 \tag{1}$$
$$2 + a - c + d = 2 \tag{2}$$
$$4 + a - d + e = 7 \tag{3}$$
$$6 - a - c - 2e + g = 3 \tag{4}$$
$$4 - a + c - d + e - 2g = 5 \tag{5}$$
$$2 - a + d + g = 0 \tag{6}$$

We solve these equations by:

1. Equation (1) can be read $c = 1 - a$.

2. Substituting $1-a$ for c in (2), that equation becomes $d = 1 - 2a$.

3. Making the new substitution (for d) in (3) we read $e = 4 - 3a$.

4. The three values (for c, d, and e in terms of a) transform (4) into $g = 6 - 6a$.

5. Making all four substitutions in (6), the equation reduces to $a = 1$.

6. Retracing the equalities above, replacing a by 1 we determine:

$$g = 6 - 6 = 0$$
$$e = 4 - 3 = 1$$
$$d = 1 - 2 = -1$$
$$c = 1 - 1 = 0$$

Recall that $b = 2$ has already been determined. Necessarily, $(x^5 + 2x^4 + 7x^3 + 3x^2 + 5x)/(x-1)^2(x^2 + x + 1)^2 = 1/(x-1) + 2/(x-1)^2 - 1/(x^2 + x + 1) + x/(x^2 + x + 1)^2$, whence $f(x) = (x^2 + 2x + 5) + 1/(x-1) + 2/(x-1)^2 - 1/(x^2 + x + 1) + x/(x^2 + x + 1)^2$.

EXERCISES 4.2

Find the partial-fraction expansion for each rational function in parts 1–10.

1. $1/(x + 1)(x - 2)$

2. $3/(x + 3)(x - 1)$

3. $x/(x^2 + 2x - 3)$

4. $x/(x^2 + 2x + 1)$

5. $(x + 2)/(x^2 + x + 1)(x - 1)$

6. $(x^2 + 3)/(x - 1)^2(x^2 - x + 1)$

7. $(3x^2 + 7x)/(x + 1)(x - 3)$

8. $(5x^5 + 2x^3)/(x^2 - 9)$

9. $1/(x + 1)^3(x - 1)$

10. $x^{10}/(x^3 - 7x - 6)$

4.3 IRRATIONAL EQUATIONS

In this section we are concerned with equations involving functions that are not rational. In particular, we are interested in functions f where $f(x)$ involves radicals enclosing the argument x. Our interest runs almost exclusively to finding solutions for equations involving such functions and zeros for such functions.

EXAMPLE 4.3.1. Find the zeros for f where $f(x) = \sqrt{x + 2} - 1$.

We wish to solve the equation $\sqrt{x + 2} - 1 = 0$ or $\sqrt{x + 2} = 1$. In the latter form of the equation, the radical is "isolated." We now make use of the fact that $a = b$ implies $a^2 = b^2$.

Squaring both sides of the above equation, we obtain $x + 2 = 1$ or $x = -1$. We conclude that $x = -1$ is the only possible zero for f. Furthermore, since $f(-1) = 0$, -1 is indeed a zero for f.

The example above was worked by isolating the single radical and squaring. The following example portrays a somewhat more difficult situation.

EXAMPLE 4.3.2. Let $f(x) = \sqrt{x + 2} - \sqrt{3 - x} - 1$. Find the zeros for f.

First, set $\sqrt{x + 2} - \sqrt{3 - x} - 1 = 0$. To make the problem (perhaps) simpler, we write $\sqrt{x + 2} = 1 + \sqrt{3 - x}$. Squaring both sides, we have $x + 2 = 1 + 2\sqrt{3 - x} + (3 - x)$ or $x + 2 = 4 - x + 2\sqrt{3 - x}$. The radical is still with us!

However, there is only one radical remaining and we may isolate it as in Example 4.3.1 and proceed accordingly. The equation is adjusted to have the appearance

$$x - 1 = \sqrt{3 - x}$$

Squaring, we find

$$x^2 - 2x + 1 = 3 - x$$

Alternatively,

$$x^2 - x - 2 = 0$$

or

$$(x - 2)(x + 1) = 0$$

The zeros of this polynomial equation are 2 and -1. Since $f(2) = 0$ and $f(-1) = -2$, 2 is the only zero for f.

Example 4.3.2 shows that additional or "extraneous" zeros may be introduced upon squaring both sides of the equation. It goes virtually without saying that all prospective zeros must be verified by substitution into the *original* form for $f(x)$.

EXAMPLE 4.3.3. Find the zeros for f where $f(x) = \sqrt{x - 2} + 1$. We know that by definition $\sqrt{x - 2} \geq 0$, implying that $\sqrt{x - 2} + 1 \geq 1$. Consequently, f cannot have a zero.

Assuming, however, that we have failed to make such an observation, we set $\sqrt{x - 2} + 1 = 0$ or $\sqrt{x - 2} = -1$. Here again we have an opportunity to observe the impossibility of the situation. Still, assuming that we do not observe the obvious, we continue by squaring both sides to obtain $x - 2 = 1$ or $x = 3$. Checking $f(3)$, though, we see that $f(3) = 2$, not zero. In any event, we conclude that f has no zeros.

EXERCISES 4.3

Find zeros for each of the following functions and give the domain for each function in 1–10.

1. $f(x) = \sqrt{x-3} - 2$

2. $f(x) = \sqrt{x^2-1} - 1$

3. $g(x) = \sqrt{x} + 2$

4. $g(x) = \sqrt{x+3} - \sqrt{x}$

5. $h(x) = \sqrt{x+6} + \sqrt{7-x} - 5$

6. $h(x) = \sqrt{x^2+2x+5} - \sqrt{x^2-2}$

7. $s(x) = \sqrt{x+4}/\sqrt{21-x} - 3/4x$

8. $s(x) = \sqrt{\sqrt{x+2}-5} - 2$

9. $r(x) = \sqrt{\sqrt{x-2}-\sqrt{x+5}+1}$

10. $r(x) = \sqrt{x+3} + \sqrt{x-3}$

11. The graph of $x^2 + y^2 = r^2$ is not the graph of a function. The upper semicircle, however, is the graph of a function. The lower semicircle is likewise the graph of a function. Describe (in equation form) these two functions.

12. The graph of $y^2 = x$ is not the graph of a function with variable x (that is, it is not the graph of a function having domain on the x-axis). Is it the graph of a function having domain on the y-axis? Sketch the graph. The upper half of the graph is the graph of a function as is the lower half. Describe the two functions.

13. Let $f(x) = \sqrt{x}$ and write $[f(x+h) - f(x)]/h$ in such a form that substitution of $h = 0$ does not yield a zero denominator.

14. Factor $x - a$ as the difference of two squares where $x > 0$ and $a > 0$.

15. Factor $x - a$ as the difference of two cubes. (The difference of two cubes has the form $y^3 - p^3$.)

4.4 THE BINOMIAL THEOREM

The polynomial f given by $f(x) = (x+2)^{25}$ is of a type of considerable interest. In particular, f is the twenty-fifth power of a *binomial*. The values of $f(x)$ are (theoretically) easy to compute and it is easily seen that $f(x)$ has the form $\Sigma_{i=0}^{25} a_i x^i$. Consider, however, the very simple question: What is the value of a_{15}? That is, what is the value of the coefficient of x^{15}?

Clearly, a_{15} can be found by multiplying the twenty-five factors. We might be able though to find a general means of calculating a_i without doing this. The development of a general theory in fact can be done in about the same length of time it would take us to do the simple, monotonous, and laborious multiplication.

The expression $(x + a)^2 = (x + a)(x + a)$ is found by multiplication to be $x^2 + xa + ax + a^2$, a sum formed by all possible products of an element from within the first parentheses and an element from within the second parentheses. There are four possibilities:

1. An x from each factor.

2. An x from the first factor and an a from the second.

3. An a from the first factor and an x from the second.

4. An a from each factor.

Similarly, $(x + a)^3 = (x + a)(x + a)(x + a)$ is calculated by finding all possible products formed by choosing either an x or an a from each of the three factors. They are then $xxx + xxa + xax + axx + xaa + axa + aax + aaa = x^3 + 3x^2a + 3xa^2 + a^3$. This process, even though bulky, might be faster than the multiplication process. Still, it only indicates the idea underlying our eventual theory.

Let n be a positive integer and examine the expression $g(x) = (x + a)^n$. The product $(x + a)(x + a) \cdots (x + a)$ can be computed by adding all possible combinations of n factors. For example, each term is formed by picking from each factor either an x or an a. The product so constructed would be of the form $x^{n-s}a^s$ (observe that $(n - s) + s = n$, the number of factors). How many terms $x^{n-s}a^s$ appear in the product?

The factor $x^{n-s}a^s$ can be thought of as being formed by "choosing s times the term a from the n factors." It can be shown that the number of ways to "choose a s times out of n attempts" is $C(n, s) = n!/s!(n - s)!$. (See Exercises B.5.7 and B.5.8 and Exercise 4.4.5.)

EXAMPLE 4.4.1. The term a_{15} of our original problem then is $C(25, 10)$ (The coefficient of x^{15} of $(x + a)^{25}$ is the coefficient of the term involving $x^{15}a^{10}$. It is the power of the constant term a that determines s.) Now $a_{15} = C(25, 10) = 25!/10!15! = 1634380$. The value $f(x) = (x + a)^{25}$ then becomes

$$f(x) = \sum_{i=0}^{25} C(25, i)x^{25-i}a^i$$

EXAMPLE 4.4.2. Write out the expansion for $(x + 2)^6$. We know that $(x + 2)^6 = \sum_{i=0}^6 C(6, i)x^{6-i}2^i = C(6, 0)x^62^0 + C(6, 1)x^52 + C(6, 2)x^42^2 + C(6, 3)x^32^3 + C(6, 4)x^22^4 + C(6, 5)x2^5 + C(6, 6)x^02^6 = x^6 + 6 \cdot 2x^5 + 15 \cdot 4x^4 + 20 \cdot 8x^3 + 15 \cdot 16x^2 + 6 \cdot 32x + 64 = x^6 + 12x^5 + 60x^4 + 160x^3 + 240x^2 + 192x + 64$.

The terms $C(n, i)$ are called *binomial coefficients* (for rather obvious reasons) and appear symmetric in some sense. The binomial coefficients appearing in Example 4.4.2 are 1, 6, 15, 20, 15, 6, and 1. They seem to be symmetric about "a center term." If this is the case, each expansion of a power of a binomial need involve the calculation of factorials for only the first one half of the terms. The symmetry of sorts results from the following.

Theorem 4.4.1. $C(n, i) = C(n, n - i)$ *for* $0 \leqslant i \leqslant n$.

Proof. $C(n, i) = n!/i!(n - i)!$ while

$$C(n, n - i) = n!/(n - i)![n - (n - i)]! = n!/(n - i)!i!.$$

EXAMPLE 4.4.3. Find the expansion for $h(x) = (x - 2)^6$. If we write $(x - 2)$ as $[x + (-2)]$, we may use the results of Example 4.4.2 by replacing 2 by -2. Thus $h(x) = C(6, 0)x^6(-2)^0 + C(6, 1)x^5(-2) + C(6, 2)x^4(-2)^2 + C(6, 3)x^3(-2)^3 + C(6, 4)x^2(-2)^4 + C(6, 5)x(-2)^5 + C(6, 6)(-2)^6 = x^6 - 12x^5 + 60x^4 - 160x^3 + 240x^2 - 192x + 64$.

Theorem 4.4.2. $(x - a)^n = \Sigma_{i=0}^{n} (-1)^i C(n, i)x^{n-i}a^i$.

Proof. We leave this as an exercise.

EXERCISES 4.4

1. Write out the expansion for each.

   a. $(x + 1)^7$ e. $(a/b - 1)^6$
   b. $(x - 1)^5$ f. $(a + b + c)^5$ [*Hint:* $a + b + c = a + (b + c)$]
   c. $(2x - 1)^4$ g. $.9^7$ [*Hint:* $.9 = 1 - 1/10$]
   d. $(x^2 - 2)^5$ h. 1.01^6 [*Hint:* $1.01 = 1 + 1/100$]

2. The value 1.1^{10} can be approximated by writing $1.1^{10} = (1 + 1/10)^{10}$ $= C(10, 0) + C(10, 1)(1/10) + C(10, 2)1/100 + C(10, 3)1/1000 + C(10, 4)1/10,000 + \cdots + C(10, 10)1/10,000,000,000$ and looking at only the first few terms. The expansion has $10 + 1 = 11$ terms. The maximum value $C(10, i)$ occurs "in the middle" as is $C(10, 5) = 252$. If we therefore use only the first four terms we have an approximation of $1 + 10 \cdot 1/10 + 45 \cdot 1/100 + 120 \cdot 1/1000 = 2.570$ and an upper bound on the error is $6(1/10,000)(252) = .1512$ (6 is the number of terms dropped; $(1/10,000)(252)$ is an upper bound for all six dropped terms). Use the first four terms of the binomial expansion to approximate each value below. Give an upper bound on the error involved.

   a. $.94^8$ b. 1.01^9 c. 2.01^{10}
   d. 101^{11} [Write $101 = 100(1 + 1/100)$.]

3. Give the coefficient of x^6 in each polynomial below.

   a. $(x + 1)^{11}$ c. $(x + 2)^8$ e. $(2 - x)^9$
   b. $(x - 1)^{11}$ d. $(3x + 2)^5$ f. $(x^2 + 2)^{51}$

4. A principle of counting says that if act A can have n possible results and if act B can have m possible results then the combined acts A and B can result in mn pairs of outcome. For example, if there are seven numbers in jar A and four in jar B, then the act of picking a number from A and a number from B results in one of twenty-eight possible pairs of numbers. Show by induction that if acts A_1, A_2, . . . , and A_n have respectively $m_1, m_2, . . . ,$ and m_n possible outcomes, then these acts may have a total of $m_1 \cdot m_2 \cdot \cdots \cdot m_n$ sets of outcomes.

5. Use the principle of counting in Exercise 4.4.4 to determine the number of ways of picking r elements from a set of s elements, $r \leq s$ obviously. [*Hint:* Upon making the first choice there are s possible choices. The second choice must be made from a set of $s - 1$ elements. There are $s(s - 1)$ outcomes in these two acts.] Use induction once you feel you have the pattern.

6. Use the principle of counting as discussed in Exercises 4.4.4 and 4.4.5 to decide how many subsets of r elements a set of $s \geq r$ elements has.

7. Use Exercise 4.4.4 to determine how many subsets a set having s elements has. [*Hint:* A subset can be formed by deciding that an element *is* or *is not* to be in the designated subset.]

8. Use induction to show that $(x + a)^n = \sum_{i=0}^{n} C(n, i)x^{n-1}a^i$. [*Hint:* Use $C(n, i) + C(n, i - 1) = C(n + 1, i)$.]

9. Prove Theorem 4.4.2.

4.5 REVIEW EXERCISES

1. A function f is *strictly increasing* if for each x and y in its domain with $x < y$, $f(x) < f(y)$. Which monomials are strictly increasing?

2. In view of Exercise 4.5.1, define: f is *strictly decreasing*. Which monomials are strictly decreasing?

3. In view of Exercise 4.5.1, define: f is *increasing*. Are any monomials increasing but not strictly increasing?

4. In view of Exercises 4.5.1 and 4.5.2, define: f is *decreasing*. Does there exist a monomial that is decreasing and not strictly decreasing?

5. Using the concepts in Exercises 4.5.1–4.5.4, answer the following:

   a. Describe conditions rendering a rational function increasing; decreasing; strictly increasing; strictly decreasing.
   b. Describe the general appearance of the graph of a (strictly) increasing function; (strictly) decreasing function.

   c. Can a function be both increasing and strictly increasing? both decreasing and strictly decreasing? both strictly decreasing and strictly increasing? both increasing and strictly decreasing? both decreasing and strictly increasing? both increasing and decreasing?

   d. If f is increasing, what can you say about $-f$?

   e. Prove that strictly increasing functions and strictly decreasing functions are 1:1. Need increasing functions be 1:1? Decreasing functions?

6. Prove that exponential and logarithmic functions are strictly increasing.

7. Can an even function be increasing? strictly increasing? decreasing? strictly decreasing?

8. Can an odd function be increasing? strictly increasing? decreasing? strictly decreasing?

9. Sketch a graph of the rational function f using the following description. The function f has vertical asymptotes $x = -1$ and $x = 2$, zeros $-2, 0, 1,$ and 3, oblique asymptote $x = y$, intersects the oblique asymptote at $x = -3$ and for some $x \in (0, 2)$, $f(x)$ is negative in $(2, 3)$.

10. Is f from Exercise 4.5.9 increasing? decreasing? even? odd? bounded? What is the implied domain for f?

11. Find the domains, zeros, poles, and asymptotic conditions (that is, vertical, horizontal, and oblique asymptotes as well as "asymptotic polynomials"). Sketch a graph of each.

   a. $(2x - 1)(x + 3)/(x^2 + 6x + 9)$
   b. $(x + 2)^4/(x + 2)^5$
   c. $(x + 2)^2/(x + 2)$
   d. $(x - 3)/(x^2 + 3)$
   e. $(x^2 + 5x + 6)/x$
   f. $(x^3 + 3x^2 + 3x + 1)/(x^2 + 7 + 6)$
   g. $x^4/(x + 1)$

12. Find intersections for the following pairs of functions. Graph each pair.

   a. $f(x) = (x^2 - 5x - 14)/(x - 2)$, $g(x) = (x - 7)/(x - 2)$
   b. $s(x) = (x^3 + 1)/(x + 1)$, $r(x) = (x^2 - 1)/5$
   c. $p(x) = (x + 2)/(x + 1)$, $1/p$
   d. $\zeta(x) = (x^2 + 5x + 6)/(x^2 + 7x + 10)$; $\xi(x) = (x^2 + 5x + 6)/(x + 3)$

13. Find the solution set for each inequality. Graph each.

   a. $f(x) \leq 0$, f as in Exercise 4.5.9.
   b. $(2x - 1)(x + 3)/(x^2 + 6x + 9) < 2$
   c. $(x + 2)^4/(x + 2)^5 < -2$
   d. $(x + 2)^2/(x + 2) < -2$
   e. $(x^2 - 3)/(x^2 + 3) < -1$
   f. $(x^3 + 3x^2 + 3x + 1)/(x^2 + 7x + 6) < x$
   g. $x^4/(x + 1) < 0$

14. Find the domain and zeros for each function.

    a. $f(x) = \sqrt{x + 3}$ c. $\eta(x) = \sqrt{(x - 3)^2}$

    b. $\mu(x) = \sqrt{x^2 + 1}$ d. $q(x) = \sqrt{x + 1} + \sqrt{x + 6} + \sqrt{x - 2} - 6$

15. Find the solution set for each inequality.

    a. $\sqrt{x} > 0$ b. $\sqrt{x + 1} < 0$ c. $\sqrt{x + 1} < 1 + \sqrt{x}$

16. Expand each of the following using the binomial theorem.

    a. $(x - 3)^5$ b. $(5 - 2x)^4$ c. $(x^{1/2} - 1)^6$ d. $(x - 2y)^4$

17. Find the coefficient of the x^8 term for each of the following.

    a. $(x - 1/2)^{10}$ b. $(2 - x)^{14}$ c. $(x^2 - 1)^{20}$

18. Use the binomial theorem to approximate each of the following numbers. Determine your maximum error.

    a. 3.1^8 b. $(-2.9)^8$ c. 11^6 d. 1010^7

4.6 QUIZ

Give yourself one and one-half hours to complete the quiz. Check your work when finished.

1. Find the domains, zeros, poles, and vertical asymptotes for the following functions. Determine any existing horizontal or oblique asymptotes and graph each function. Determine if either function is bounded, even, or odd.

    a. $f(x) = x/(x + 8)x^2$ b. $g(x) = x^2/(x^4 + 1)$

2. Find all points of intersection for f and g from 1.

3. Determine the solution set for $f(x) \leq g(x)$ for f and g as in 1.

4. Write f in its partial fraction expansion.

5. Find the zeros for s, where

$$s(x) = \sqrt{\sqrt{x - 1} + \sqrt{x + 1}} + 1$$

6. Find the term involving y^5 in the polynomial $(2 - y)^{12}$.

7. Approximate 2.01^{11} to within a tolerance of 10^{-4}.

4.7 ADVANCED EXERCISES

1. Let us introduce a new concept in regard to our "space" of real numbers. For $a \in R$ define $N_r(a) = \{x \in R : |x - a| < r\}$ $(r > 0)$ and call $N_r(a)$ the *neighborhood* of radius r about a. Recalling that $|x - a| < r$ is equivalent to $a - r < x < a + r$, we see that $N_r(a) = (a - r, a + r)$, an open interval "centered at a." Neighborhoods describe a condition of "closeness" or proximity.

a. Show that if $a \neq b$, there is a neighborhood $N_r(a)$ and a neighborhood $N_s(b)$ such that $N_r(a) \cap N_s(b) = \phi$.

b. Given any two points a and b, some neighborhood contains both points. Prove this conjecture.

c. Show that a set $A \subset R$ is bounded if and only if $A \subset N_r(a)$ for some r and some a.

d. Show that the intersection of two neighborhoods is empty or contains a neighborhood.

e. Show that if $0 < r < s$, $N_r(x) \subset N_s(x)$.

f. Show that a is the only point common to all neighborhoods in $\{N_{1/n}(a) : n \in N\}$. That is, show that $\cap_{n=1}^{\infty} N_{1/n}(a) = \{a\}$.

2. Let $f : A \to R$ be a real function and suppose that each neighborhood about a contains a point of A. Define $\lim_{x \to a} f(x) = L$ to mean that given any neighborhood $N_r(L)$, there is a neighborhood $N_s(a)$ such that $f[\underline{N_s(a)} \cap A] \subset N_r(L)$. [$\underline{N_s(a)}$ is $N_s(a) - \{a\}$ and is called a *deleted neighborhood* of a.] We read $\lim_{x \to a} f(x) = L$: the *limit of $f(x)$ as x approaches a is L*. The definition is given geometric significance in Figure 4.7.1. The neighborhood $N_r(L)$ is seen to contain $f[\underline{N_s(a)} \cap A]$. We get the feeling that "f maps everything close to a close to L."

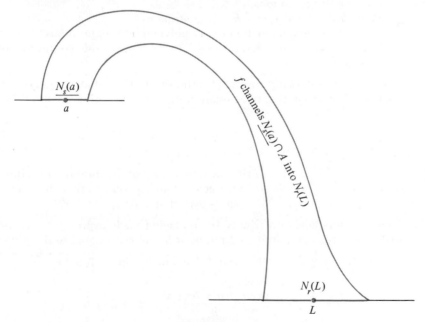

Figure 4.7.1

a. Show that $\lim_{x \to a} i_R(x) = a$. [*Hint:* For $N_r(a)$, pick a neighborhood $N_s(a)$ with $i_R[N_s(a)] \subset N_r(a)$.]

b. Show that $\lim_{x \to a} f(x) = a^2$ if $f(x) = x^2$.

   c. Show that $\lim_{x \to 2} 1/x = 1/2$. [*Hint:* Given $N_r(1/2)$, show that $f[N_r(2)] \subset N_r(1/2)$ where $f(x) = 1/x$. You may assume $r < 1/2$.]

   d. Show that $\lim_{x \to 0} x/x = 1$. [Observe that 0 is not in the domain for this function.]

   e. If $\lim_{x \to a} g(x) = K$ and $\lim_{x \to a} f(x) = L$, f and g having the same domain, then

$$\lim_{x \to a} (f + g)(x) = K + L$$
$$\lim_{x \to a} (kf)(x) = kL \text{ for } kR$$
$$\lim_{x \to a} (fg)(x) = KL$$
$$\lim_{x \to a} (f/g)(x) = K/L \text{ if } L \neq 0$$

Show the above statements are true in the case $f + g$ and kf.

3. We say that f is *continuous at a point* a in its domain if $\lim_{x \to a} f(x) = f(a)$. That is, as x approaches a, $f(x)$ approaches $f(a)$. Exercise 4.7.2 shows that i_R is continuous at every point of its domain (such functions are said to be *continuous*) as is $f(x) = x^2$. If f is given by $f(x) = 1/x$, f is continuous at $x = 2$. In fact that function is continuous for all $x \neq 0$.

   a. Use mathematical induction and Exercise 4.7.2e to show that $f(x) = x^n$ is continuous.

   b. Show from a. that $f(x) = kx^n$ is continuous.

   c. Use induction to show that every polynomial is continuous.

   d. Use parts a.–c. and Exercise 4.7.2e to discuss the continuity of rational functions.

4. Calculate the following limits. [Continuity tells us that $\lim_{x \to a} f(x) = f(a)$ which is to say that to evaluate $\lim_{x \to a} f(x)$ we "plug in" the value a for x in $f(x)$.]

   a. $\lim\limits_{x \to 1} (3x^2 + 2)$ c. $\lim\limits_{x \to -1} (3x^3 - x^2 + 1)$

   b. $\lim\limits_{x \to 2} (x + 1)/(x + 2)$ d. $\lim\limits_{x \to 1} (x + 2)/(x - 3)$

   e. $\lim\limits_{x \to 1} (x - 1)^2/(x - 1)$ (Replace this function by one that is continuous at $x = 1$ and is equal to this one everywhere except at $x = 1$.)

5. Calculate the following limits by replacing each expression by one that represents a function continuous at $h = 0$ but is equal to the given expression everywhere other than $h = 0$. For example, $\lim\limits_{h \to 0} \dfrac{(x + h)^2 - x^2}{h}$

$$\frac{(x + h)^2 - x^2}{h} = \frac{x^2 + 2xh + h^2 - x^2}{h} = 2x + h \text{ if } h \neq 0.$$

Thus

$$\lim_{h \to 0} \frac{(x + h)^2 - x^2}{h} = 2x$$

since $2x + h$ goes to $2x$ as h approaches zero.

a. $\lim\limits_{h \to 0} \dfrac{i_R(x + h) - i_R(x)}{h}$

b. $\lim\limits_{h \to 0} \dfrac{m(x + h) + b - (mx + b)}{h}$

c. $\lim\limits_{h \to 0} \dfrac{f(x + h) - f(x)}{h}$ where $f(x) = x^3$

d. $\lim\limits_{h \to 0} \dfrac{f(x + h) - f(x)}{h}$ where $f(x) = x^n$, $n \in N$

e. $\lim\limits_{h \to 0} \dfrac{g(x + h) - g(x)}{h}$ where $g(x) = 1/x$

f. $\lim\limits_{h \to 0} \dfrac{g(x + h) - g(x)}{h}$ where $g(x) = (x + 1)(x + 2)$

g. $\lim\limits_{h \to 0} \dfrac{\mu(x + h) - \mu(x)}{h}$ where $\mu(x) = f(x) + g(x)$ and

$$\lim\limits_{h \to 0} \dfrac{f(x + h) - f(x)}{h} = Df(x) \quad \text{and} \quad \lim\limits_{h \to 0} \dfrac{g(x + h) - g(x)}{h} = Dg(x).$$

h. $\lim\limits_{h \to 0} \dfrac{\eta(x + h) - \eta(x)}{h}$ where f is as in g. above and

$\eta(x) = kf(x)$, $k \in R$.

i. $\lim\limits_{h \to 0} \dfrac{r(x + h) - r(x)}{h}$ where f and g are as in g. and $r(x) = f(x)g(x)$.

j. $\lim\limits_{h \to 0} \dfrac{s(x + h) - s(x)}{h}$ where f and g are as in g. above and

$s(x) = f(x)/g(x)$, $g(x) \neq 0$ and $Dg(x) \neq 0$.

$Df(x)$ is called the *derivative* of f at x. Observe that $f(x + h) - f(x)$ represents the change in the functional value of f generated by the change from x to $x + h$. $Df(x)$, then, might be thought of as a rate of change of f with respect to the variable at x. The quantity $\dfrac{f(x + h) - f(x)}{h}$

represents the slope of the line connecting $(x, f(x))$ and $(x + h, f(x + h))$. (See Figure 4.7.2.) The derivative then might also be thought of as the slope of the limiting position of that line as $x + h$ approaches x. What do you think might be geometric significance of the line through $(x, f(x))$ with slope $Df(x)$? (See Figure 4.7.3.)

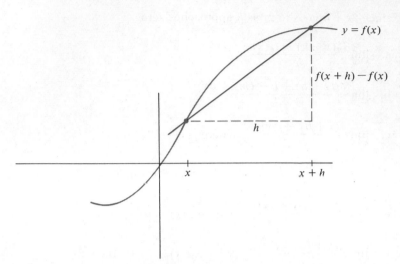

Figure 4.7.2

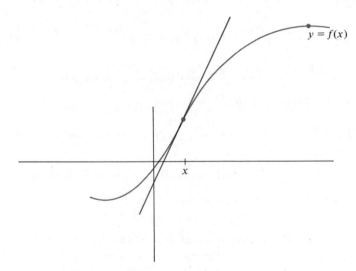

Figure 4.7.3

CHAPTER 5

The Trigonometric Functions

5.1 TWO TRIGONOMETRIC FUNCTIONS

Enough material has now been extended to allow the definition of the trigonometric functions. The definitions of two basic *trigonometric functions* will be given in this section.

$$\text{sine} = p_2 \circ \alpha : R \to R$$
$$\text{cosine} = p_1 \circ \alpha : R \to R$$

Note: The two functions just defined are real valued functions of a real variable.

Sine (a) and cosine (a) will be abbreviated to sin a and cos a respectively.

What are some of the properties of these functions, and how does one set about evaluating them? The second question will be answered first.

Let $a \in R$. Then $\alpha(a)$ is a point (x, y) on the unit circle and it follows from the definitions above that:

$$\text{sine } (a) = \sin a = p_2 \circ \alpha(a) = p_2(\alpha(a)) = p_2((x, y)) = y$$
$$\text{cosine } (a) = \cos a = p_1 \circ \alpha(a) = p_1(\alpha(a)) = p_1((x, y)) = x$$

From the above analysis, we may say that the *value* of the sine function for any real number a is merely the second coordinate of the point which is the image of a under α. The cosine of a is the first coordinate of the same point. Thus, to determine the sine and cosine functional values, we must know the correspondence α.

What are the ranges of sine and cosine? We note only that $\alpha[R]$ is C and $p_1[C] = p_2[C] = [-1, 1]$. Necessarily, sin $[R] = \cos [R] = [-1, 1]$. Hence we may write

$$\text{sine}: R \to [-1, 1]$$
$$\text{cosine}: R \to [-1, 1]$$

These statements imply that for each real number a,

$$-1 \leqslant \sin a \leqslant 1$$
$$-1 \leqslant \cos a \leqslant 1$$

Figures 5.1.1 and 5.1.2 show graphs of the sine and cosine functions (recall Exercise 2.5.19). The graph portrays the periodic nature of these functions. The periodicity is seen by the following [using the fact that $\alpha(a) = \alpha(a + 2\pi)$]:

$$\sin a = p_2 \circ \alpha(a) = p_2(\alpha(a)) = p_2(\alpha(a + 2\pi)) = \sin (a + 2\pi)$$
$$\cos a = p_1 \circ \alpha(a) = p_1(\alpha(a)) = p_1(\alpha(a + 2\pi)) = \cos (a + 2\pi)$$

The graphs are then sketched by tracing $\alpha(a)$ once around the unit circle, starting at $(1, 0)$ (that is, following $\alpha(a)$ as a goes from 0 to 2π).

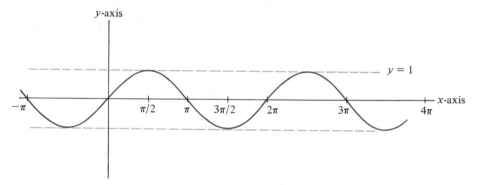

Figure 5.1.1 $y = \sin x$.

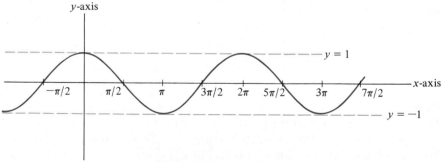

Figure 5.1.2 $y = \cos x$.

EXERCISES 5.1

1. Find the value of the image of each of the following under sine and cosine.

   a. All the special values of α listed in Section 2.5.
   b. The negatives of all special values of α listed in Section 2.5.

2. Determine the polarity (positiveness or negativeness) of sin a and cos a where $\alpha(a)$ is in quadrant I. II. III. IV.

3. Describe the following sets relative to the special values. For example, $\{x: \sin x = 0\} = \{x: \alpha(x) = (\pm 1, 0)\} = \{n\pi: n \in Z\}$.

   a. $\{x: \sin x = \frac{1}{2}\}$ b. $\{x: \sin x = -\frac{1}{2}\}$
   c. $\{x: \sin x = \sqrt{3}/2\}$ d. $\{x: \sin x = -\sqrt{3}/2\}$
   e. $\{x: \sin x = 1/\sqrt{2}\}$ f. $\{x: \sin x = -1/\sqrt{2}\}$
   g. $\{x: \sin x = 1\}$ h. $\{x: \sin x = -1\}$
   i. $\{x: \sin x = 2\}$ j. $\{x: \sin^2 x + \cos^2 x = 1\}$
   k. Do problems a.–i. with cosine replacing sine.

4. If sin $a = y$ and cos $a = x$, compute $\alpha(a)$.

5. If sin $a = y$ and cos $a = x$, compute sin $(-a)$ and cos $(-a)$.

6. If sin $a = y$ and cos $a = x$, compute sin $(\pi/2 - a)$ and cos $(\pi/2 - a)$. (See Exercise 2.5.16.)

7. Show that if $(x, y) \neq (0, 0)$ there is a real number a such that $(x, y) =$ (rcos a, rsin a) where $r = (x^2 + y^2)^{1/2}$. (See Exercise 2.5.17.)

5.2 SOME PROPERTIES OF SINE AND COSINE

In the next few paragraphs a number of characteristics of the sine and cosine functions will be portrayed and developed algebraically through the use of (1) properties of the α function and (2) some ingenuity in the use of manipulative techniques of algebra.

Let $\alpha(a) = (x, y)$. By property 5 of α, $\alpha(-a) = (x, -y)$. Thus

$$\sin (a) = y$$
$$\cos (a) = x$$
$$\sin (-a) = p_2 \circ \alpha(-a) = p_2(\alpha(-a))$$
$$= p_2((x, -y)) = -y = -\sin a$$
$$\cos (-a) = p_1 \circ \alpha(-a) = p_1(\alpha(-a))$$
$$= p_1((x, -y)) = x = \cos a$$

Restating these relationships, we have

$$\sin (-a) = -\sin a$$
$$\cos (-a) = \cos (a)$$

It is seen here that changing the *polarity* (sign) of *a* changes the polarity of the value of the image under the sine function while causing no change in the value of the image under the cosine. This pair of results will be used several times in the work that follows.

EXAMPLE 5.2.1. Compute $\sin (-\pi/4)$ and $\cos (-7\pi/6)$.
    Since $\sin \pi/4 = 1/\sqrt{2}$, $\sin (-\pi/4) = -1/\sqrt{2}$. Moreover, the statement $\cos 7\pi/6 = -\sqrt{3}/2$ shows that $\cos (-7\pi/6) = -\sqrt{3}/2$ also. Note also that:

$$(\sin a)^2 + (\cos a)^2 = x^2 + y^2 = 1 \quad ((x, y) \text{ is on } C)$$

In the future, $(\sin a)^2$ will be written $\sin^2 a$. In general, we will write $f^2(x)$ for $(f(x))^2$. Thus, we have

$$\sin^2 a + \cos^2 a = 1$$

Are the two trigonometric functions periodic? Indeed they are:

$$\sin (a + 2\pi) = p_2(\alpha(a + 2\pi)) = p_2(\alpha(a)) = \sin a$$

using only the fact that α has period 2π. Similarly,

$$\cos (a + 2\pi) = p_1(\alpha(a + 2\pi)) = p_1(\alpha(a)) = \cos a$$

Hence the sine and cosine functions have as a period the number 2π. It also happens to be the case that 2π is the *primitive period* for each of the two functions. Since induction shows that $2n\pi$ is a period for any $n \in Z$,

$$\sin (a + 2n\pi) = \sin a$$
$$\cos (a + 2n\pi) = \cos a$$

Necessarily, adding an *even* multiple of π to the argument *a* of the sine or cosine function does not change the image value.
    Consider any point (x, y) on the unit circle. Since the image of α is all of the circle, it follows that $(x, y) = \alpha(a)$ for some real number *a*. (Note again that $\sin a = y$ and $\cos a = x$.) We now see

$$(x, y) = (\cos a, \sin a) \quad \text{where} \quad (x, y) = \alpha(a)$$

That is, every point on the unit circle can be written in terms of the sine and cosine functions. (See Exercise 5.1.7 for a generalization of this idea.)

This can also be interpreted as: if x and y are real numbers so that $x^2 + y^2 = 1$, (x, y) is on the unit circle and $(x, y) = (\cos a, \sin a)$ for some real number a.

If we couple this fact with property 6 of α, we derive many interesting and important functional properties of sine and cosine. Again, property 6 states that:

$$d(\alpha(a), \alpha(b)) = d(\alpha(a - b), \alpha(0))$$

However, from what has just been said, $\alpha(a) = (\cos a, \sin a)$ while $\alpha(b) = (\cos b, \sin b)$.

Writing out the formula for $d(\alpha(a), \alpha(b))$, we see:

$$\begin{aligned}
d(\alpha(a), \alpha(b)) &= d((\cos a, \sin a), (\cos b, \sin b)) \\
&= [(\cos a - \cos b)^2 + (\sin a - \sin b)^2]^{1/2} \\
&= [(\cos^2 a + \sin^2 a) + (\cos^2 b + \sin^2 b) \\
&\quad - 2\,(\cos a \cos b + \sin a \sin b)]^{1/2} \\
&= [2 - 2(\cos a \cos b + \sin a \sin b)]^{1/2}
\end{aligned}$$

In like manner,

$$d(\alpha(a - b), \alpha(0)) = [2 - 2\cos(a - b)]^{1/2}$$

Equating the two expressions as related by property 6, and squaring both sides, we have

$$2 - 2(\cos a \cos b + \sin a \sin b) = 2 - 2\cos(a - b)$$

Alternatively,

$$\cos(a - b) = \cos a \cos b + \sin a \sin b$$

We can now determine functional values at $(a - b)$ in terms of the functional values at a and b.

EXAMPLE 5.2.2. For example, find $\cos \pi/12$.

The number $\pi/12$ is not one of the special values. However, since $\pi/12 = \pi/4 - \pi/6$,

$$\begin{aligned}
\cos(\pi/12) &= \cos(\pi/4 - \pi/6) \\
&= \cos(\pi/4) \cos(\pi/6) + \sin(\pi/4) \sin(\pi/6) \\
&= (1/\sqrt{2})(\sqrt{3}/2) + (1/\sqrt{2})(1/2) \\
&= (\sqrt{3} + 1)/2\sqrt{2}
\end{aligned}$$

If we make use of specific values for a (say, $\pi/2$, π, and so on), various useful equalities result. For instance, if $a = \pi/2$, then

$$\cos(a - b) = \cos(\pi/2 - b) = \cos \pi/2 \cos b + \sin \pi/2 \sin b$$
$$\cos(\pi/2 - b) = \sin b$$

Moreover, if in this equality we replace b by $\pi/2 - c$, we see that

$$\cos c = \sin (\pi/2 - c).$$

The last two results above may be stated in the following manner: if two numbers a and b are such that $a + b = \pi/2$, $\sin a = \cos b$ and $\sin b = \cos a$. The numbers a and b with this characteristic (that is, $a + b = \pi/2$) are called *complementary numbers;* if functions f and g are such that whenever a and b are complementary numbers, $f(a) = g(b)$ and $g(a) = f(b)$, f and g are called *cofunctions.* Examples of cofunctions other than sine and cosine will be given at a later time.

EXAMPLE 5.2.3. Suppose $\cos \pi/24 = .9914$. Compute $\sin \pi/24$, $\cos 11\pi/24$ and $\sin 11\pi/24$.

The relation $\sin^2 a + \cos^2 a = 1$ together with the fact that $\alpha(\pi/24)$ is in the first quadrant forces $\sin \pi/24 = [1 - (\cos^2 \pi/24)]^{1/2}$ or approximately .1305. Since $11\pi/24 = \pi/2 - \pi/24$, $\cos 11\pi/24 = \sin \pi/24 = .1305$, approximately, and $\sin 11\pi/24 = \cos \pi/24 = .9914$.

Having been successful in computing $\cos (a - b)$, we might feel that it is possible to formulate such an expression for $\cos (a + b)$. This we indeed can do, since we can write $a + b$ as $a - (-b)$. Making use of previous knowledge, we write

$$\cos (a + b) = \cos (a - (-b))$$
$$\cos (a + b) = \cos a \cos (-b) + \sin a \sin (-b)$$
$$\cos (a + b) = \cos a \cos b - \sin a \sin b$$

The final form follows from $\cos (-b) = \cos b$ while $\sin (-b) = -\sin b$.

EXAMPLE 5.2.4. Using the information from Example 5.2.3 and this new identity, we see that $\cos 13 \pi/24 = \cos (\pi/2 + \pi/24) = \cos \pi/2 \cos \pi/24 - \sin \pi/2 \sin \pi/24 = -\sin \pi/24 = -.1305$.

Our previous knowledge, together with our last achievements, will now be used to develop "formulas" for $\sin (a + b)$ and $\sin (a - b)$.

$$\sin (a + b) = \cos (\pi/2 - (a + b)) = \cos ((\pi/2 - a) - b)$$
$$= \cos (\pi/2 - a) \cos b + \sin (\pi/2 - a) \sin b$$
$$\sin (a + b) = \sin a \cos b + \cos a \sin b$$

Similar to our tactics employed above, we write $a - b$ as $a + (-b)$ so that

$$\sin (a - b) = \sin a \cos (-b) + \cos a \sin (-b)$$
$$\sin (a - b) = \sin a \cos b - \cos a \sin b$$

These formulas, of course, are analogous to those for the cosine of the sum and difference of two numbers. It has been mentioned that we can now find sine and cosine values for some numbers not found among the special values given for α. However, in retrospect it also allows us to increase the list of values under α.

EXAMPLE 5.2.5. We know that $\alpha(\pi/12) = (\cos \pi/12, \sin \pi/12)$ $= ((\sqrt{3} + 1)/2\sqrt{2}, (\sqrt{3} - 1)/2\sqrt{2})$ since

$$\sin \pi/12 = \sin (\pi/4 - \pi/6) = \sin \pi/4 \cos \pi/6 - \cos \pi/4 \sin \pi/6$$
$$= 1/\sqrt{2} \cdot \sqrt{3}/2 - 1/\sqrt{2} \cdot 1/2 = (\sqrt{3} - 1)/2\sqrt{2}$$

Taking special values for a, we again are capable of deriving even more relationships:

$$\cos (\pi/2 + b) = \cos \pi/12 \cos b - \sin \pi/2 \sin b = -\sin b$$
$$\cos (\pi + b) = \cos \pi \cos b - \sin \pi \sin b = -\cos b$$
$$\cos (b + (2n + 1)\pi) = \cos ((b + \pi) + 2n\pi) = \cos (b + \pi)$$
$$= -\cos b$$
$$\sin (\pi + b) = \sin \pi \cos b + \cos \pi \sin b = -\sin b$$
$$\sin (b + (2n + 1)\pi) = \sin ((b + \pi) + 2n\pi) = \sin (b + \pi)$$
$$= -\sin b$$

The above few lines show that adding an odd multiple of π to the argument of either the sine or cosine function merely changes the polarity of the image value. Coupled with the fact that adding an even multiple of π changes nothing insofar as the image value is concerned, we may put the two statements together into a *reduction formula* for each of the functions. Let n be any integer.

$$\sin (b + n\pi) = (-1)^n \sin b$$
$$\cos (b + n\pi) = (-1)^n \cos b$$

EXAMPLE 5.2.6. Calculate $\sin (25\pi/4)$ and $\cos (34\pi/3)$. Now $25\pi/4 = 6\pi + \pi/4$ so that $\sin 25\pi/4 = \sin (6\pi + \pi/4) = (-1)^6 \sin \pi/4$ $= 1/\sqrt{2}$. Likewise, $34\pi/3 = 11\pi + \pi/3$ so that $\cos 34\pi/3 = \cos (\pi/3 + 11\pi) = (-1)^{11} \cos \pi/3 = -\frac{1}{2}$.

EXAMPLE 5.2.7. Write $\cos 8\pi/3$ as $\pm \cos a$ or $\pm \sin a$ for some a satisfying $0 \leqslant a \leqslant \pi/4$.

Since $8\pi/3 = 2\pi + 2\pi/3$, we may write (using the property of cofunctions) $\cos 8\pi/3 = \cos 2\pi/3 = \sin (\pi/2 - 2\pi/3) = \sin (-\pi/6)$ $= -\sin \pi/6$. Alternatively, we could have written $\cos 8\pi/3 = \cos (3\pi - \pi/3) = -\cos \pi/3 = -\sin \pi/6$ using the reduction formula at another point.

The following is a list of the relationships developed in this section.

$$\alpha(a) = (\cos a, \sin a)$$
$$\sin (-a) = -\sin a$$
$$\cos (-a) = \cos a$$
$$\sin^2 a + \cos^2 a = 1$$
$$\sin (a + n\pi) = (-1)^n \sin a; \, n \in Z$$
$$\cos (a + n\pi) = (-1)^n \cos a; \, n \in Z$$
$$\cos (a - b) = \cos a \cos b + \sin a \sin b$$
$$\cos (a + b) = \cos a \cos b - \sin a \sin b$$
$$\cos (\pi/2 - b) = \sin b$$
$$\sin (\pi/2 - b) = \cos b$$
$$\sin (a + b) = \sin a \cos b + \cos a \sin b$$
$$\sin (a - b) = \sin a \cos b - \cos a \sin b$$

EXERCISES 5.2

1. Given that $\cos a = 3/5$, $\sin b = 8/17$, $\cos c = -5/13$, $\sin d = -24/25$, $\alpha(a) \in$ quadrant IV, $\alpha(b) \in$ quadrant I, $\alpha(c) \in$ quadrant II, and $\alpha(d) \in$ quadrant III, compute:

   a. $\sin a$
   b. $\cos b$
   c. $\sin c$
   d. $\cos d$
   e. $\sin (-a)$
   f. $\cos (-b)$
   g. $\sin (-c)$
   h. $\cos (-d)$
   i. $\sin (36\pi + a)$
   j. $\cos (b - 15\pi)$
   k. $\cos (c + 4367\pi)$
   l. $\sin (d + 19\pi)$
   m. $\cos (397\pi/2 + a)$
   n. $\sin (43\pi/6 + b)$
   o. $\cos (c + 945\pi/4)$
   p. $\sin (1631\pi/3 - d)$

2. Using a, b, c, and d as in Exercise 5.2.1 compute:

   a. $\sin (a + b)$
   b. $\cos (a + b)$
   c. $\sin (a - b)$
   d. $\cos (a - b)$
   e. $\sin (a + c)$
   f. $\cos (a + c)$
   g. $\sin (-c)$
   h. $\cos (a - c)$
   i. $\sin (a + d)$
   j. $\cos (a + d)$
   k. $\sin (a - d)$
   l. $\cos (a - d)$
   m. $\sin (b + c)$
   n. $\cos (b + c)$
   o. $\sin (b - c)$
   p. $\cos (b - c)$
   q. $\sin (b + d)$
   r. $\cos (b + d)$
   s. $\sin (b - d)$
   t. $\cos (b - d)$
   u. $\sin (c + d)$
   v. $\cos (c + d)$
   w. $\sin (c - d)$
   x. $\cos (c - d)$

3. Using a, b, c, and d as in Exercise 5.2.1 compute:

   a. $\sin 2a = \sin (a + a)$
   b. $\cos 2a$
   c. $\sin 2b$
   d. $\cos 2b$
   e. $\sin 2c$
   f. $\cos 2c$
   g. $\sin 2d$
   h. $\cos 2d$
   i. $\alpha(a)$
   j. $\alpha(b)$
   k. $\alpha(c)$
   l. $\alpha(d)$
   m. $\alpha(a + b)$
   n. $\alpha(b + c)$
   o. $\alpha(c + d)$
   p. $\alpha(a - c)$
   q. $\alpha(b - c)$
   r. $\alpha(c - d)$
   s. $\alpha(b - d)$

4. Given that $\sin \pi/5 = .5878$, $\sin \pi/8 = .3827$, and $\cos \pi/12 = .9511$, calculate each of the following:

   a. $\cos \pi/5$
   b. $\cos \pi/8$
   c. $\sin \pi/10$
   d. $\cos (-\pi/5)$
   e. $\cos (-\pi/8)$
   f. $\sin (-\pi/10)$
   g. $\sin (-\pi/5)$
   h. $\sin (-\pi/8)$
   i. $\cos (-\pi/10)$

5. Using the answers and information in Exercise 5.2.4 calculate:

    a. $\cos 3\pi/10$ [*Hint:* $3\pi/10 = \pi/2 - \pi/5$]   g. $\cos (-3\pi/10)$
    b. $\sin 3\pi/10$ h. $\sin (-3\pi/10)$
    c. $\cos 3\pi/8$ i. $\cos (-3\pi/8)$
    d. $\sin 3\pi/8$ j. $\sin (-3\pi/8)$
    e. $\cos 2\pi/5$ k. $\cos (-2\pi/5)$
    f. $\sin 2\pi/5$ l. $\sin (-2\pi/5)$

6. Given that a and b are complementary numbers, find $\sin a$, $\cos a$, $\sin b$, $\cos b$, and $\alpha(b)$ in each of the following.

    a. $\alpha(a) = (3/5, 4/5)$ d. $\alpha(a) = (-3/5, 4/5)$
    b. $\alpha(a) = (5/13, 12/13)$ e. $\alpha(a) = (8/17, 15/17)$
    c. $\alpha(a) = (1/\sqrt{2}, -1/\sqrt{2})$ f. $\alpha(a) = (x, y)$

7. Using the answers and information from Exercises 5.2.4 and 5.2.5, find the values for:

    a. $\cos 23\pi/10$ f. $\sin 46\pi/5$ k. $\cos \pi/20$
    b. $\sin 53\pi/10$ g. $\cos 83\pi/8$ l. $\sin 9\pi/40$
    c. $\cos 25\pi/8$ h. $\sin 502\pi/5$ m. $\cos 3\pi/5$
    d. $\sin 33\pi/8$ i. $\cos 13\pi/40$ n. $\sin 3\pi/5$
    e. $\cos 21\pi/5$ j. $\sin 3\pi/40$

8. Find each of the following values using some of the identities developed. (See Exercise 5.2.1 for some values needed.)

    a. $\sin 5\pi/12$ e. $\sin (-7\pi/12)$ i. $\sin 17\pi/60$
    b. $\cos 5\pi/12$ f. $\cos (-7\pi/12)$ j. $\cos 3\pi/20$
    c. $\sin 7\pi/12$ g. $\sin (-5\pi/12)$ k. $\sin \pi/8$
    d. $\cos 7\pi/12$ h. $\cos (-5\pi/12)$ l. $\cos 13\pi/10$

9. Compute $\alpha(a)$ for each a below. [*Hint:* Use Exercise 5.2.8.]

    a. $a = 5\pi/12$ b. $a = 7\pi/12$ c. $-7\pi/12$ d. $a = -5\pi/12$

10. Show that $\sin (a + \pi/2) = \cos a$. Then calculate $\alpha(a + \pi/2)$ in each of the following.

    a. $\alpha(a) = (3/5, 4/5)$ d. $\alpha(a) = (7/25, -24/25)$
    b. $\alpha(a) = (5/13, 12/13)$ e. $\alpha(a) = (x, y)$
    c. $\alpha(a) = (-8/17, 15/17)$

11. Write each of the following as $\pm \sin a$ or $\pm \cos a$ for a satisfying $0 \leqslant a \leqslant \pi/4$.

    a. $\cos 21\pi/4$ d. $\cos 89\pi/4$ g. $\sin 58\pi/3$
    b. $\sin 23\pi/6$ e. $\sin 72\pi/5$ h. $\cos 3163\pi/16$
    c. $\sin 5\pi/6$ f. $\cos 23\pi/5$

12. Write the equation (see Exercise 5.1.7) of the line passing through the origin and

    a. $(\cos \pi/4, \sin \pi/4)$ c. $(\cos \pi/10, \sin \pi/10)$ e. $(\cos 5\pi/3, \sin 5\pi/3)$
    b. $(\cos \pi/5, \sin \pi/5)$ d. $(\cos 3\pi/4, \sin 3\pi/4)$ f. $(\cos 7\pi/6, \sin 7\pi/6)$

If A is a ray emanating from the origin, what do all the points $(x, y) \neq (0, 0)$ on A have in common when written in the form $(r \cos a, r \sin a)$ for $0 \leqslant a \leqslant 2\pi$?

13. How do we know what the entire graph of $y = \sin x$ looks like if we only graph the curve for $0 \leqslant x \leqslant 2\pi$?

14. Which of the sine and cosine is an even function? Odd function?

5.3 SOME FURTHER RELATIONSHIPS INVOLVING SINE AND COSINE

In this section we will develop more identities using properties derived in the preceding section. Let us develop some *double-number* formulas.

$$\sin 2a = \sin (a + a) = \sin a \cos a + \cos a \sin a$$
$$\sin 2a = 2 \sin a \cos a$$
$$\cos 2a = \cos (a + a) = \cos a \cos a - \sin a \sin a$$
$$\cos 2a = \cos^2 a - \sin^2 a$$
$$\cos 2a = 2 \cos^2 a - 1$$
$$\cos 2a = 1 - 2 \sin^2 a$$

The latter two forms of the double-number formulas for the cosine come as a result of writing the identity $\sin^2 a + \cos^2 a = 1$ as

$$\cos^2 a = 1 - \sin^2 a$$

or

$$\sin^2 a = 1 - \cos^2 a$$

Having double-number formulas, we might feel that *half-number* formulas should arise in a quite natural way. (We can view a as half of $2a$ as well as we think of $2a$ as being twice a.)

Replacing $2a$ by b (and hence a by $b/2$) in the double-number formula for the cosine, we see that they become (looking at the latter forms only):

$$\cos b = 2 \cos^2 (b/2) - 1$$
$$\cos b = 1 - 2 \sin^2 (b/2)$$

Solving each equation for the expression involving $b/2$, we successfully arrive at the half-number formulas:

$$\cos b/2 = \pm [(1 + \cos b)/2]^{1/2}$$
$$sin\ b/2 = \pm [(1 - \cos b)/2]^{1/2}$$

The ambiguous signs in front of the radical must be determined by observing the quadrant location of $\alpha(b/2)$.

EXAMPLE 5.3.1. Compute $\cos \pi/8$ and $\sin 8\pi/5$ knowing that $\cos 16\pi/5 = -.81$.

Since $\pi/8 = \frac{1}{2}(\pi/4)$ and $\alpha(\pi/8)$ is in quadrant I,

$$\cos \pi/8 = \left[\frac{1 + \cos \pi/4}{2}\right]^{1/2} = \left[\frac{1 + 1/\sqrt{2}}{2}\right]^{1/2}$$

or approximately .92. Now, $8\pi/5 = (1/2)(16\pi/5)$ and $3\pi/2 < 8\pi/5 < 2\pi$, whence $\alpha(8\pi/5)$ is in quadrant IV and

$$\sin 8\pi/5 = -\left[\frac{1 - \cos 16\pi/5}{2}\right]^{1/2} = -\left[\frac{1 + .81}{2}\right]^{1/2}$$

or approximately $-.95$.

Expressions involving products of functional values for the sine and cosine are easily generated. For example, if we add the identities for $\cos (a - b)$ and $\cos (a + b)$ we see that

$$\cos (a - b) = \cos a \cos b + \sin a \sin b$$
$$\cos (a + b) = \cos a \cos b - \sin a \sin b$$
$$\overline{\cos (a - b) + \cos (a + b) = 2 \cos a \cos b}$$

The expression may be rewritten as

$$\cos a \cos b = \tfrac{1}{2}[\cos (a - b) + \cos (a + b)]$$

If instead of adding (in the above derivation) we subtract, the result is

$$\sin a \sin b = \tfrac{1}{2}[\cos (a - b) - \cos (a + b)]$$

Similar results are obtained by using $\sin (a + b)$ and $\sin (a - b)$. Add the two equations below.

$$\sin (a + b) = \sin a \cos b + \cos a \sin b$$
$$\sin (a - b) = \sin a \cos b - \cos a \sin b$$
$$\overline{\sin (a + b) + \sin (a - b) = 2 \sin a \cos b}$$

Solving as done above, we see that

$$\sin a \cos b = \tfrac{1}{2}[\sin (a + b) + \sin (a - b)]$$

Interchanging a and b we find that

$$\sin b \cos a = \tfrac{1}{2}[\sin (a + b) - \sin (a - b)]$$

EXAMPLE 5.3.2. Compute $\cos 5\pi/12$.

By the first of the identities developed above, $\cos 5\pi/12 \cos \pi/12$ $= \tfrac{1}{2}(\cos (5\pi/12 - \pi/12) + \cos (5\pi/12 + \pi/12)) = \tfrac{1}{2}(\cos \pi/3 + \cos \pi/2)$ $= 1/4$.

The form of the above identities is easily changed by replacing $(a - b)$ by y and $(a + b)$ by x. We solve for a and b to observe that

$$a = \tfrac{1}{2}(x + y)$$
$$b = \tfrac{1}{2}(x - y)$$

Substituting these expressions into the aforementioned equations, we have:

$$\cos y + \cos x = 2 \cos \tfrac{1}{2}(x + y) \cos \tfrac{1}{2}(x - y)$$
$$\cos y - \cos x = 2 \sin \tfrac{1}{2}(x + y) \sin \tfrac{1}{2}(x - y)$$
$$\sin x + \sin y = 2 \sin \tfrac{1}{2}(x + y) \cos \tfrac{1}{2}(x - y)$$
$$\sin x - \sin y = 2 \sin \tfrac{1}{2}(x - y) \cos \tfrac{1}{2}(x + y)$$

EXAMPLE 5.3.3. Compute $\sin 3\pi/10 - \sin \pi/5$, knowing that $\sin \pi/20 = .156$.

Since $\tfrac{1}{2}(3\pi/10 + \pi/5) = \pi/4$ and $\tfrac{1}{2}(3\pi/10 - \pi/5) = \pi/20$, we can calculate the desire of quantity using the special values and the given information.

$$\sin 3\pi/10 - \sin \pi/5 = 2 \sin \tfrac{1}{2}(3\pi/10 - \pi/5) \cos \tfrac{1}{2}(3\pi/10 + \pi/5)$$
$$= 2 \sin \pi/20 \cos \pi/4$$
$$= 2(.156)(1/\sqrt{2})$$

or approximately .221.

The following is a list of the identities derived in this section.

$$\sin 2a = 2 \sin a \cos a$$
$$\cos 2a = \cos^2 a - \sin^2 a$$
$$= 2 \cos^2 a - 1$$
$$= 1 - 2 \sin^2 a$$

$$\sin a/2 = \pm \left[\frac{1 - \cos a}{2}\right]^{1/2}$$

$$\cos a/2 = \pm \left[\frac{1 + \cos a}{2}\right]^{1/2}$$

$$\cos a \cos b = \tfrac{1}{2}(\cos (a - b) + \cos (a + b))$$
$$\sin a \sin b = \tfrac{1}{2}(\cos (a - b) - \cos (a + b))$$
$$\sin a \cos b = \tfrac{1}{2}(\sin (a - b) + \sin (a + b))$$
$$\cos a + \cos b = 2 \cos \tfrac{1}{2}(a + b) \cos \tfrac{1}{2}(a - b)$$
$$\cos a - \cos b = 2 \sin \tfrac{1}{2}(a + b) \sin \tfrac{1}{2}(b - a)$$
$$\sin a + \sin b = 2 \sin \tfrac{1}{2}(a + b) \cos \tfrac{1}{2}(a - b)$$
$$\sin a - \sin b = 2 \sin \tfrac{1}{2}(a - b) \cos \tfrac{1}{2}(a + b)$$

EXERCISES 5.3

1. Compute $\sin 2a$ and $\cos 2a$ in each of the following cases.

   a. $\alpha(a) = (3/5, 4/5)$
   b. $\alpha(a) = (-3/5, 4/5)$
   c. $\sin a = 15/17$, $\cos a = 8/17$
   d. $\sin a = -15/17$, $\alpha(a)$ is in quadrant III
   e. $\cos a = 5/13$, $\alpha(a)$ is in quadrant IV.
   f. $\cos a = 12/13$, $\alpha(a)$ is in quadrant I.

2. Using half-number formulas, compute $\sin a/2$ and $\cos a/2$ in each case below.

   a. $\alpha(a) = (3/5, 4/5)$, $a \in [0, 2\pi)$ [*Hint:* $\alpha(a) \in$ quadrant I, and $a \in [0, 2\pi)$ implies $a \in [0, \pi/2)$, whence $a/2 \in [0, \pi/4)$.]
   b. $\alpha(a) = (-3/5, 4/5)$, $a \in [0, 2\pi)$
   c. $\alpha(a) = (8/17, 15/17)$, $a \in [2\pi, 4\pi)$
   d. $\sin a = -15/17$, $a \in [\pi, 3\pi/2)$
   e. $\cos a = 5/13$, $a \in [7\pi/2, 2\pi)$
   f. $\cos a = 12/13$, $a \in [50\pi, 101\pi/2)$
   g. $\sin a = 7/24$, $a \in (0, 2\pi)$, $\alpha(a) \notin$ quadrant I.

3. Write each of the following products as a sum.

   a. $\sin 2a \cos 3a$ c. $\cos 5a \cos 2a$ e. $\sin 3a \sin 2a$
   b. $\sin 3a \cos 3a$ d. $\sin 2a \cos 5a$ f. $2 \sin 7a \cos 2a$

4. Compute each of the following (writing each product as a sum).

   a. $\cos \pi/8 \cos 3\pi/8$ b. $\sin \pi/12 \sin 7\pi/12$ c. $\sin \pi/16 \cos 3\pi/16$
   d. $\cos 7\pi/5 \sin 2\pi/5$ (use $\sin \pi/5 = .59$)

5. Write each of the following sums as a product.

   a. $\sin a + \sin 2a$ c. $\sin 5a/3 - \sin 5a/6$ e. $\cos 3a - \cos 5a$
   b. $\cos 3a + \cos 5a$ d. $\sin 2a - \sin a$ f. $\cos 6a + \cos 7a$

6. Compute each of the following (writing each of the sums as a product).

   a. $\cos 7\pi/12 + \cos \pi/12$ c. $\sin 13\pi/16 + \sin 3\pi/16$
   b. $\cos \pi/8 - \cos 5\pi/8$ d. $\sin 3\pi/5 - \sin 2\pi/5$ (use $\sin \pi/10 = .31$)

7. Use the sine and cosine to describe the slope of a line through $(0, 0)$ and $(x, y) \neq (0, 0)$. [*Hint:* Write (x, y) in the form $(r \cos a, r \sin a)$.]

5.4 WORKING WITH IDENTITIES

An equation is a (not necessarily true) mathematical statement of equality. Whenever variable terms are involved in such statements, we have seen that there is, a priori, a set called the *substitution set* (that is, the set from which values of the variable are chosen. In our encounters the substitution set has been $R - A$ where A is the *restriction set* (that is, the set of values for which some expression in the statement of the equality is not well defined).

EXAMPLE 5.4.1. In the statement $1/x = 1/x$, the restriction set is $\{0\}$ for only zero renders any expression undefined. The substitution set is then $R - \{0\}$, the set of all nonzero real numbers.

Any equation that is a valid statement for every value of the substitution set is said to be an *identity*. All other equations are said to be *conditional*. The statement $1/x = 1/x$ examined above is an identity since $1/x = 1/x$ is valid for all x in $R - \{0\}$. That is, if $x \neq 0$, $1/x = 1/x$. (We cannot say that $1/0 = 1/0$ since $1/0$ is not defined.)

EXAMPLE 5.4.2. The equation $x = 2x$ must be conditional. The substitution set is R, but if $x = 1$ $x \neq 2x$. In fact, only $x = 0$ gives validity to the equation.

EXAMPLE 5.4.3. The equation $\sqrt{x} = -2$ is also a conditional equation. By definition, $\sqrt{x} \geq 0$ and necessarily $\sqrt{x}$ cannot be negative. Consequently the equation fails to be valid for *any* element of the substitution set $\{x \in R : x \geq 0\}$.

The relationships derived in Sections 5.1 through 5.3 were derived *independently of the choices of values for the variables.* (The variables were usually a, b, c, x, or y.) Thus, these relationships are indeed *trigonometric identities*.

You might suspect that countless new identities can be developed from those already given. In order to do so, however, familiarity with the identities already given is a must. The following is a complete list of our previous results.

$$\sin^2 a + \cos^2 a = 1$$
$$\cos (-a) = \cos a$$
$$\sin (-a) = -\sin a$$
$$\cos (a + n\pi) = (-1)^n \cos a$$
$$\sin (a + n\pi) = (-1)^n \sin a$$
$$\cos a = \sin (\pi/2 - a)$$
$$\sin a = \cos (\pi/2 - a)$$
$$\cos (a + b) = \cos a \cos b - \sin a \sin b$$
$$\cos (a - b) = \cos a \cos b + \sin a \sin b$$
$$\sin (a + b) = \sin a \cos b + \cos a \sin b$$
$$\sin (a - b) = \sin a \cos b - \cos a \sin b$$
$$\cos 2a = \cos^2 a - \sin^2 a = 2 \cos^2 a - 1 = 1 - 2 \sin^2 a$$
$$\sin 2a = 2 \sin a \cos a$$

$$\cos\ (a/2) = \pm(\tfrac{1}{2}(1 + \cos\ a))^{1/2}$$
$$\sin\ (a/2) = \pm(\tfrac{1}{2}(1 - \cos\ a))^{1/2}$$
$$\cos\ a \cos\ b = \tfrac{1}{2}(\cos\ (a - b) + \cos\ (a + b))$$
$$\sin\ a \sin\ b = \tfrac{1}{2}(\cos\ (a - b) - \cos\ (a + b))$$
$$\sin\ a \cos\ b = \tfrac{1}{2}(\sin\ (a - b) + \sin\ (a + b))$$
$$\cos\ a \sin\ b = \tfrac{1}{2}(\sin\ (a + b) - \sin\ (a - b))$$
$$\cos\ a + \cos\ b = 2 \cos\ \tfrac{1}{2}(a + b)\ \cos\ \tfrac{1}{2}(a - b)$$
$$\cos\ a - \cos\ b = 2 \sin\ \tfrac{1}{2}(a + b)\ \sin\ \tfrac{1}{2}(b - a)$$
$$\sin\ a + \sin\ b = 2 \sin\ \tfrac{1}{2}(a + b)\ \cos\ \tfrac{1}{2}(a - b)$$
$$\sin\ a - \sin\ b = 2 \sin\ \tfrac{1}{2}(a - b)\ \cos\ \tfrac{1}{2}(a + b)$$

Several problems will be investigated in order to exhibit some uses of a few of the derived identities. Some of the work is repetitious and will serve as a review.

EXAMPLE 5.4.4. Suppose $\alpha(a) = (3/5, 4/5)$ and $\alpha(b) = (15/17, 8/17)$. Find each of the following:

1. $\cos\ a$	7. $\sin\ (a + b)$	13. $\cos\ (a + 3\pi)$
2. $\cos\ b$	8. $\sin\ (a - b)$	14. $\sin\ (a + 4\pi)$
3. $\sin\ a$	9. $\cos\ 2a$	15. $\alpha(a + b)$
4. $\sin\ b$	10. $\sin\ 2a$	16. $\alpha(a - b)$
5. $\cos\ (a + b)$	11. $\cos\ a/2$	
6. $\cos\ (a - b)$	12. $\sin\ a/2$	

Solutions:

1. $\cos\ a = 3/5$
2. $\cos\ b = 15/17$
3. $\sin\ a = 4/5$
4. $\sin\ b = 8/17$
5. $\cos\ (a + b) = \cos\ a \cos\ b - \sin\ a \sin\ b = 3/5 \cdot 15/17 - 4/5 \cdot 8/17 = 13/85$
6. $\cos\ (a - b) = \cos\ a \cos\ b + \sin\ a \sin\ b = 3/5 \cdot 15/17 + 4/5 \cdot 8/17 = 77/85$
7. $\sin\ (a + b) = \sin\ a \cos\ b + \cos\ a \sin\ b = 4/5 \cdot 15/15 + 3/5 \cdot 8/17 = 84/85$
8. $\sin\ (a - b) = \sin\ a \cos\ b - \cos\ a \sin\ b = 4/5 \cdot 15/17 - 3/5 \cdot 8/17 = 36/85$
9. $\cos\ 2a = \cos^2\ a - \sin^2\ a = (3/5)^2 - (4/5)^2 = 9/25 - 16/25 = -7/25$
10. $\sin\ 2a = 2 \sin\ a \cos\ a = 2(4/5)(3/5) = 24/25$
11. $\cos\ a/2 = \pm\left[\dfrac{1 + \cos\ a}{2}\right]^{1/2} = \pm\left[\dfrac{1 + 3/5}{2}\right]^{1/2} = \pm(8/10)^{1/2}$
    $= \pm(4/5)^{1/2} = \pm 2/\sqrt{5}$

There is ambiguity as to the $+$ and $-$ since the value of a is not given to us.

12. $\quad \sin a/2 = \pm \left[\dfrac{1 - \cos a}{2} \right]^{1/2} = \pm \left[\dfrac{1 - 3/5}{2} \right]^{1/2} = \pm (2/10)^{1/2}$

$\qquad = \pm (1/5)^{1/2} = \pm 1/\sqrt{5}$

The $\pm$ is ambiguous for the same reason as in Solution 11.

13. $\quad \cos (a + 3\pi) = (-1)^3 \cos a = -\cos a$
14. $\quad \sin (a + 4\pi) = (-1)^4 \sin a = \sin a$
.15. $\quad \alpha(a + b) = (\cos (a + b), \sin (a + b)) = (13/85, 84/85)$
16. $\quad \alpha(a - b) = (\cos (a - b), \sin (a - b)) = (77/85, 36/85)$

The following example illustrates a technique for justifying identities. The process used is that of showing that one of the two expressions can be transformed into the other by (1) substitution of identical expressions and (2) the use of reversible algebraic manipulations.

EXAMPLE 5.4.5. Show that

$$\cos a + \frac{\sin^2 a}{\cos a} = \frac{1}{\cos a}$$

is an identity.

The more complicated expression will be simplified. The statement of equality rules out the possibility that $\cos a = 0$ since $1/\cos a$ would in that situation be meaningless. Since $\cos a \neq 0$ implies that $a \neq (2n + 1) \pi/2$ for some integer n, a will not be allowed to take on any such value during this discussion.

To add the two expressions on the left, we need a common denominator. Thus

$$\cos a + \frac{\sin^2 a}{\cos a} = \frac{\cos^2 a}{\cos a} + \frac{\sin^2 a}{\cos a} = \frac{\cos^2 a + \sin^2 a}{\cos a}$$

However, one of the basic identities developed shows that

$$\frac{\cos^2 a + \sin^2 a}{\cos a} = \frac{1}{\cos a}$$

Thus the left-hand expression (in the original statement of the problem) has been simplified to look exactly like the right. This procedure could have been reversed. (In fact, if the identity is valid, the process *must* be reversible.)

$$\frac{1}{\cos a} = \frac{\cos^2 a + \sin^2 a}{\cos a} = \frac{\cos^2 a}{\cos a} + \frac{\sin^2 a}{\cos a} = \cos a + \frac{\sin^2 a}{\cos a}$$

If reversible steps are used, we *need not* show both procedures.

EXAMPLE 5.4.6. Examine the expression $\sin a + \sqrt{3} \cos a$, which may be rewritten $1 \cdot \sin a + \sqrt{3} \cos a$. Since $1^2 + (\sqrt{3})^2 = 4$, the point $(1/2, \sqrt{3}/2)$ is on the unit circle. In particular, $(1/2, \sqrt{3}/2) = \alpha(\pi/3)$. Thus $\sin a + \sqrt{3} \cos a = 2(\frac{1}{2} \sin a + \sqrt{3}/2 \cos a) = 2(\sin a \cos \pi/3 + \cos a \sin \pi/3) = 2 \sin (a + \pi/3)$.

Alternatively, the point $(\sqrt{3}/2, 1/2) = \alpha(\pi/6)$ is on the unit circle and $\sin a + \sqrt{3} \cos a = 2(\frac{1}{2} \sin a + (\sqrt{3}/2) \cos b) = 2(\sin a \sin \pi/6 + \cos a \cos \pi/6) = 2 \cos (a - \pi/6)$. Observe that the two numbers $\pi/6$ and $\pi/3$ brought into the discussion are complementary numbers.

EXERCISES 5.4

Using established identities, algebraic maneuvering, and substitution, verify that the following are trigonometric identities, giving the substitution set and restriction set for each statement.

1. $\sin (\pi/2 + a) = \sin (\pi/2 - a)$

2. $1 + \dfrac{\cos^2 a}{\sin^2 a} = \dfrac{1}{\sin^2 a}$ (Caution: $\sin a$ cannot be zero whence $a \neq n\pi$ for $n \in Z$.)

3. $1 + \dfrac{\sin^2 a}{\cos^2 a} = \dfrac{1}{\cos^2 a}$ (What values of a are to be eliminated?)

4. $\dfrac{\sin a}{\cos a} + \dfrac{\cos a}{\sin a} = \dfrac{1}{\cos a \sin a}$

5. $\dfrac{\dfrac{\cos a}{\sin a}}{1 + \dfrac{\cos^2 a}{\sin^2 a}} = \sin a \cos a$

6. $\dfrac{\sin a}{1 + \cos a} = \dfrac{1}{\sin a} - \dfrac{\cos a}{\sin a}$

7. $\dfrac{1}{\sin a} - \dfrac{\cos^2 a}{\sin a} = \sin a$

8. $1 - \dfrac{\cos^2 a}{1 + \sin a} = \sin a$

9. $\dfrac{1}{\cos^2 a \sin^2 a} = \dfrac{1}{\cos^2 a} + \dfrac{1}{\sin^2 a}$

10. $\cos a = \sin (a + \pi/6) + \cos (a + \pi/3)$

11. $\sin (a + \pi/6) + \cos (\pi/3 - a) = 2 \sin (a + \pi/6)$

12. $\sin (a + \pi/6) + \cos (\pi/3 - a) = 2 \cos (\pi/3 - \pi)$

13. $\sin (a + \pi/6) = \cos (\pi/3 - a)$

14. $\sin (\pi/4 - a) \sin (\pi/4 + a) = \frac{1}{2} \cos 2a$

15. $3 \sin a + 4 \cos a = 5 \sin (a + b)$ where $\alpha(b) = (3/5, 4/5)$

16. $5 \sin a + 12 \cos a = 13 \sin (a + b)$ where $\alpha(b) = (5/13, 12/13)$

17. $5 \sin a + 12 \cos a = 13 \cos (a - b)$ where $\alpha(b) = (12/13, 5/13)$

18. $(1/\sqrt{2}) \sin a + (1/\sqrt{2}) \cos a = \sin (a + \pi/4)$

19. $x \sin a + y \cos a = (x^2 + y^2)^{1/2} \sin (a + b)$ where $\alpha(b) = (x/(x^2 + y^2)^{1/2}, y/(x^2 + y^2)^{1/2})$.

20. $\dfrac{\sin^3 a + \cos^3 a}{\sin a + \cos a} = 1 - \sin a \cos a$

21. $\dfrac{1 + \sin a}{1 - \sin a} - \dfrac{1 - \sin a}{1 + \sin a} = \dfrac{4 \sin a}{\cos^2 a}$

22. $\dfrac{\sin 2a}{\cos a} = 2 \sin a$

23. $\dfrac{2 \cos 2a}{\sin 2a} = \dfrac{\cos a}{\sin a} - \dfrac{\sin a}{\cos a}$

24. $\sin (a + b) \sin (a - b) = \sin^2 a - \sin^2 b$

25, $\cos (a + b) \cos (a - b) = \cos^2 a - \sin^2 b$

26. $\sin a = 2 \sin a/2 \cos a/2$

27. $\dfrac{1}{1 + \sin a} + \dfrac{1}{1 - \sin a} = \dfrac{2}{\cos^2 a}$

28. $\dfrac{1 - \cos a}{\sin a} = \dfrac{\sin a}{1 + \cos a}$

29. $\sin a = \sin (a - b) \cos b + \cos (a - b) \sin b$

30. $\dfrac{\sin a}{1 + \cos a} + \dfrac{1 + \cos a}{\sin a} = \dfrac{2}{\sin a}$

31. $\cos^4 a - \sin^4 a = \cos^2 a - \sin^2 a$

32. $\sin 3a = 3 \sin a - 4 \sin^3 a$ [*Hint:* $\sin 3a = \sin (2a + a)$.]

33. $(\cos a - \sin a)^2 + 2 \sin a \cos a = 1$

34. $\dfrac{\sin^3 a + \cos^3 a}{\sin a + \cos a} = 1 - \tfrac{1}{2} \sin 2a$

35. $\dfrac{\sin 3a}{\sin a} - \dfrac{\cos 3a}{\cos a} = 2$

5.5 FOUR ADDITIONAL TRIGONOMETRIC FUNCTIONS

Now that the sine and cosine functions have been discussed, it is possible, and profitable, to look at four other trigonometric functions. They are:

$$\text{tangent} = \text{sine/cosine} = \frac{p_2 \circ \alpha}{p_1 \circ \alpha}$$

$$\text{cotangent} = \text{cosine/sine} = \frac{p_1 \circ \alpha}{p_2 \circ \alpha}$$

$$\text{secant} = 1/\text{cosine} = \frac{1}{p_1 \circ \alpha}$$

$$\text{cosecant} = 1/\text{sine} = \frac{1}{p_2 \circ \alpha}$$

The first problem at hand is to determine a domain and a range for each of the four. Furthermore, it is of interest to know the mechanics of obtaining the image of any given value from the domain. Let $\alpha(a) = (x, y)$. Then $\cos a = x$ and $\sin a = y$. We conclude:

$$\tan a = y/x \qquad \cot a = x/y$$
$$\sec a = 1/x \qquad \csc a = 1/y$$

The abbreviations are obvious.

Now when do each of the above make sense (when are they well defined)? Clearly, y/x makes sense as long as x is different from zero. The next logical question is: When does $x = 0$? The only points (on the unit circle) having first coordinate zero are the points $(0, 1)$ and $(0, -1)$. The points form the images of α under all odd multiples of $\pi/2$. That is to say the first coordinate value x of a point $(x, y) \in C$ is zero if and only if (x, y) is $\alpha(a)$ for a some odd multiple of $\pi/2$. Thus we have

tangent: $R - \{a : a \text{ is an odd multiple of } \pi/2\} \to R$
secant: $R - \{a : a \text{ is an odd multiple of } \pi/2\} \to R$

By a like analysis, $y = 0$ implies that $\alpha(a)$ is $(1, 0)$ or $(-1, 0)$. But $\alpha(a)$ is one of these points if and only if $a = n\pi$ for some integer. Necessarily, the cotangent and cosecant functions are not well defined for such values a.

cotangent: $R - \{a : A = n\pi \text{ for some } n \in Z\} \to R$
cosecant : $R - \{a : A = n\pi \text{ for some } n \in Z\} \to R$

It can be shown that the images of the domains of the tangent, cotangent, secant, and cosecant are respectively R, R, $R - A$, and $R - A$, where $A = \{x \in R : -1 < x < 1\} = (-1, 1)$.

EXAMPLE 5.5.1. An illustration that the image of the secant function is $R - A$ will be given.

First, $\sec a$ cannot be in A (that is, $-1 < \sec a < 1$ is impossible). Let $\alpha(a) = (x, y)$, $x \neq 0$. Then $\sec a = 1/x$. Since $-1 \leq x \leq 1$, $|x| \leq 1$. Then, $|\sec a| = |1/x| = 1/|x| \geq 1$. The inequality $-1 < \sec a < 1$ is equivalent to the inequality $|\sec a| < 1$ and since the two inequalities are contradictory, $R - A$ certainly contains the range of the secant.

Let $a \in R - A$. Then $|a| \geqslant 1$ and $|1/a| \leqslant 1$. We can show (and indeed should show) that $(1/a, (1 - 1/a^2)^{1/2})$ is a point on the unit circle. Necessarily there is an $x \in R$ with $\alpha(x)$ this given point. Consequently $\sec x = 1/(1/a) = a$, and a is an element of the image of the secant. This argument illustrates that all elements of R except those in A are image elements under this trigonometric function. Thus the range of the secant is R–A.

The mechanics of finding values of images under each of the new functions has actually been pointed out. For instance, if $\alpha(a) = (3/5, 4/5)$, then $\sin (a) = 4/5$ and $\cos a = 3/5$, giving

$$\tan a = (4/5)/(3/5) = 4/3$$
$$\cot a = (3/5)/(4/5) = 3/4$$
$$\sec a = 1/(3/5) = 5/3$$
$$\csc a = 1/(4/5) = 5/4$$

The secant and cosecant are obviously seen as the reciprocals of the cosine and sine functions respectively. In other words, when each makes sense,

$$\cos a \sec a = 1,$$

and

$$\sin a \csc a = 1$$

Furthermore, it is noticeable that the tangent and cotangent share this same relationship. That is,

$$\tan a \cot a = 1$$

Many other relationships exist in the form of identities. (Again, it must be kept in mind that identities must be valid for all values of the variables where each individual expression makes sense.) For instance,

$$\tan (a + b) = \sin (a + b)/\cos (a + b) = \frac{\sin a \cos b + \cos a \sin b}{\cos a \cos b - \sin a \sin b}$$

$$= \frac{\dfrac{\sin a \cos b}{\cos a \cos b} + \dfrac{\cos a \sin b}{\cos a \cos b}}{\dfrac{\cos a \cos b}{\cos a \cos b} - \dfrac{\sin a \sin b}{\cos a \cos b}}$$

$$\tan (a + b) = \frac{\tan a + \tan b}{1 - \tan a \tan b}$$

In general, division is not a reversible operation. How do we know we did not divide the numerator and the denominator by zero? The answer is, we do not know!

If cos a cos b is zero, then either cos $a = 0$ or cos $b = 0$, which is to say that either a or b is an odd multiple of $\pi/2$. Necessarily, either tan a or tan b fails to be defined. The above relationship holds for all a and b *as long as neither is an odd multiple of $\pi/2$*. The expression is seen to be an identity.

It should not take you long to convince yourself that 2π is a period for each of the new trigonometric functions. It is also the primitive period for the secant and cosecant. However,

$$\tan (a + \pi) = \sin (a + \pi)/\cos (a + \pi) = (-\sin a)/(-\cos a)$$
$$= \sin a/\cos a = \tan a$$

Likewise,

$$\cot (a + \pi) = \cot a$$

Because of the property just displayed, π is seen to be a period for the tangent and cotangent functions. The number π is also the primitive period for both. No proof of this is offered.

Graphs of the four new trigonometric functions are given in Figures 5.5.1, 5.5.2, 5.5.3, and 5.5.4 with a comparison of the mutually reciprocal functions shown in Figures 5.5.5, 5.5.6, and 5.5.7. The fact that the sine and cosine functions "differ by $\pi/2$" [that is, sin $(a + \pi/2) = \cos a$ and cos $(a - \pi/2) = \sin a$] is seen in Figure 5.5.8.

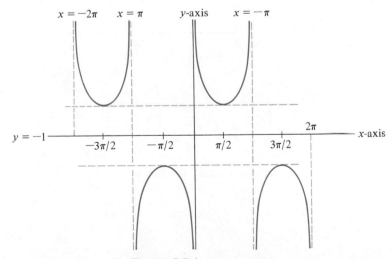

Figure 5.5.1 $y = \csc x$.

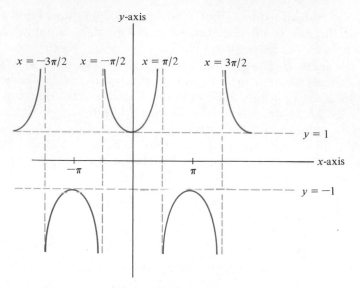

Figure 5.5.2 $y = \sec x$.

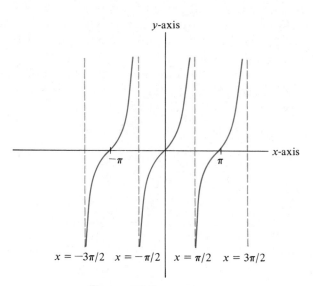

Figure 5.5.3 $y = \tan x$.

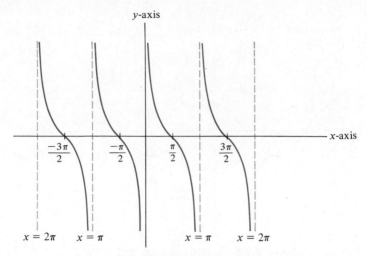

Figure 5.5.4 $y = \cot x$.

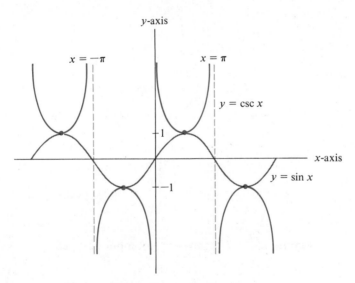

Figure 5.5.5 Sine and cosecant curves.

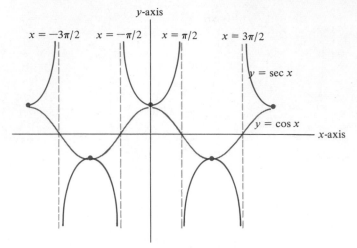

Figure 5.5.6 Cosine and secant curves.

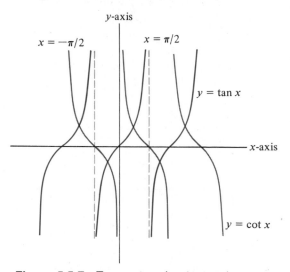

Figure 5.5.7 Tangent and cotangent curves.

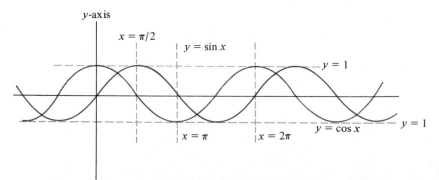

Figure 5.5.8 Sine and cosine curves.

Still other identities may be developed. The identity for tan $(a + b)$ yields an identity for tan $2a$ (*a double-number formula*).

$$\tan 2a = \tan (a + a) = \frac{\tan a + \tan a}{1 - \tan a \tan a}$$

$$\tan 2a = \frac{2 \tan a}{1 - \tan^2 a}$$

Furthermore,

$$\tan a/2 = \frac{\sin a/2}{\cos a/2} = \pm \left[\frac{\dfrac{1 - \cos a}{2}}{\dfrac{1 + \cos a}{2}} \right]^{1/2}$$

$$\tan a/2 = \pm \left(\frac{1 - \cos a}{1 + \cos a} \right)^{1/2}$$

As is implied by their names, tangent and cotangent are cofunctions. Let a and b be complementary numbers. Then $\sin a = \cos b$ and $\cos a = \sin b$. Consequently,

$$\tan a = \frac{\sin a}{\cos a} = \frac{\cos b}{\sin b} = \cot b$$

and

$$\tan b = \frac{\sin b}{\cos b} = \frac{\cos a}{\sin a} = \cot a$$

Similarly, the secant and cosecant are cofunctions. The exercises ask you to show many more relationships.

EXERCISES 5.5

1. Fill in the following table.

	0	$\pi/6$	$\pi/4$	$\pi/3$	$\pi/2$	$2\pi/3$	$3\pi/4$	$5\pi/6$	π	$7\pi/6$	$5\pi/4$	$4\pi/3$	$3\pi/2$	$5\pi/3$	$7\pi/4$	$11\pi/6$	2π
sin																	
cos																	
tan																	
cot																	
sec																	
csc																	

2. Give the polarity of tan a, cot a, sec a, and csc a, where $\alpha(a)$ is in quadrant I, II, III, IV. Which of the four trigonometric functions of this chapter are even? odd?

3. Develop an identity for cot $(a + b)$ in a manner analogous to that for tan $(a + b)$.

4. Give a "double-number formula" the cotangent.

5. Develop a "half-number formula" the cotangent.

6. Show that the secant and cosecant are cofunctions.

7. Verify that the following are identities.

   a. tan $(-a) = -$tan a b. cot $(-a) = -$cot a
   c. sec $(-a) = $sec a d. csc $(-a) = -$csc a

8. Develop an identity for tan $(a - b)$ in terms of tan a and tan b.

9. Let n be in **Z** and prove the following are identities.

   a. tan $(a + n\pi) = $ tan a
   b. cot $(a + n\pi) = $ cot a
   c. tan $(\pi/2 + a) = -$cot a
   d. cot $(2/\pi + a) = -$tan a

10. Verify the following identities.

   a. sec $(\pi/2 + a) = -$csc a
   b. csc $(\pi/2 + a) = -$sec a
   c. tan $a + $ cot $a = $ sec a csc a
   d. $1 + $ tan^2 $a = $ sec^2 a
   e. $1 + $ cot^2 $a = $ csc^2 a

   f. $\dfrac{\tan x + 1}{\sin x + \cos x} = \sec x$

   g. csc $a - $ cos a cot $a = $ sin a

   h. $\dfrac{\cot a}{1 + \cot^2 a} = \sin a \cos a$

   i. $\dfrac{\sin t + \sin q}{\cot t - \cos q} = -\cot \tfrac{1}{2}(t - q)$

   j. $\dfrac{\sin x - \sin y}{\cos x + \cos y} = \tan \tfrac{1}{2}(x - y)$

   k. $\dfrac{\cos 6x + \cos 4x}{\sin 6x + \sin 4x} = \cot x$

   l. $\dfrac{\sin (2x - y) + \sin y}{\cos (2x - y) + \cos y} = \tan x$

   m. sin $t - $ sin $3t = 2$ sin^3 $t - 2$ sin t cos^2 t

   n. $\dfrac{\tan a - \cot a}{\tan a + \cot a} = 2$ sin^2 $a - 1$

o. $\cos^2 a - \sin^2 a = \dfrac{1 - \tan^2 a}{1 + \tan^2 a}$

p. $\cot a + \tan a = \cot a \sec^2 a$

q. $(\sec a - \tan a)^2 = \dfrac{1 - \sin a}{1 + \sin a}$

r. $\dfrac{1 + \csc a}{\csc a - 1} = \dfrac{1 + \sin a}{1 - \sin a}$

11. Prove the following identities.

a. $\dfrac{\sin a}{1 + \cos a} = \csc a - \cot a$

b. $2 \cot 2a = \cot a - \tan a$

c. $\dfrac{\sin a}{1 + \cos a} + \dfrac{1 + \cos a}{\sin a} = 2 \csc a$

5.6   REVIEW EXERCISES

1. If a circle has radius r and one of its arcs is 1/8 the circle, what is the associated length of the arc? If the arc is 1/6 of the circle? If the arc is 1/12 of the circle?

2. Answer the questions of Exercise 5.6.1, where $r = 1$. Where $r = 5$.

3. What is the length of the arc (counterclockwise traversal) from $(1, 0)$ to each indicated point?

a.	$(\sqrt{3}/2, 1/2)$	f.	$(-1/\sqrt{2}, 1/\sqrt{2})$	k.	$(-1/2, -\sqrt{3}/2)$
b.	$(1/\sqrt{2}, 1/\sqrt{2})$	g.	$(-\sqrt{3}/2, 1/2)$	l.	$(0, -1)$
c.	$(1/2, \sqrt{3}/2)$	h.	$(-1, 0)$	m.	$(1/2, -\sqrt{3}/2)$
d.	$(0, 1)$	i.	$(-\sqrt{3}/2, -1/2)$	n.	$(1/\sqrt{2}, -1/\sqrt{2})$
e.	$(-1/2, \sqrt{3}/2)$	j.	$(-1/\sqrt{2}, -1/\sqrt{2})$	o.	$(\sqrt{3}/2, -1/2)$

4. In which quadrant is each of the following?

a.	$\alpha(1)$	f.	$\alpha(-1)$	k.	$\alpha(16)$	p.	$\alpha(-1/3)$
b.	$\alpha(3)$	g.	$\alpha(-3)$	l.	$\alpha(-3\pi/8)$	q.	$\alpha(2095\pi/3)$
c.	$\alpha(5)$	h.	$\alpha(-5)$	m.	$\alpha(27\pi/10)$	r.	$\alpha(34\pi/7)$
d.	$\alpha(7)$	i.	$\alpha(-7)$	n.	$\alpha(-3\pi/5)$	s.	$\alpha(-2000)$
e.	$\alpha(9)$	j.	$\alpha(-9)$	o.	$\alpha(1/2)$		

5. Find each of the following points.

a.	$\alpha(2700\pi)$	e.	$\alpha(195\pi/4)$	h.	$\alpha(1191\pi/6)$
b.	$\alpha(-51\pi)$	f.	$\alpha(-85\pi/5)$	i.	$\alpha(75\pi/2)$
c.	$\alpha(\pi/3)$	g.	$\alpha(95\pi/6)$	j.	$\alpha(-3491\pi/2)$
d.	$\alpha(2702\pi/3)$				

6. Find the values of sine and cosine at the values of Exercise 5.6.5.

7. Find the values of secant, cosecant, tangent, and cotangent of each value in Exercise 5.6.5.

8. Given that sin 1 = .84, find the following.

a. cos 1	h. tan 2	n. tan 3	y. $\alpha(1/2)$
b. sec 1	i. sin 1/2	o. cot 3	t. sin (-1)
c. csc 1	j. cos 1/2	p. sin 3/2	u. tan (-1)
d. tan 1	k. tan 1/2	q. cos 5/2	v. $\alpha(1)$
e. cot 1	l. sin 3	r. tan 7/2	w. $\alpha(-1)$
f. sin 2	m. cos 3	s. cos (-1)	x. $\alpha(2)$
g. cos 2			

9. Show that for $0 < a < \pi/2$, $0 < \sin a < \tan a$

10. Show that the following are identities.

a. $\sin a \cot a - \cos a$

b. $\cos a \tan a = \sin a$

c. $\tan a \csc a = \sec a$

d. $\cot a \sec a = \csc a$

e. $\dfrac{\sec a}{\csc a} = \tan a$

f. $\dfrac{\csc a}{\sec a} = \cot a$

g. $\dfrac{\sin a + \cos a}{\sec a + \csc a} = \sin a \cos a$

h. $\cot a + \csc a + 2 = (1 + \sin a) + \cot a(1 + \cos a)$

i. $2 \sin a + \sec a = (\sec a)(\sin a + \cos a)$

j. $1/2 \csc a \sec a = \csc 2a$

k. $\sin 4a = 8 \cos^3 a \sin a - 4 \cos a \sin a$

l. $\cos 4a = \cos^4 a - 8 \cos^2 a + 1$

m. $\dfrac{\tan a - \sin a}{\cos a} + \dfrac{\sin a \cos a}{1 + \cos a} = \dfrac{\tan a}{1 + \cos a}$

n. $\dfrac{\sec a + \tan a}{\cos a - \tan a - \sec a} = -\csc a$

o. $\sin a + \cos a + \dfrac{\sin a}{\cot a} = \sec a + \csc a - \dfrac{\cos a}{\tan a}$

p. $\tan a/2 = \dfrac{\sin a}{1 + \cos a}$

q. $\tan a/2 = \csc a - \cot a$

r. $\dfrac{1 - \cos 2a}{\sin 2a} = \tan a$

s. $\dfrac{\sin 2a}{\sin a} = \sec a + \dfrac{\cos 2a}{\cos a}$

t. $\dfrac{\sin 3a}{\sin a} - \dfrac{\cos 3a}{\cos a} = 2$

u. $\dfrac{\sin 5a - \sin 3a}{\cos 5a - \cos 3a} = \tan a$

11. Write the following products as sums and the sums as products.

    a. sin 6a + sin 10a   e. sin 10a sin 6a
    b. sin 6a + cos 10a   f. sin 10a cos 6a
    c. cos 6a + cos 10a   g. cos 10a cos 6a
    d. sin (x + h) − sin x   h. cos (x − h) cos x

12. Write each of the following in the form $(r \cos a, r \sin a)$. (See Exercise 5.1.7.)

    a. All those points of Exercise 2.4.9.   j. $(3x, x)$, $x > 0$
    b. $(1, 1)$ k. $(x, -3x)$, $x < 0$
    c. $(-1, -1)$ l. $(x, -3x)$, $x > 0$
    d. (x, x) where $x > 0$ m. $(3x, -x)$, $x < 0$
    e. (x, x) where $x < 0$ n. $(3x, -x)$, $x > 0$
    f. $(x, -x)$ where $x > 0$ o. $(x, 0)$, $x < 0$
    g. $(x, -x)$ where $x < 0$ p. $(x, 0)$, $x > 0$
    h. $(x, 3x)$, $x > 0$ q. $(0, x)$, $x < 0$
    i. $(x, 3x)$, $x < 0$ r. $(0, x)$, $x > 0$

13. Show that $x \sin a + y \cos a = (x^2 + y^2)^{1/2} \sin (a + b)$, where $\alpha(b) = (x/(x^2 + y^2)^{1/2}, y/(x^2 + y^2)^{1/2})$.

14. Figure 5.6.1 shows a line passing through $(0, 0)$ and $\alpha(\theta)$. The perpendicular dropped from (a, b) to the x-axis forms a right triangle. Describe sin θ, cos θ, tan θ, and so forth in terms of a and b. Prove that your conclusions are correct.

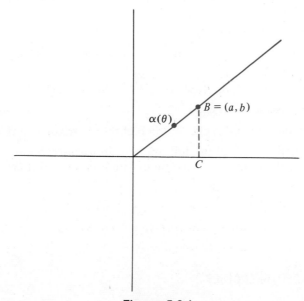

Figure 5.6.1

15. The triangle of Figure 5.6.1 is shown separately in Figure 5.6.2. The vertex A is that placed over the vertex in the coordinate system of Figure 5.6.1. We say that an *angle* is formed at A and that the angle has measure θ radians, $0 < \theta < \pi/2$. [The line from A to B passes through $\alpha(\theta)$.] Using Exercise 5.6.14, determine the values among a, b, c, A, and B that are missing from the list in each case. (We write "A has measure θ radians" in the form $A = \theta$ radians.)

    a. $\theta = \pi/3$, $a = 3$ b. $\theta = \pi/4$, $b = 6$ c. $\theta = \pi/6$, $c = 2$

How do you think the measure of B is related to the measure of A? Why?

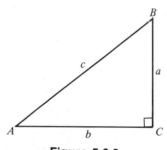

Figure 5.6.2

5.7 QUIZ

Work this quiz in one hour and verify your answers.

1. Compute the six trigonometric functional values for a where $\alpha(a) = (8/17, -15/17)$.

2. Compute $\alpha(2a)$ and $\alpha(a/2)$ for a as in 1.

3. Let $\sin a = y$ and $\cos b = x$ with $\alpha(a)$ in quadrant III. Compute:

    a. $\cos (a + b)$ c. $\cos (a + n\pi)$
    b. $\sin (a - b)$ d. $\frac{1}{2}[\sin (a + b) + \sin (a - b)]$.

4. Compute: $\cos \pi/8 \cos 3\pi/8$.

5. Give the domain and range for each of the six trigonometric functions.

Determine whether each of the following is an identity. Show by example any that are not identities and give justification to those that are.

6. $\sin^4 a - \cos^4 a = -\cos 2a$

7. $(\tan^4 a + \tan^2 a)/(\sec^2 a - \sec^2 a \cos^2 a) = \tan^2 a$

8. $(\sin^2 a - \cos^2 a)(\sin a + \cos a)/(\sin a - \cos a) = 1$

9. $(\tan a + \sec a)(\sin a - 1) \sec^2 a = -1$

5.8 ADVANCED EXERCISES

1. $\alpha(\pi/8)$ can be calculated from half-angle formulas for the sine and

cosine functions. However, knowing that equal chords determine equal angles, find $\alpha(\pi/8)$ by

a. writing the equation of the line l passing through $(1, 0)$ and $(1/\sqrt{2}, 1/\sqrt{2})$.

b. finding the equation of l', the perpendicular bisector of that segment of l from $(1, 0)$ to $(1/\sqrt{2}, 1/\sqrt{2})$.

c. finding the point of intersection of l' and C.

Calculate $\alpha(\pi/8)$ by this method.

2. Exercise 5.1.7 showed that all points could be written in the form $(r \cos a, r \sin a)$ for some $r > 0$ and some $a \in [0, 2\pi)$, provided, of course, the point is not $(0, 0)$. We can then symbolize the "location" of such a point by specifying r and a. Such a manner of depicting points is known as the *polar form*. That is, $(r \cos a, r \sin a)$ can be written as $(r, a)^*$. The ordered-pair notation here is different from the original *rectangular system* introduced earlier. The first element r denotes distance from the origin while a determines a ray starting at $(0, 0)$ and passing through $\alpha(a)$. The intersection of this half-line and the circle $x^2 + y^2 = r^2$ determines a single point (the point in question). Write in polar form those points in Exercise 5.5.12.

    Note: Polar form may be more general than this description. For example, the restriction of a to $[0, 2\pi)$ need not hold. Some do not prefer to restrict r to being non-negative.

3. Show that tan $[(-\pi/2, \pi/2)] = R$ and that tan $| (-\pi/2, \pi/2)$ is $1:1$. In particular, this shows that $(-\pi/2, \pi/2) \sim R$. [*Hint:* Let r be any real number. Then, let $x = \dfrac{1}{1+r^2}$ and $y = \pm(1 - x^2)^{1/2}$ the $+$ or $-$ agreeing with whether $r > 0$ or $r < 0$.] Show $(x, y) \in C$ and is on the arc from $(0, -1)$ to $(0, 1)$ when traveling in a counterclockwise direction. Then show that if $\alpha(a) = (x, y)$, $-\pi/2 < a < \pi/2$ and tan $a = r$, then $x = \left[\dfrac{1}{1+r^2}\right]^{1/2}$ and $y = \pm(1 - x^2)^{1/2}$. This latter shows tan $| (-\pi/2, \pi/2)$ is $1:1$, while the former argument shows that tan $[(-\pi/2, \pi/2)] = R$.

    Previously (in other advanced exercises) we saw that $(a, b) \sim (c, d)$ for nonempty, open intervals. Thus, our result here shows that $(a, b) \sim R$ for each nonempty, open interval (a, b).

4. Verify the following.

a. sin $| [-\pi/2, \pi/2]: [-\pi/2, \pi/2] \rightarrow [-1, 1]$ is $1:1$ and onto.

b. cos $| [0, \pi]: [0, \pi] \rightarrow [-1, 1]$ is $1:1$ and onto.

c. cot $|(0, \pi): (0, \pi) \rightarrow R$ is $1:1$ and onto. [*Hint:* Only need show tan $(a + \pi/2) = $ cot a.]

5. Using induction and the formulas for cos $(a \pm b)$ and sin $(a \pm b)$, show:

a. sin $(a + n\pi) = (-1)^n \sin a$, $n \in Z$.

b. cos $(a + n\pi) = (-1)^n \cos a$, $n \in Z$.

c. sin $(a + (2n + 1)\pi/2) = (-1)^n \cos a$, $n \in Z$.

d. cos $(a + (2n + 1)\pi/2) = (-1)^n \sin a$, $n \in Z$.

CHAPTER 6

Inverse Trigonometric Functions

6.1 SETS DETERMINED BY TRIGONOMETRIC FUNCTIONS

Various equations involving trigonometric functions have been encountered throughout the material presented in the text. These equations have for the most part been identities rather than conditional equations. Since conditional trigonometric equations have much value in mathematics, we will take time to examine a few.

One simple example of a conditional trigonometric equation is given by

$$\sin x = k, \ k \in R$$

EXAMPLE 6.1.1. Find the solution set for

$$\sin x = 1/2$$

You should recognize by inspection that $x = \pi/6$ is a possible solution to the latter equation. The value $5\pi/6$ is another candidate. These two numbers represent the only values in $[0, 2\pi)$ satisfying the conditions. (Why?) However, $\pi/6 + 2n\pi$ and $5\pi/6 + 2n\pi$ are also solutions for n any integer (2π is a period for the sine function). The set of solutions (solution set) for the equation $\sin x = 1/2$ is

$$\{x: \sin x = 1/2\}$$
$$= \{x: x = \pi/6 + 2n\pi \quad \text{or} \quad 5\pi/6 + 2n\pi, \ n \in Z\}$$
$$= \{x: x = n\pi + (-1)^n \ \pi/6 \text{ for } n \in Z\}$$

218

Figure 6.1.1 shows a geometric interpretation of this solution. See Exercise 5.1.3 for more of this type of analysis.

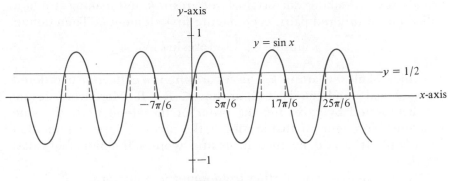

Figure 6.1.1

EXAMPLE 6.1.2. Find the set of solutions for the equation $\cos x = 3$.

This problem is easily answered. We recall from a previous discussion the fact that $-1 \leq \cos x \leq 1$. Thus $\cos x = 3$ is impossible and there exist no solutions. Consequently, the solution set is seen to be ϕ.

We will introduce at this time some new symbolism. The set **arcsin** k $(k \in R)$ is given by

$$\text{arcsin } k = \{x \in R : \sin x = k\}.$$

The symbol **arcsin** k is actually an abbreviation of **arcsine** k. The pronunciation is obvious. We will read the statement "$x \in$ **arcsin** k" in the following way: x is a number whose image under the sine function is k. This may be shortened (by an abuse of the language) to: x is a number whose sine is k.

In a similar manner, we define:

> **arccosine** $k = $ **arccos** $k = \{x : \cos x = k\}$
> **arctangent** $k = $ **arctan** $k = \{x : \tan x = k\}$
> **arcsecant** $k = $ **arcsec** $k = \{x : \sec x = k\}$
> **arccosecant** $k = $ **arccsc** $k = \{x : \csc x = k\}$
> **arccotangent** $k = $ **arccot** $k = \{x : \cot x = k\}$

We could have defined arcsine instead of **arcsin** k. This idea lends itself to the functional notation (for example, compare the meaning of sine as opposed to sin a). Our concept of **arcsin** k would not be greatly altered if we were to consider

$$\text{arcsin } k = \{(k, x): \sin x = k\}.$$

We simply are taking our original set **arcsin** k and making it into a collection of ordered pairs, each having first element k. Then define

$$\text{arcsine} = \bigcup_{k \in R} (\text{arcsin} k)$$

the union of all sets **arcsin** k. The set arcsine is a collection of ordered pairs of real numbers. That is, arcsine $\subset R \times R$. Moreover, arcsine can be formed by reversing the order of the elements of the sine function. We might say that arcsine is the *inverse* of the sine relation (function). It would be nice if arcsine were a function, but, alas, *it is not.*

Now examine some further trigonometric equations.

EXAMPLE 6.1.3. Solve the equation

$$2 + 3 \sin x = 6 - \sin x$$

Adding sin x to both sides and subtracting 2, we reveal an alternate form of the equation to be

$$4 \sin x = 4$$

Division of both sides by 4 yields

$$\sin x = 1$$

The indicated solution set is therefore **arcsin** 1. The (expected) solution set can also be described as $\{x: x = \pi/2 + 2n\pi, \ n \in Z\}$. Whether or not this is the solution set must be checked against the original equation.

To verify that each element of **arcsin** 1 is a solution of the given equation, let us suppose that $x \in$ **arcsin** 1. Does x satisfy

$$2 + 3 \sin x = 6 - \sin x ?$$

Seeing that sin $x = 1$, we realize that the left-hand expression of the original equation becomes $2 + 3 = 5$ while the right-hand one is $6 - 1 = 5$. Necessarily, each element of the set **arcsin** 1 is a solution of (satisfies) the stated equation.

EXAMPLE 6.1.4. Find the solution set for

$$\cos^2 x - \sin x = -1$$

Since $\cos^2 x = 1 - \sin^2 x$, the original equation becomes

$$1 - \sin^2 x - \sin x = -1$$

or

$$\sin^2 x + \sin x - 2 = 0$$

The left-hand side of the latter form is a *quadratic* equation in the variable $\sin x$. Substituting $y = \sin x$, we observe that the equation takes on the appearance of the algebraic equation $y^2 + y - 2 = 0$, which in turn factors into the form

$$(y + 2)(y - 1) = 0$$

The trigonometric equation then must factor likewise into

$$(\sin x + 2)(\sin x - 1) = 0$$

Now if $a \cdot b = 0$, $a = 0$ or $b = 0$. As a result we conclude from above that $\sin x + 2 = 0$ or $\sin x - 1 = 0$. This is to say, $\sin x = -2$ or $\sin x = 1$. However, $\sin x = -2$ is not possible since $-1 \leqslant \sin x \leqslant 1$. Thus the only possible solutions arise from the equation $\sin x = 1$. Consequently the supposed solution set is **arcsin 1**.

To investigate this supposed solution, note that if $\sin x = 1$, then $\cos x = \pm(1 - \sin^2 x)^{1/2} = 0$. Substituting these values into the original form of the equation, we have

$$\cos^2 x - \sin x = 0 - 1 = -1$$

Hence the solution set is as predicted.

EXAMPLE 6.1.5. Solve the equation

$$\tan x \sin x - \tan x = 0$$

Since $\tan x$ is a factor of each expression on the left, we can "factor it out" to obtain:

$$\tan x(\sin x - 1) = 0$$

What conclusion remains?

$$\tan x = 0 \quad \text{or} \quad \sin x - 1 = 0$$

The latter result is to say that $\sin x = 1$. Remembering that **arcsin 1** $= \{x : x = \pi/2 + 2n\pi \text{ for } n \in Z\}$, we remark that for $x \in$ **arcsin 1**, $\tan x$ is not defined. (Review again the domain of the tangent function.) It is then clear that *no* member of **arcsin 1** can be a solution for $\tan x \sin x - \tan x = 0$.

Consider *arctan 0*. If $x \in$ arctan 0, is x a solution of our equation? Given that tan $x = 0$, tan $x = $ sin $x/$cos x implies that sin $x = 0$ (cos $x \neq 0$). Thus, tan x sin $x -$ tan $x = 0 \cdot 0 - 0 = 0$. Clearly, the solution set is **arctan** 0.

The last problem emphasizes the need for checking tentative results to see whether or not they represent genuine solutions.

EXAMPLE 6.1.6. Compute the solution set for 5 sin $x + 12$ cos x $= 13$. By a previous identity, 5 sin $x + 12$ cos $x = 13$ sin $(x + a)$ where sin $a = 12/13$ and cos $a = 5/13$. The equation now becomes

$$13 \sin (x + a) = 13$$

or

$$\sin (x + a) = 1$$

giving

$$x + a \in \textbf{arcsin } 1$$

The latter statement implies that $x + a = \pi/2 + 2n\pi$ for some $n \in Z$. Alternatively, $x = (\pi/2 - a) + 2n\pi$. The proposed solution set $\{x: x = (\pi/2 - a) + 2n\pi, \; n \in Z, \; \sin a = 12/13 \; \cos a = 5/13\}$ should be verified by check.

EXAMPLE 6.1.7. As a final example, find the solution set for

$$\sin^2 2x = 1$$

The equation is similar to the algebraic equation $y^2 = 1$, having roots 1 and -1. Thus, $2x \in$ **arcsin** $1 \cup$ **arcsin** $(-1) = \{a: a = (2n + 1)\pi/2$ for $n \in Z\}$. Since $2x = (2n + 1) \pi/2$, $x = (2n + 1) \pi/4$. This is merely a means of describing the set of all *odd multiples* of $\pi/4$. Alternatively, the set is given as (arcsin $1/\sqrt{2}) \cup$ (arcsin $(-1/\sqrt{2}))$.

The solution set, like that of the previous example, needs verification.

EXERCISES 6.1

1. Describe each of the following sets in terms of the special values. $0, \pi/6, \pi/4$, and so forth.

    a. arcsin 0
    b. arcsin 1/2
    c. arcsin $(-1/2)$
    d. arcsin $(\sqrt{3}/2)$
    e. arcsin $(-\sqrt{3}/2)$
    f. arcsin 1
    g. arcsin (-1)
    h. arcsin $(1/\sqrt{2})$
    i. arcsin $(-1/\sqrt{2})$

2. Redo Exercise 6.1.1 with **arccos** replacing **arcsin**.

3. Follow the instructions of Exercise 6.1.1.

   a. arctan 0   d. arctan $(\sqrt{3})$ f. arctan $(1/\sqrt{3})$
   b. arctan 1   e. arctan $(-\sqrt{3})$ g. arctan $(-1/\sqrt{3})$
   c. arctan (-1)

4. Do Exercise 6.1.3 with **arccot** replacing **arctan**.

5. Follow the instruction of Exercise 6.1.1.

   a. arcsec 1   d. arcsec $(-\sqrt{2})$ g. arcsec 2
   b. arcsec (-1) e. arcsec $(2/\sqrt{3})$ h. arcsec (-2)
   c. arcsec $(\sqrt{2})$ f. arcsec $(-2/\sqrt{3})$

6. Do Exercise 6.1.5 with **arccsc** replacing **arcsec**.

7. Describe each of the sets below (in terms of special values):

   a. arcsin 1 ∩ arccos 0   f. arcsin $\sqrt{3}/2$ ∪ arcsin $(-\sqrt{3}/2)$
   b. arcsin 1/2 ∩ arccos $\sqrt{3}/2$ g. arccos $1/\sqrt{2}$ ∪ arccos $(-1/\sqrt{2})$
   c. arcsin 0 ∩ arccot 0   h. arctan 1 ∪ arctan (-1)
   d. arccos $(-1/\sqrt{2})$ ∩ arctan (-1) i. arctan 0 ∪ arccos 0
   e. arctan $(-\sqrt{3})$ ∩ arcsin $\sqrt{3}/2$

8. List the elements in each set.

   a. arcsin 0 ∩ $[-\pi/2, \pi/2]$ n. arccos 1 ∩ $[0, \pi]$
   b. arcsin $1/\sqrt{2}$ ∩ $[-\pi/2, \pi/2]$ o. arccos (-1) ∩ $[0, \pi]$
   c. arcsin 1/2 ∩ $[-\pi/2, \pi/2]$ p. arccos $(-1/2)$ ∩ $[0, \pi]$
   d. arcsin $\sqrt{3}/2$ ∩ $[-\pi/2, \pi/2]$ q. arccos $(-\sqrt{3}/2)$ ∩ $[0, \pi]$
   e. arcsin 1 ∩ $[-\pi/2, \pi/2]$ r. arccos $(-1/\sqrt{2})$ ∩ $[0, \pi]$
   f. arcsin (-1) ∩ $[-\pi/2, \pi/2]$ s. arctan 0 ∩ $(-\pi/2, \pi/2)$
   g. arcsin $(-1/2)$ ∩ $[-\pi/2, \pi/2]$ t. arctan $1/\sqrt{3}$ ∩ $(-\pi/2, \pi/2)$
   h. arcsin $(-\sqrt{3}/2)$ ∩ $[-\pi/2, \pi/2]$ u. arctan 1 ∩ $(-\pi/2, \pi/2)$
   i. arcsin $(-1/\sqrt{2})$ ∩ $[-\pi/2, \pi/2]$ v. arctan $\sqrt{3}$ ∩ $(-\pi/2, \pi/2)$
   j. arccos 0 ∩ $[0, \pi]$ w. arctan $(-\sqrt{3})$ ∩ $(-\pi/2, \pi/2)$
   k. arccos $1/\sqrt{2}$ ∩ $[0, \pi]$ x. arctan (-1) ∩ $(-\pi/2, \pi/2)$
   l. arccos 1/2 ∩ $[0, \pi]$ y. arctan $(-1/\sqrt{3})$ ∩ $(-\pi/2, \pi/2)$
   m. arccos $\sqrt{3}/2$ ∩ $[0, \pi]$

9. Evaluate each of the following.

   a. tan a, $a \in$ arctan 3   e. cos a, $a \in$ arcsin 8/17
   b. sin a, $a \in$ arcsin 1/3   f. tan a, $a \in$ arcsin 5/13
   c. cos a, $a \in$ arccos 1/4   g. cos a, $a \in$ arctan 3/4
   d. sin a, $a \in$ arccos 3/5   h. tan a, $a \in$ arccos 0

10. Describe each set in terms of special values.

    a. arcsin $(\sin \pi/6)$ c. arctan $(\tan \pi/3)$
    b. arccos $(\cos \pi/4)$ d. arccos $(\tan \pi/4)$

11. Solve the following trigonometric equations using the "arc" notation.
    Where possible, also write the solution set in terms of the special values,
    0, $\pi/6$, $\pi/4$, etc.

a. $2 \sin \theta = \sqrt{2}$
b. $3 \tan \theta = \sqrt{3}$
c. $5 \cos \theta = 3\sqrt{3} - \cos \theta$
d. $\sin^2 a = 1$
e. $\cos^2 a = 1$
f. $\tan^2 a = 3$
g. $\tan^2 a = 1$
h. $\sin a + \dfrac{\sin 2a}{\cos a} = 3/\sqrt{2}$
i. $2 \sin^2 a - \sin a - 1 = 0$
j. $2 \cos^2 a - \cos a = 0$
k. $\sin^2 a - \cos^2 a = 0$
l. $\cos a \, (\sin^2 a - 1) = 0$
m. $\tan a = \sin a$
n. $\tan a = \cos a$

o. $\cos a = \sin a$
p. $2 \sin^2 a + \cos a = 2$
q. $2 \cos^2 a + \sin a = 1$
r. $\sin a \tan a = 0$
s. $\sin a \cos a = 2$
t. $\tan^2 a + \sin^2 a = \sec a - \cos^2 a$
u. $\tan a - \sec^2 a = 1$
v. $\sin^2 a + \cos a = 0$
w. $\sin^2 a - \sin a - 2 = 1$
x. $\sin 2\theta = \cos \theta$
y. $\tan \theta = \sin 2\theta$
z. $\sin 2\theta = 1$
a'. $2 \cos 3\theta = 2$
b'. $\theta \cos \theta - \theta - \cos \theta + 1 = 0$
c'. $3 \sin a + 4 \cos a = 5$

12. Find the possible choices for $\alpha(a)$ where

a. $a \in \arcsin 1$ d. $a \in \operatorname{arccsc} (-\sqrt{2})$ g. $a \in \arccos (-5/13)$
b. $a \in \arccos \sqrt{3}/2$ e. $a \in \operatorname{arcsec} (2/\sqrt{3})$ h. $a \in \arctan 8/15$
c. $a \in \arctan (-\sqrt{3})$ f. $a \in \arcsin 3/5$ i. $a \in \arcsin -7/25$

6.2 INVERSE TRIGONOMETRIC FUNCTIONS

The "arc" sets given in Section 6.1 show an important characteristic of the trigonometric functions, a property that should have
been evident as a result of their periodicity.

Consider, for instance, the sine function. Reiterating, $\sin: R \to$
$[-1, 1]$ is an onto function. The sine function relates to each x, the
number $\sin x$. Reversing the correspondence does *not* give rise to a
function (the sine function is not 1:1). For each $y \in [-1, 1]$, the
inverse correspondence relates to y every element of **arcsin** y. Thus
the inverse correspondence is not a function; **arcsin** y is an "infinite"
set. Figure 6.1.1 shows that 1/2, for example, is mated to many real
numbers x.

The boldface portion of the curve in Figure 6.2.1 shows that
section of the sine function whose domain is restricted to the closed
interval $[-\pi/2, \pi/2]$. This portion of the curve is seen to be a graph
of $\sin |[-\pi/2, \pi/2]$. Alternatively, this curve segment is a graph of
$y = \sin x$ where $x \in [-\pi/2, \pi/2]$.

Every horizontal line (of the form $y = k$) strikes the colored part
of the curve *at most once*. If $-1 \leq k \leq 1$, the line strikes this segment
exactly once. Such a situation is geometric indication that the restriction is 1:1. In addition, the graph also points to the fact that for the

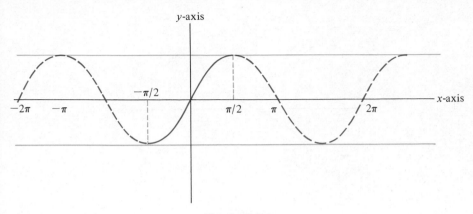

Figure 6.2.1

colored portion, sin x makes an excursion through *all* the values of $[-1, 1]$. Hence the restriction (restricted function) has an inverse. The inverse will be symbolized via

$$\sin^{-1}: [-1, 1] \rightarrow [-\pi/2, \pi/2]$$

The correspondence is given by $y \rightarrow a \in [-\pi/2, \pi/2]$ where $\sin a = y$ (that is, $\sin^{-1} y = a$ where $\sin a = y$).

The inverse mating may also be displayed in the following manner:

$$\sin^{-1} y \in [-\pi/2, \pi/2] \quad \text{and} \quad \sin(\sin^{-1} y) = y$$

Furthermore, $\sin^{-1} y \in \arcsin y \cap [-\pi/2, \pi/2]$.

The symbolism $\sin^{-1} k$ is not the (-1) power (reciprocal) of the $\sin k$. $1/\sin k = (\sin k)^{-1}$ is the symbolism to be used for an exponent of (-1).

The *inverse sine function* ($\sin^{-1}$) is really the inverse of the given restriction and not of the sine function itself.

Figure 6.2.2 shows a graph of $x = \sin y$ with $y \in [-\pi/2, \pi/2]$. Figure 6.2.1 shows $y = \sin x$. The two figures differ only because x and y are interchanged. Figure 6.2.3 illustrates a graph of $y = \sin^{-1} x$. We note that Figures 6.2.2 and 6.2.3 are identical. This can be shown to be correct from a set-theoretic analysis.

EXAMPLE 6.2.1. As examples of the $\sin^{-1}$ correspondence, compute: (1) $\sin^{-1} 1/2$, (2) $\sin^{-1}(-1/2)$, and (3) $\sin^{-1} 1$.

By definition, $\sin^{-1} 1/2 \in \arcsin 1/2$ and $\sin^{-1} 1/2 \in [-\pi/2, \pi/2]$. Elements of $\arcsin 1/2$ are of the form $n\pi + (-1)^n \pi/6$. (See Figure

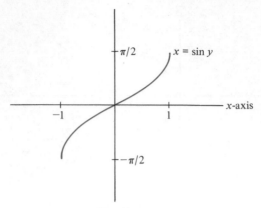

Figure 6.2.2

6.1.1.) The only element from this collection belonging to $[-\pi/2, \pi/2]$ is $\pi/6$. Thus $\sin^{-1} 1/2 = \pi/6$.

We cannot overemphasize the fact that $\sin^{-1} 1/2$ is a real number, not a set.

Now **arcsin** $(-1/2)$ includes the numbers $7\pi/6$, $11\pi/6$, $19\pi/6$, . . . and none of these is in $[-\pi/2, \pi/2]$. However, $-\pi/6$, $-5\pi/6$, $-13\pi/6$, $-17\pi/6$, . . . are also in **arcsin** $(-1/2)$. Only the element $-\pi/6$ is in the necessary interval. Problem 2 is then answered by $\sin^{-1}(-1/2) = -\pi/6$.

To solve (3) we only need note that arcsin 1 contains the number $\pi/2$. Necessarily, we conclude that $\sin^{-1} 1 = \pi/2$.

For each of the trigonometric functions there are many restrictions giving rise to an inverse. It is important that you be aware that the ranges of the restrictions and the ranges of the corresponding unrestricted functions coincide.

Agreement: The restrictions to be used (when discussing inverse trigonometric functions) are listed below.

$$\text{cosine} \mid [0, \pi]$$
$$\text{tangent} \mid (-\pi/2, \pi/2)$$
$$\text{cosecant} \mid (0, \pi/2] \cup (-\pi, -\pi/2]$$
$$\text{secant} \mid [0, \pi/2) \cup [\pi, 3\pi/2)$$
$$\text{cotangent} \mid (0, \pi)$$

The corresponding inverse functions are:

$$\cos^{-1}: [-1, 1] \to [0, \pi]$$
$$\tan^{-1}: R \to (-\pi/2, \pi/2)$$
$$\csc^{-1}: R - (-1, 1) \to (0, \pi/2] \cup (-\pi, -\pi/2]$$
$$\sec^{-1}: R - (-1, 1) \to [0, \pi/2) \cup [\pi, 3\pi/2)$$
$$\cot^{-1}: R \to (0, \pi)$$

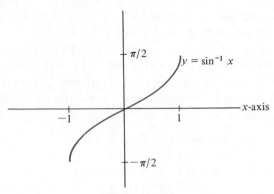

Figure 6.2.3

These functions are called respectively the *inverse cosine, inverse tangent, inverse cosecant, inverse secant,* and *inverse cotangent.* In each case, the domain of the inverse function agrees with the image of the trigonometric function for which it is named. The range of each inverse function is the restricted domain of the corresponding function.

Examples will be used to illustrate the new inverse functions further.

EXAMPLE 6.2.2. Find (1) $\cos^{-1}\frac{1}{2}$, (2) $\cos^{-1}(-\frac{1}{2})$, (3) $\tan^{-1} 1$, (4) $\csc^{-1} 2$, (5) $\sec^{-1}(-2)$, and (6) $\cot^{-1}(-1)$.

As in the case of $\sin^{-1}\frac{1}{2}$, there are two items to note upon examination of $\cos^{-1}\frac{1}{2}$. First, $\cos^{-1}\frac{1}{2}$ belongs to **arccos** $\frac{1}{2}$. Secondly $\cos^{-1}\frac{1}{2}$ $\in [0, \pi]$. The set **arccos** $\frac{1}{2} = \{x: \cos x = \frac{1}{2}\} = \{x: x = \pm\pi/3 + 2n\pi, n \in Z\}$. The one element from this set that is also in $[0, \pi]$ is $\pi/3$. In set symbolism, **arccos** $\frac{1}{2} \cap [0, \pi] = \{\pi/3\}$. By definition, $\cos^{-1}\frac{1}{2} = \pi/3$.

Now

arccos $(-\frac{1}{2}) = \{x: x = 2\pi/3 + 2n\pi \quad \text{or} \quad 4\pi/3 + 2n\pi, n \in Z\}$
$= \{x: x = (2n + 1)\pi \pm \pi/3\}$.

Since **arccos** $(-\frac{1}{2}) \cap [0, \pi] = \{2\pi/3\}$, $\cos^{-1}(-\frac{1}{2}) = 2\pi/3$.

In order to answer (3) we must see that **arctan** $1 = \{x: x = \pi/4 + n\pi$ for $n \in Z\}$. Since $\pi/4 \in (-\pi/2, \pi/2)$, $\tan^{-1} 1 = \pi/4$.

Problem (4) can be solved by observing that **arccsc** $2 = $ **arcsin** $\frac{1}{2}$ $= \{x: x = n\pi + (-1)^n \pi/6\}$ ($x \in$ **arccsc** 2 implies $\csc x = 2$ and $\sin x = \frac{1}{2}$ whence $x \in$ **arcsin** $\frac{1}{2}$ and conversely). Since $\pi/6$ belongs to the set, $\pi/6 = \csc^{-1} 2$.

Similarly, **arcsec** $(-2) = $ **arccos** $(-\frac{1}{2})$. Because $4\pi/3 \in (\pi, 3\pi/2)$, $\sec^{-1}(-2) = 4\pi/3$.

The set **arccot** $(-1) =$ **arctan** $(-1) = \{x: x = 5\pi/4 + n\pi, n \in Z\}$. It follows by definition that $\cot^{-1} (-1) = 3\pi/4$.

The following list affords a check as to whether or not an inverse function has been correctly evaluated. The list makes use of (1) the "arc" sets and (2) the ranges of the individual inverse functions. The intersections will either contain a single element or will be empty (depending on k).

$$\sin^{-1} k \in (\textbf{arcsin } k) \cap [-\pi/2, \pi/2]$$
$$\cos^{-1} k \in (\textbf{arccos } k) \cap [0, \pi]$$
$$\tan^{-1} k \in (\textbf{arctan } k) \cap (-\pi/2, \pi/2)$$
$$\csc^{-1} k \in (\textbf{arccsc } k) \cap [(0, \pi/2] \cup (-\pi, -\pi/2]]$$
$$\sec^{-1} k \in (\textbf{arcsec } k) \cap [[0, \pi/2) \cup [\pi, 3\pi/2)]$$
$$\cot^{-1} k \in (\textbf{arccot } k) \cap (0, \pi)$$

EXAMPLE 6.2.3. As one more example, compute $\tan^{-1} 0$. Now,

$$(\textbf{arctan } 0) \cap (-\pi/2, \pi/2) = \{0\}$$

Hence, $\tan^{-1} 0 = 0$.

EXERCISES 6.2

1. Find the value of each of the following.

   a. $\sin^{-1} (0)$ h. $\cos^{-1} (-1)$ n. $\sin^{-1} (-1/\sqrt{2})$
   b. $\sin^{-1} (\sqrt{3}/2)$ i. $\tan^{-1} (1/\sqrt{3})$ o. $\cos^{-1} (-\sqrt{3}/2)$
   c. $\sin^{-1} (1/\sqrt{2})$ j. $\tan^{-1} (\sqrt{3})$ p. $\cos^{-1} (-1/\sqrt{2})$
   d. $\sin^{-1} (-1)$ k. $\tan^{-1} (-1)$ q. $\cos^{-1} (-1/\sqrt{2})$
   e. $\cos^{-1} (\sqrt{3}/2)$ l. $\tan^{-1} (\sin 0)$ r. $\tan^{-1} (-1/\sqrt{3})$
   f. $\cos^{-1} (1/\sqrt{2})$ m. $\sin^{-1} (-\sqrt{3}/2)$ s. $\tan^{-1} (-3)$
   g. $\cos^{-1} (1)$

2. Evaluate each of the following:

   a. $\sin (\sin^{-1} 1/4)$ e. $\cos (\sin^{-1} (-4/5))$ i. $\cos^{-1} (\cos (-\pi/5))$
   b. $\cos (\cos^{-1} 1/5)$ f. $\sin^{-1} (\sin \pi/10)$ j. $\tan^{-1} (\tan \pi/7)$
   c. $\tan (\tan^{-1} 6)$ g. $\sin^{-1} (\sin -\pi/7)$ k. $\tan^{-1} (\tan 11\pi/10)$
   d. $\sin (\tan^{-1} 3/4)$ h. $\cos^{-1} (\cos 3\pi/4)$ l. $\tan^{-1} (\tan 9\pi/10)$

3. Sketch a graph of each inverse trigonometric function.

4. Determine the following points.

   a. $\alpha(\sin^{-1} 1/2)$ f. $\alpha(\sin^{-1} (\sin 5\pi/4))$
   b. $\alpha(\cos^{-1} (-1/2))$ g. $\alpha(\sin^{-1} (\cos (-\pi/6)))$
   c. $\alpha(\tan^{-1} (-1))$ h. $\alpha(\cos^{-1} (\cos 5\pi/6))$
   d. $\alpha(\sec^{-1} 5/4)$ i. $\alpha(\cos^{-1} (\sin \pi/3))$
   e. $\alpha(\csc^{-1} (-17/8))$ j. $\alpha(\tan^{-1} (\cot \pi/4))$

6.3 HANDLING INVERSE TRIGONOMETRIC FUNCTIONS

Certain problems of evaluation (see Exercise 6.2.2) become interesting whenever the trigonometric functions and the inverse functions appear together.

EXAMPLE 6.3.1. Find the values of: (1) sin (sin^{-1} a), (2) sin^{-1} (sin $5\pi/6$), (3) tan (sin^{-1} 3/5), and (4) cos (tan^{-1} 5/12).

Problem (1) has been examined previously. It was seen that sin (sin^{-1} a) = a if $-1 \leqslant a \leqslant 1$. This is really the age old question, "What is the name of the man who's name is Jones?" Alternatively, it is a question of noting that f ($f^{-1}(a)$) = a.

On the other hand, sin^{-1} (sin a) is *not necessarily a*. Computation of a solution to (2) shows this truth: sin^{-1} (sin $5\pi/6$) = sin^{-1} $(-\frac{1}{2})$ = $-\pi/6$. The answer sin^{-1} (sin $5\pi/6$) = $5\pi/6$ is *invalid!* This comes about because of the use of a restricted trigonometric function.

The solution for (3) requires further thought. Since 3/5 > 0, the value of sin^{-1} 3/5, say x, must be such that $0 < x < \pi/2$. The identity sin^2 x + cos^2 x = 1 implies that cos x = $\pm$4/5 (sin x = 3/5 by definition). The fact that $0 < x < \pi/2$ demands that cos x = 4/5. Rewriting sin^{-1} 3/5 in place of x, we see:

$$\cos (\sin^{-1} 3/5) = 4/5$$

Then,

$$\tan (\sin^{-1} 3/5) = \frac{\sin (\sin^{-1} 3/5)}{\cos (\sin^{-1} 3/5)}$$

$$= \frac{3/5}{4/5} = 3/4$$

Problem 4 is somewhat like 3. However, the calculation may not be so clear. Two techniques will be used, each having a special significance.

Since we wish to know the value of cos (tan^{-1} 5/12), we may just as well examine sec (tan^{-1} 5/12) and use the reciprocal. Let tan^{-1} 5/12 = x. It is seen that $0 < x < \pi/2$ and tan x = 5/12. Using the identity tan^2 x + 1 sec^2 x, we quickly observe that sec x = 13/12, and cos (tan^{-1} 5/12) = cos x = 1/sec x = 12/13.

An alternative approach utilizes the identity tan x = sin x/cos x. Now tan x = 5/12, and it is clear that sin x $\neq$ 5 and cos x $\neq$ 12. (Why?) However, it is a known algebraic truth that there is a number r so that $5/r$ = sin x and $12/r$ = cos x, whence $5r/12r$ = 5/12 = sin x/cos x

as before. Recalling that $\sin^2 x + \cos^2 x = 1$, $1 = (5/r)^2 + (12/r)^2 = 169/r^2$ which is to say that $r^2 = 169$ and $r = \pm 13$. Since $\sin x > 0$, $r = 13$, and $\sin x = 5/13$. Clearly, $\cos x = 12/13$.

The latter also extends a technique for finding $\sin x$ and $\cos x$ (up to $\pm$) whenever $\tan x$ is known. The r is a square root of the sum of the squares of the numerator and denominator of the expression for the value of the tangent.

As we might expect, equations arise during the use of the inverse trigonometric functions. The following illustrates how such an equation might be solved (using known identities).

EXAMPLE 6.3.2. Show that $\tan^{-1} 1/2 + \tan^{-1} 1/3 = \pi/4$. Since $0 < 1/3 < 1/2 < 1$, it follows that $0 < \tan^{-1} 1/3 < \tan^{-1} 1/2 < \pi/4$. (Why?) Furthermore, $0 < \tan^{-1} 1/2 + \tan^{-1} 1/3 < \pi/2$. Thus, $\pi/4$ and $\tan^{-1} 1/2 + \tan^{-1} 1/3$ are both in $(0, \pi/2)$. If we can show that $\tan(\tan^{-1} 1/3) = \tan \pi/4 = 1$, we are finished. (*Note:* $\tan a = \tan b$ does not imply that $a = b$ unless both a and b are between the same two multiples of π.)

$$\tan\ (\tan^{-1} 1/2 + \tan^{-1} 1/3) = \frac{\tan\ (\tan^{-1} 1/2) + \tan\ (\tan^{-1} 1/3)}{1 - \tan\ (\tan^{-1} 1/2)\ \tan\ (\tan^{-1} 1/3)}$$

$$= \frac{1/2 + 1/3}{1 - 1/2\ 1/3} = \frac{5/6}{5/6} = 1$$

The equation is seen to be valid.

EXERCISES 6.3

1. Find the value of each of the following.

    a. $\sin\ (\sin^{-1}\ (1/2))$
    b. $\cos\ (\cos^{-1}\ (1/3))$
    c. $\tan\ (\tan^{-1}\ (65))$
    d. $\sin^{-1}\ (\sin 17\pi/6)$
    e. $\cos^{-1}\ (\cos\ (-\pi/12))$

    f. $\tan^{-1}\ (\tan 5\pi/6)$
    g. $\sin\ (\sin^{-1} 2)$
    h. $\cos\ (\cos^{-1}\ (-3/2))$
    i. $\tan^{-1}\ (\tan \pi/2)$

2. Find the value of each of the following.

    a. $\tan^{-1}\ (\sin \pi/2)$
    b. $\cos^{-1}\ (\sin 9\pi/4)$
    c. $\sin^{-1}\ (\cos \pi/6)$
    d. $\sin^{-1}\ (\tan 26\pi)$
    e. $\sin\ (\cos^{-1} 1/2)$
    f. $\tan\ (\sin^{-1}\ (-1/2))$
    g. $\cos\ (\tan^{-1} 1)$
    h. $\csc\ (\sin^{-1} 1/2)$
    i. $\sec\ (\sin^{-1}\ (-1/2))$

    j. $\cos\ (\sin^{-1} 3/5)$
    k. $\sin\ (\cos^{-1} 3/5)$
    l. $\tan\ (\cos^{-1} 3/5)$
    m. $\sin\ (\tan^{-1} 12/5)$
    n. $\cos\ (\tan^{-1} 8/15)$
    o. $\cos\ (\cot^{-1}\ (-7/24))$
    p. $\sin\ (\cos^{-1}\ (-7/25))$
    q. $\cos\ (\sin^{-1}\ (-5/13))$
    r. $\tan\ (\cot^{-1}\ (-4/3))$

3. Use identities where necessary to find the value of each of the following.

    a. $\sin (\tan^{-1} (1) + \cos^{-1} (-\frac{1}{2}))$
    b. $\sin (\sin^{-1} (3/4) + \cos^{-1} (+7/25))$
    c. $\cos (\tan^{-1} (3/4) + \sin^{-1} (8/17))$
    d. $\cos (\cos^{-1} (\frac{1}{2}) - \sin^{-1} (1))$
    e. $\tan (\sin (5/13) + \cos^{-1} (7/25))$
    f. $\tan (\sin^{-1} (5/13) = \sin^{-1} (4/5))$
    g. $\sin (2 \tan^{-1} (4/3))$
    h. $\cos (2 \sin^{-1} (3/5))$
    i. $\cos (\frac{1}{2} \tan^{-1} (1))$
    j. $\sin (\frac{1}{2} \cos^{-1} (3/5))$
    k. $\tan (2 \sin^{-1} (24/25))$
    l. $\sin (2 \sin^{-1} (\frac{1}{2}) + \cos^{-1} (4/5))$
    m. $\sin (\pi/2 - \cos^{-1} (1/3))$
    n. $\cos (\pi/2 - \sin^{-1} (-2/3))$

4. Verify that the following equations hold.

    a. $\tan^{-1} 6 + \tan^{-1} 1/6 = \pi/2$
    b. $\sin^{-1} (\frac{1}{2}) + \sin^{-1} (3/2) = \pi/2$
    c. $\sin^{-1} (a) + \sin^{-1} (-a) = 0$ if $a \leqslant 1$
    d. $\cos^{-1} (\frac{1}{2}) + \cos^{-1} (-\frac{1}{2}) = \pi$
    e. $\cos^{-1} 12/13 + \cos^{-1} 24/25 = \cos^{-1} 253/325$
    f. $\sin^{-1} (3/5) + \sin^{-1} (4/5) = \pi/2$
    g. $\tan^{-1} (3/4) + \tan^{-1} 1/3 = \tan^{-1} 13/19$
    h. $\tan^{-1} 1/5 + \tan^{-1} 1/6 = \tan^{-1} 11/30$
    i. $\tan^{-1} 1/6 - \tan^{-1} 1/7 = \tan^{-1} 1/41$
    j. $\tan^{-1} 1/4 + 2 \tan^{-1} 1/5 = \tan^{-1} 32/43$
    k. $\tan^{-1} (1/5) + \tan^{-1} (3/2) = \pi/4$
    l. $\tan^{-1} (1/8) + \tan^{-1} (\frac{1}{2}) = \tan^{-1} (2/3)$

6.4 GEOMETRIC PROPERTIES OF THE TRIGONOMETRIC FUNCTIONS

We have, on previous occasions, seen graphs of $y = \sin x$. The outstanding characteristics of this curve are:

1. a maximum value of $+1$ 3. periodicity

2. a minimum value of -1 4. $(0, 0)$ a point on the curve

The minimum and maximum values have the same absolute value, namely 1. This number is called the *amplitude* or peak value of the curve $y = \sin x$. The amplitude of a real function f might be defined as the least upper bound of $\{y: y = |f(x)|$ for some x in the domain of $f\}$. Figure 6.4.1 illustrates.

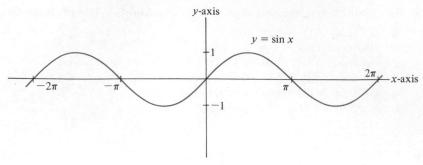

Figure 6.4.1

Examine the curve $y = a \sin x$ with $a > 0$ ($a > 0$ is not a necessary condition but will be employed). Since the excursions of the expression $\sin x$ carry it between -1 and $+1$, it follows that $a \sin x$ must have minimum and maximum values $-a$ and a respectively. In a manner analogous to the above, we define the amplitude of $y = a \sin x$ to be a.

EXAMPLE 6.4.1. The example $y = 3 \sin x$ is shown in Figure 6.4.2. The student should compare Figures 6.4.1 and 6.4.2 to see that the curves are identical except for minimum and maximum excursions. If we are not concerned with whether or not both axes have the same scale of measurement, both $\sin x$ and $a \sin x$ can be given by the same curve (simply relabel the graduations on the y-axis).

The curve $y = \sin (x + b)$ (in Figure 6.4.3) resembles $y = \sin x$ as given in Figure 6.4.1. However, one difference appears. The point

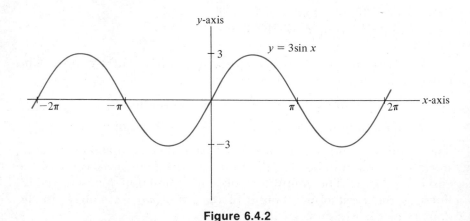

Figure 6.4.2

$(0, 0)$ is not on the curve $y = \sin (a + b)$. The argument of the sine curve given here is $x + b$, not x. The value $\sin 0 = 0$ still holds but when $x + b = 0$, $x = -b$. Thus the point $(-b, 0)$ on $y = \sin (x + b)$ is analogous to the point $(0, 0)$ for the curve $y = \sin x$. The curve $y = \sin (x + b)$ appears to be the curve $y = \sin x$ after a slight "shift" to the left. The only visible difference between Figures 6.4.1 and 6.4.3 is the placement of the y-axis. The value $|b|$ is called the *phase displacement*.

Let $k > 0$ be any real number. The expression $\sin kx$ is periodic of primitive period 2π. That is, if kx_0 and kx_1 differ by 2π, $\sin kx_0 = \sin kx_1$. Furthermore, $kx_0 - kx_1 = 2\pi$ implies that $x_0 - x_1 = 2\pi/k$. Thus the graph (in Figure 6.4.4) of $\sin kx$ shows the apparent periodicity of the expression to be $2\pi/k$. We will call $2\pi/k$ the *period* of $\sin kx$ *relative to* x.

EXAMPLE 6.4.2. Figure 6.4.5 illustrates the curve $y = \sin 2x$. In this figure, $k = 2$, whence $2\pi/k = 2\pi/2 = \pi$. The period of $\sin 2x$ relative to x is π.

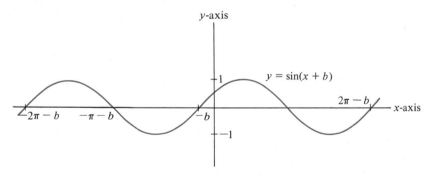

Figure 6.4.3

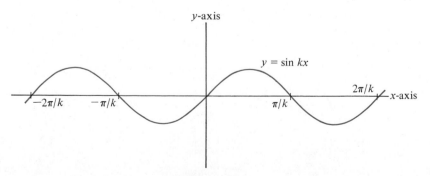

Figure 6.4.4

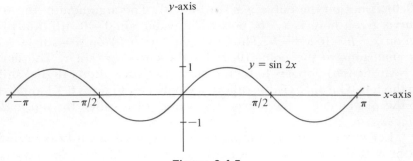

Figure 6.4.5

In the expression sin kx, $k/2\pi$ is sometimes called the *frequency*. That part of a sine curve from some point x to $x + 2\pi/k$ (the difference of $x + 2\pi/k$ and x is one period of sin kx relative to x) is called a *cycle*. The curve goes through all values of its range during a cycle.

The most general form of the sine curve is

$$y = a \sin k(x + b), \quad k > 0$$

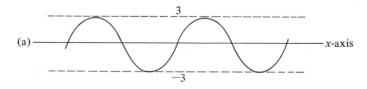

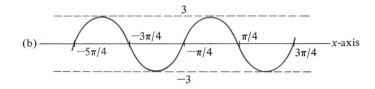

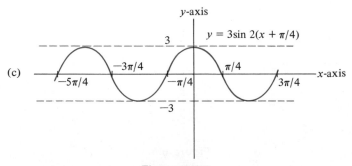

Figure 6.4.6

The following items are to be noted:

1. The *amplitude* of $y = a \sin k(x + b)$ is a
2. The *phase* of $y = a \sin k(x + b)$ is $b \, [(-b, 0)$ is on the curve$]$
3. The *period* of $y = a \sin k(x + b)$ relative to x is $2\pi/k$

EXAMPLE 6.4.3. Figure 6.4.6 shows a step-by-step construction of a sketch of $y = a \sin k(x + b)$ for the special case $y = 3 \sin 2(x + \pi/4)$.

An example will be used to summarize the material of this section.

EXAMPLE 6.4.4. Write $\sqrt{2} \sin 4x - \sqrt{2} \cos 4x$ in the form $a \sin k(x + b)$ and state the amplitude, period, and phase.

Using a previous identity,

$$\sqrt{2} \sin 4x - \sqrt{2} \cos 4x = 2 \sin (4x - \pi/4)$$
$$= 2 \sin 4(x - \pi/16)$$

The amplitude is 2, the period relative to x is $2\pi/4 = \pi/2$, and the phase is $\pi/16$. Figure 6.4.7 shows the sketch of this function.

A similar analysis exists for $a \cos k(x + b)$.

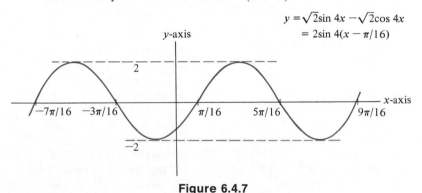

Figure 6.4.7

EXERCISES 6.4

1. Give the period relative to x, the phase, and the amplitude of each of the following. (*Hint:* Use identities to rewrite h.–j.)

    a. $3 \sin 5x$

    b. $2 \sin (3x + 1)$

    c. $\frac{1}{2} \cos 2(x - \pi/6)$

    d. $(3/4) \cos (2/3x - \pi/8)$

    e. $\sin (\pi - \pi/2)$

    f. $6 \cos (x - \pi/3)$

    g. $(5/3) \sin (5/3x + \pi)$

    h. $\sin x + \cos x$

    i. $5 \sin 2x + 12 \cos 2x$

    j. $\sin 3x + \sqrt{3} \cos 3x$

2. Give the frequency in each part of Exercise 6.4.1.

3. Sketch each curve (3 cycles) from Exercise 6.4.1 above.

4. Find a period for f in each case. Example: Let $f(x) = \sin (x/7) + \sin (x/5)$. The component parts of $f(x)$ have period 14π and 10π respectively. The least common multiple of 14 and 10 is 70. It can be shown that 70 is a period for both $\sin x/7$ and $\sin x/5$ and hence a period for f.

    a. $f(x) = \sin x/3 + \sin x/4$
    b. $f(x) = \sin 3x/4 + \sin x$
    c. $f(x) = \cos 2x/3 + \cos x/6$

6.5 REVIEW EXERCISES

1. Describe the element of each of the following sets.

    a. $\arcsin 0 \cap \arctan 0$
    b. $\arcsin 0 \cap \arccos 0$
    c. $\arcsin 1/\sqrt{2} \cap \arccos 1/\sqrt{2}$
    d. $\arcsin (-1/\sqrt{2}) \cap \arccos 1/\sqrt{2}$
    e. $\arccos 0 \cap \arctan 0$
    f. $\arccos 0 \cap \text{arccot } 0$
    g. $\arcsin 3 \cap \arccos \frac{1}{2}$
    h. $\arcsin 1 \cap \text{arccsc } 1$
    i. $\arctan (-1) \cap \arccos 1/\sqrt{2}$
    j. $\arcsin (-\sqrt{3}/2) \cap \arctan (-\sqrt{3})$

    k. $\arctan (\sin \pi/2)$
    l. $\arcsin 1 \cup \arccos 1$
    m. $\arcsin 0 \cup \arccos 0$
    n. $\arcsin 1/\sqrt{2} \cup \arccos 1/\sqrt{2}$
    o. $\arctan 1 \cup \arcsin 1/\sqrt{2}$
    p. $\arctan 1 \cup \arctan (-1)$
    q. $\arcsin 1/\sqrt{2} - \arccos (-1/\sqrt{2})$
    r. $\arcsin 1/\sqrt{2} - \arccos 1/\sqrt{2}$
    s. $\arctan (-1) - \arccos (-1/\sqrt{2})$

2. List the elements in each set.

    a. $\arcsin 0 \cap [\pi/2, 3\pi/2]$
    b. $\arcsin 1/\sqrt{2} \cap [\pi/2, 3\pi/2]$
    c. $\arcsin 1/2 \cap [\pi/2, 3\pi/2]$
    d. $\arcsin \sqrt{3}/2 \cap [\pi/2, 3\pi/2]$
    e. $\arcsin 1 \cap [\pi/2, 3\pi/2]$
    f. $\arcsin (-1) \cap [\pi/2, 3\pi/2]$
    g. $\arcsin (-\frac{1}{2}) \cap [\pi/2, 3\pi/2]$
    h. $\arcsin (-\sqrt{3}/2) \cap [\pi/2, 3\pi/2]$
    i. $\arcsin (-1/\sqrt{2}) \cap [\pi/2, 3\pi/2]$
    j. $\arccos 0 \cap [-\pi, 0]$
    k. $\arccos 1/\sqrt{2} \cap [-\pi, 0]$
    l. $\arccos \frac{1}{2} \cap [-\pi, 0]$
    m. $\arccos \sqrt{3}/2 \cap [-\pi, 0]$

    n. $\arccos 1 \cap [-\pi, 0]$
    o. $\arccos (-1) \cap [-\pi, 0]$
    p. $\arccos (-\frac{1}{2}) \cap [-\pi, 0]$
    q. $\arccos (-\sqrt{3}/2) \cap [-\pi, 0]$
    r. $\arccos (-1/\sqrt{2}) \cap [-\pi, 0]$
    s. $\arctan 0 \cap (\pi/2, 3\pi/2)$
    t. $\arctan 1/\sqrt{3} \cap (\pi/2, 3\pi/2)$
    u. $\arctan 1 \cap (\pi/2, 3\pi/2)$
    v. $\arctan \sqrt{3} \cap (\pi/2, 3\pi/2)$
    w. $\arctan (-\sqrt{3}) \cap (\pi/2, 3\pi/2)$
    x. $\arctan (-1) \cap (\pi/2, 3\pi/2)$
    y. $\arctan (-1/\sqrt{3}) \cap (\pi/2, 3\pi/2)$

3. Each of the elements in the sets resulting in Exercise 6.1.8 can be described in terms of inverse functions. Do so.

4. Solve the following trigonometric equations.

    a. $\sin x \cos x = 0$
    b. $\sin x \cos x = \sin x$
    c. $\tan x + \cot x = 2$

    d. $\sin x \cos x = 1$
    e. $\sin^2 x - \cos x + 4 = -1$
    f. $3 \csc^2 2x - 5 = \cot 2x$

5. Evaluate the following.

    a. $\sin(\sin^{-1} a)$
    b. $\cos(\cos^{-1} a)$
    c. $\tan(\tan^{-1} a)$
    d. $\sin^{-1}(\sin 5\pi/6)$
    e. $\sin^{-1}(\sin 11\pi/6)$
    f. $\cos^{-1}(\cos 2\pi/3)$
    g. $\cos^{-1}(\cos(-\pi/4))$
    h. $\tan^{-1}(\tan \pi/7)$
    i. $\tan^{-1}(\tan 8\pi/7)$

    j. $\sin(\cos^{-1}(-5/12))$
    k. $\cos(\sin^{-1}(-8/17))$
    l. $\tan(\sin^{-1}(3/5))$
    m. $\tan(\cos^{-1}(12/13))$
    n. $\sin(\tan^{-1}(3/4))$
    o. $\cos(\tan^{-1}(-8/15))$
    p. $\sin^{-1}(\tan 95\pi/4)$
    q. $\cos^{-1}(\tan 95\pi/4)$
    r. $\tan^{-1}(\sin 3\pi/4)$

6. Prove or disprove the validity of the following statements.

    a. $\sin^{-1} 3/5 + \cos^{-1} 5/13 = \cos^{-1}(-16/65)$
    b. $\sin^{-1} 8/17 + \cos^{-1} 12/13 = \sin^{-1} 21/221$
    c. $\sin^{-1} 3/5 + \cos^{-1} 5/13 + \sin^{-1} 63/65$
    d. $\tan^{-1} 1/8 + \tan^{-1} 1/7 = \tan^{-1} 3/11$
    e. $\tan^{-1} 1/7 - \tan^{-1} 1/8 = \tan^{-1} 1/55$
    f. $\tan^{-1} 4/3 + \tan^{-1} 4/5 = \tan^{-1}(-32)$

7. Give the amplitude, phase, and period with respect to x in each example below.

    a. $3 \sin(7x + 1)$
    b. $(1/5) \cos(3x - 2)$
    c. $(3/2) \sin 5(x - 1/3)$
    d. $5 \cos 3(x - \pi/7)$
    e. $3 \sin x + \cos x$
    f. $5 \sin x + 12 \cos x$
    g. $3 \sin 2x + 4 \cos 2x$

8. Give the frequency of each expression in Exercise 6.5.7.

9. Sketch 4 cycles of each expression in Exercise 6.5.7.

10. Refer to Figure 6.5.1 and Exercises 5.6.14 and 5.6.15. Compute those values missing from among a, b, c, A and B granted that $A + B = \pi/2$ radians.

    a. $A = \sin^{-1} 3/5$ radians, $a = 1$
    b. $A = \cos^{-1} 5/13$ radians, $b = 5$
    c. $A = \tan^{-1} 8/15$ radians, $c = 2$
    d. $B = \cot^{-1} 8/15$ radians, $c = 2$

    e. $a = 6$, $b = 8$
    f. $a = 7$, $c = 25$
    g. $b = 1$, $c = 2$

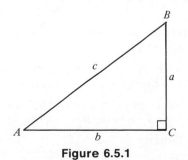

Figure 6.5.1

6.6 QUIZ

Finish this self-test in one hour and check your efforts.

1. Find the solution sets for the following equations:

    a. $\sec^2 a + \tan a = 6$
    b. $\cot a + \cot a \cos a = 0$
    c. $\sin a \cos^2 a = 1$

2. Sketch 3 cycles of $3 \sin 2x + 4 \cos 2x$

3. Calculate:

    a. $\sin (\sin^{-1} .61)$ c. $\sin^{-1} (\sin 285\pi/9)$
    b. $\sin^{-1} (\cos \pi/5)$ d. $\tan^{-1} (\cos 0)$

4. Determine:

    a. $\alpha(\sin^{-1} (-\sqrt{3}/2))$
    b. $\alpha(\cos^{-1} (-\sqrt{3}/2))$
    c. $\alpha(\tan^{-1} 1)$

5. Show whether each of the following is true.

    a. $\tan^{-1} 3 + \tan^{-1} (-2) = \tan^{-1} (-1/7)$
    b. $\tan^{-1} 2 + \tan^{-1} 3 = -\pi/4$
    c. $\sin^{-1} 3/5 + \cos^{-1} 12/13 = \tan^{-1} 56/33$

6.7 ADVANCED EXERCISES

1. Find 3 closed intervals other than $[-\pi/2, \pi/2]$ which would make appropriate restrictions for the sine function in order that an inverse might be defined.

2. Find 3 closed intervals other than $[0, \pi]$ which would make appropriate restrictions for the cosine function in order that an inverse might be defined.

3. Find 3 open intervals other than $(-\pi/2, \pi/2)$ which would make appropriate restrictions for the tangent function in order that an inverse might be defined.

4. Why is the restriction for the tangent function the open interval $(-\pi/2, \pi/2)$ instead of $[-\pi/2, \pi/2]$?

5. Why does the restriction $\cos | [-\pi/2, \pi/2]$ prove unsatisfactory for purposes of defining an inverse cosine function?

6. Let m, n, p, and q be positive integers. What is the primitive period of $f(x) = \sin (m/n)x + \sin (p/q)x$ with respect to x?

7. Write $\cos x$ in the form $\sin (x + a)$.

8. Consider $a \sin k(x + b)$. Let f be the frequency and p the period relative to x. Show that $pf = 1$.

9. Write the equation of the line through $(0, 0)$ and the point:

    a. $\alpha(\sin^{-1} \frac{1}{2})$ c. $\alpha(\cos^{-1} 1/\sqrt{2})$ e. $\alpha(\tan^{-1} 0)$
    b. $\alpha(\sin^{-1} (-\sqrt{3}/2))$ d. $\alpha(\cos^{-1} (-1))$ f. $\alpha(\tan^{-1} (-1/\sqrt{3}))$

10. Note that $\sin^{-1} = \alpha^{-1} \circ p_2^{-1}$ with appropriate restrictions. Suppose $f: X \to Y$ and $g: Y \to Z$ are $1:1$ and onto functions. Show that $(g \circ f)^{-1} = f^{-1} \circ g^{-1}$.

11. Let b arcsin $k = \{x: \sin x/b = k\}$. Show that if $cx \in$ arcsin k, $x \in (1/c)$ arcsin k; $c \neq 0$.

12. Let $G \subset R \times R$ be given by

    $$G = \{(x, y): x, y \in R, \quad \sin x = \sin y\}$$

    Show that G is an equivalence relation on $R \times R$.

13. Let $G \subset [-\pi/2, \pi/2] \times [-\pi/2, \pi/2]$ be given as in Exercise 6.7.12. Describe G completely in terms of the types of ordered pairs it has.

14. Describe the relations on R determined by each of the six "arc" sets (that is, determine arccos, arctan, etc.). Describe also the relations for each "arc" set determined analogous to the determination of G in Exercise 6.7.12; G in Exercise 6.7.13. What Cartesian product is used in each case?

CHAPTER 7

Extensions of Previous Material

7.1 POLAR COORDINATES

Previous discussions and exercises have drawn attention to the fact that each point $(p, q) \neq (0, 0)$ of the plane can be written in the form $(r \cos a, r \sin a)$ where:

1. a is a real number, and

2. $r = (p^2 + q^2)^{1/2} = d((p, q), (0, 0))$.

Since the line l passing through (p, q) and $(0, 0)$ has the equation $py = qx$ (why?) the unit circle C and l intersect at $(p/r, q/r)$ and $(-p/r, -q/r)$. The point $(p/r, q/r)$, lying on the unit circle, has the representation $\alpha(a) = (\cos a, \sin a)$ for some real number a ($0 \leq a < 2\pi$ if we desire). Thus $(p, q) = (r \cos a, r \sin a)$.

The point $(-p, -q)$ is given by $(-r \cos a, -r \sin a)$ or by $(r \cos (a + \pi), r \sin (a + \pi))$. Moreover, (p, q) itself may be given by $(-r \cos (a + \pi), -r \sin (a + \pi))$. That is, since r need not be positive, more than one such representation (for each point) exists. In fact, an infinite variety of forms exists since a may be replaced by $a + 2n\pi$ for any $n \in Z$.

Since the r and a, in this sense, completely describe (or "locate") the point (p, q) it seems reasonable that we might develop a system of identification for points of the plane using this type of information. If (p, q) is a point of the plane with $(p, q) = (r \cos a, r \sin a)$, *the polar* (coordinate) *form* for (p, q) is given by the ordered-pair notation (r, a)*. To insure that no confusion arises because of the similarity between the notations of the rectangular coordinate and polar coordinate systems, the polar coordinates will always be given with an asterisk.

EXAMPLE 7.1.1. Sketch a graph showing the location of the points $(6, \pi/3)^*$ and $(-6, \pi/3)^*$.

From the special value list, we determine that the line through the pole (origin) and $(6, \pi/3)^*$ intersects the unit circle at $(\frac{1}{2}, \sqrt{3}/2)$ the rectangular coordinates of $(6, \pi/3)^*$ are $(3, 3\sqrt{3}/2))$.

The location of $(6, \pi/3)^*$ is shown in Figure 7.1.1. The point $(-6, \pi/3)^*$ is the same point as $(6, \pi/3 + \pi)^*$ or $(6, 4\pi/3)^*$ (verify this) so that $(-6, \pi/3)^*$ can be located as in Figure 7.1.2.

EXAMPLE 7.1.2. Give four polar (coordinate) representations for $(3, 4)$. Since $5^2 = 3^2 + 4^2$, $(3, 4)$ may be given by

$$(5, \sin^{-1} 4/5)^*$$
$$(-5, \pi + \sin^{-1} 4/5)^*$$
$$(5, 2\pi + \sin^{-1} 4/5)^*$$
$$(5, -2\pi + \sin^{-1} 4/5)^*$$

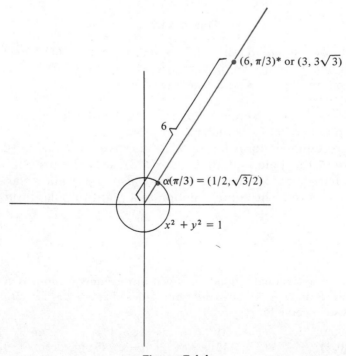

$(6, \pi/3)^*$ or $(3, 3\sqrt{3})$

6

$\alpha(\pi/3) = (1/2, \sqrt{3}/2)$

$x^2 + y^2 = 1$

Figure 7.1.1

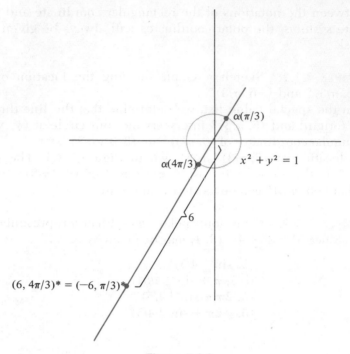

Figure 7.1.2

This follows since $\alpha(\sin^{-1} 4/5) = \alpha(2\pi + \sin^{-1} 4/5) = \alpha(-2\pi + \sin^{-1} 4/5)$ and if $\alpha(\sin^{-1} 4/5) = (x, y)$, $\alpha(\pi + \sin^{-1} 4/5) = (-x, -y)$. (What is $\alpha(\sin^{-1} 4/5)$? $\alpha(\sin^{-1} a)$?) (See Figure 7.1.3.)

EXAMPLE 7.1.3. Suppose $(3, \pi/2)^*$ is a polar-form representation of a point. What is its rectangular representation?

The rectangular form for $(3, \pi/2)^*$ is given by $(0, 3)$: $(0, 3)$ is 3 units from the pole and $(0, 1)$ is $\alpha(\pi/2)$. (See Figure 7.1.4.) We observe that $(3, 5\pi/2)^*$ and $(3, -3\pi/2)^*$ also represent polar forms of the same point. The representation of a point in polar form is not unique.

EXERCISES 7.1

1. Each of the following points is given in rectangular form. Write each in polar form $(r, a)^*$ for two different values of a, $0 \leqslant a < 2\pi$. (*Hint:* Use a negative value for r.)

   a. $(3, 0)$ e. $(3, 4)$ i. $(-5, 12)$ m. $(-2, 3)$
   b. $(0, 1)$ f. $(5, 12)$ j. $(-8, -15)$ n. $(-1, \sqrt{3})$
   c. $(-1, 0)$ g. $(8, 15)$ k. $(6, 5)$ o. $(-5\sqrt{3}, 5)$
   d. $(0, -2)$ h. $(3, -4)$ l. $(-2, 3)$ p. (x, y)

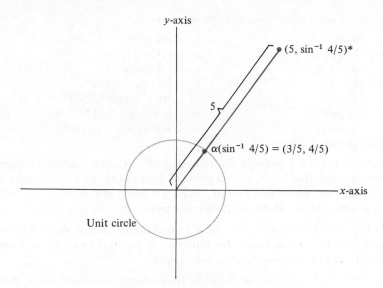

Figure 7.1.3

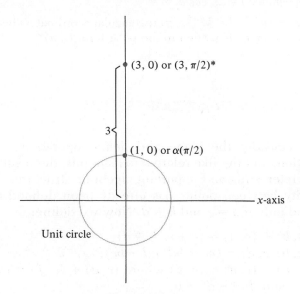

Figure 7.1.4

2. Do Exercise 7.1.1 with $-2\pi \leq a < 0$.

3. Let $(r, a)^*$ be a point written in polar form. Show that:

    a. $p_1((r, a)^*) = r \cos a$ b. $p_2((r, a)^*) = r \sin a$

4. If $(x, y) = (r \cos a, r \sin a)$, $x \neq 0$, $r > 0$, and $a \in \arctan k$; find k.

5. Convert each polar form to rectangular form.

a. $(3, 0)$ *   i. $(-2, \pi)$ *
b. $(2, \pi)$ *   j. $(-3, \pi/2)$ *
c. $(3, -\pi/2)$ *   k. $(-5, 5\pi/3)$ *
d. $(5, -\pi)$ *   l. $(20, \tan^{-1} 8/15)$ *
e. $(1, \pi/2)$ *   m. $(10, \pi + \tan^{-1} 5/12)$ *
f. $(6, 5\pi/6)$ *   n. $(8, \pi - \sin^{-1} \frac{1}{2})$ *
g. $(2, 3\pi/4)$ *   o. $(3, \pi + \cos^{-1} (-\frac{1}{2}))$ *
h. $(10, \tan^{-1} 3/4)$ *

6. Let $y = mx + b$ represent the equation of a straight line in rectangular coordinates (that is, (x, y) is a point of the line, (x, y) a rectangular representation). Show that this becomes $a = \tan^{-1} m$ if $b = 0$ ($\tan^{-1} m$ is a constant). [*Hint:* Write (x, y) in polar form (in terms of r and a).]

7. Let $x^2 + y^2 = a^2$ represent a circle in rectangular form. Show this equation becomes $r = a$ when the equation is written in terms of r and a [(r, a) * a polar representation for (x, y)].

8. Let $(x - h)^2 + y^2 = h^2$ be a circle (in rectangular form). Show that in polar form [(x, y) converted to the polar form (r, a)*], the equation has the appearance $r = 2h \cos a$.

9. Show $x^2 + (y - k)^2 = k^2$ [(x, y) rectangular coordinates] becomes $r = 2k \sin a$ when (x, y) is written in the polar form (r, a)*.

7.2 THE PLANE AS AN ALGEBRAIC SYSTEM

If we consider the plane and define operations of addition, multiplication, and the like relative to its points, the resulting system has a very interesting and important algebraic structure.

Equality between points has already been defined by $(a, b) = (c, d)$ if and only if $a = c$ and $b = d$. Now we define:

$$(a, b) + (c, d) = (a + c, b + d);$$
$$(a, b)(c, d) = (ac - bd, ad + bc);$$
$$(a, b) - (c, d) = (e, f) \text{ where } (c, d) + (e, f) = (a, b);$$
$$\text{and if } c^2 + d^2 \neq 0,$$
$$(a, b) \div (c, d) = (g, h) \text{ where } (c, d)(g, h) = (a, b).$$

Addition is perhaps the easiest operation to perform. We merely add the respective first elements and then the respective second elements.

EXAMPLE 7.2.1. From the above, we see that $(2, 1) + (3, 5) = (2 + 3, 1 + 5) = (5, 6)$.

Multiplication is somewhat more complicated. The following example illustrates.

EXAMPLE 7.2.2. $(2, 1)(3, 5) = (2 \cdot 3 - 1 \cdot 5,\ \ 2 \cdot 5 + 1 \cdot 3) = (6 - 5,\ 10 + 3) = (1, 13)$.

Our definition of subtraction does not provide a procedure for calculating a difference; it merely gives a means for verifying a correct answer. Now, it is claimed that $(a - c, b - d)$ is a solution to $(a, b) - (c, d)$. That is, the answer is found in a manner analogous to the process for addition. That the answer is valid is seen by

$$(c, d) + (a - c, b - d) = (c + a - c, d + b - d) = (a, b)$$

EXAMPLE 7.2.3. From the above,

$$(2, 1) - (3, 5) = (2 - 3, 1 - 5) = (-1, -4)$$

As in the case of subtraction, a proposed answer will be given for the division shown above. (*Note:* If both c and d are zero, the division is not defined.) This proposed answer for $(a, b) \div (c, d)$ is

$$\left(\frac{ac + bd}{c^2 + d^2},\ \frac{bc - ad}{c^2 + d^2} \right)$$

We verify this proposal via the following.

$$(c, d) \left(\frac{ac + bd}{c^2 + d^2},\ \frac{bc - ad}{c^2 + d^2} \right)$$

$$= \left(\frac{ac^2 + bcd - bcd + ad^2}{c^2 + d^2},\ \frac{dac + bd^2 + bc^2 - cad}{c^2 + d^2} \right)$$

$$= (a, b)$$

EXAMPLE 7.2.4. The following shows a computation of a solution to a division process.

$$(2, 1) / (3, 5) = (2, 1) \div (3, 5) = \left(\frac{2 \cdot 3 + 1 \cdot 5}{3^2 + 5^2},\ \frac{1 \cdot 3 - 2 \cdot 5}{3^2 + 5^2} \right)$$

$$= \left(\frac{11}{34},\ \frac{-7}{34} \right)$$

Let us now introduce another multiplication called *scalar multiplication*. Let k be any real number. We define:

$$k(a, b) = (ka, kb)$$

Scalar multiplication is simply a multiplication of each coordinate by a common real number.

EXAMPLE 7.2.5. $6(2, 1) = (12, 6)$.

Examining this definition in a sort of reverse manner, we see that we may in effect *factor out* any real number appearing as a factor common to both elements of the ordered pair.

A geometric interpretation can be given to the operations of addition, subtraction, and scalar multiplication.

EXAMPLE 7.2.6. The point (a, b) (Figure 7.2.1) can be used to describe a "line segment" between $(0, 0)$ and (a, b). By placing an arrowhead in the line segment at (a, b), we give rise to the intuitive feeling that we are considering a "vector" from $(0, 0)$ to (a, b). The line segment has become a directed line segment in some sense.

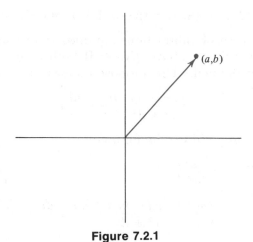

Figure 7.2.1

EXAMPLE 7.2.7. The point $(-a, -b) = -(a, b)$ is seen to determine a vector of the same length as that determined by (a, b) but the direction indicated is opposite that of Figure 7.2.1. (See Figure 7.2.2.)

EXAMPLE 7.2.8. The point $6(a, b) = (6a, 6b)$ denotes a vector in the same direction as that of Figure 7.2.1 but of length six times as great (see Figure 7.2.3). On the other hand, $-6(a, b) = (-6a, -6b) = -(6a, 6b)$ gives a vector point in the opposite direction with similar length (Figure 7.2.4).

EXAMPLE 7.2.9. Since $(a, b) + (c, d) = (a + c, b + d)$ the addition of the two is seen geometrically as the "head-to-tail" addition

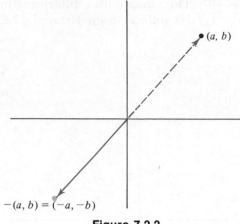

Figure 7.2.2

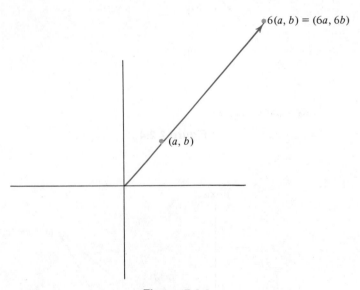

Figure 7.2.3

of the two vectors involved (Figure 7.2.5). The vector for (c, d) is moved so that its initial point [originally $(0,0)$] is superimposed with (a, b) and so that the vector is parallel (parallel in the sense that straight lines are parallel and the oriented direction coincides with that of the original vector). The resulting vector is that from $(0, 0)$ to the terminal point of the displaced vector. The vector sum is often called the *resultant vector*.

EXAMPLE 7.2.10. The geometric interpretation of $(a, b) - (c, d) = (a, b) + (-(c, d))$ follows from Example 7.2.9 and is given in Figure 7.2.6.

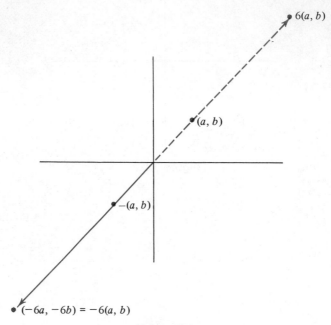

Figure 7.2.4

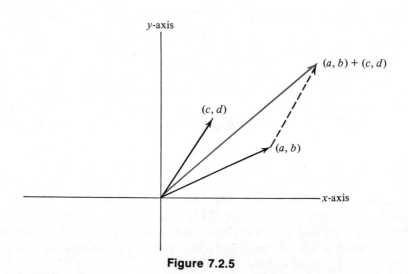

Figure 7.2.5

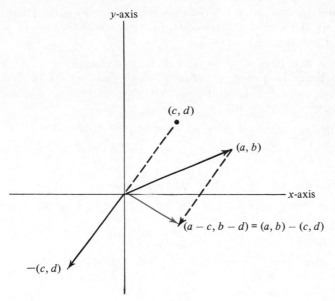

Figure 7.2.6

EXERCISES 7.2

1. Compute each of the following.

a. $(3, 2) + (1, 5)$
b. $(2, -1) + (3, 2)$
c. $(-3, -1) + (-2, 1)$
d. $(3, 5) + (-1, -1)$
e. $(3, 4) + (0, -2)$
f. $(6, 2) - (1, 4)$
g. $(3, 1) - (2, -3)$
h. $(-2, 5) - (-3, 2)$
i. $(-2, -1) - (-1, -2)$
j. $(5, -6) - (5, 6)$

k. $(1, 2)(2, 1)$
l. $(1, 3)(3, 5)$
m. $(1, 2)(6, 3)$
n. $(2, 5) + (-1, -1)$
o. $(2, 4)(-1, -1)$
p. $(12, 2)(\frac{1}{2}, \frac{1}{2})$
q. $(3, 0)(0, 1)$
r. $(5, 6)(-1, -2)$
s. $(-2, 1)(1, -2)$
t. $(2, 0)(0, 3)$

u. $(1, 2) \div (3, 1)$
v. $(4, 4) \div (3, 5)$
w. $(6, 2) \div (1, 2)$
x. $(6, 2) \div (2, 0)$
y. $(2, -3) \div (-4, 1)$
z. $(5, 2) \div (-3, -2)$
a'. $(16, 0) \div (0, 1)$
b'. $(3, -2) \div (1, 1)$
c'. $(5, -6) \div (2, 3)$
d'. $(0, 5) \div (5, 0)$

2. Find each of the following $((a, b)^n = (a, b)(a, b) \cdots (a, b)$ where n factors (a, b) appear).

a. $(a, b) + (0, 0)$
b. $(a, b)(1, 0)$
c. $(a, b) \div (1, 0)$
d. $(a, b) - (0, 0)$
e. $(a, b)(0, 0)$
f. (a, b)

g. $(a, b)(a, -b)$
h. $(a, b)(0, 1)$
i. $(0, 1)^2$
j. $(0, 1)(0, -1)$
k. $(0, 1)^3$
l. $(0, 1)^4$

m. $(0, 1)^{4n}, n \in N$
n. $(0, 1)^{4n+1}, n \in N$
o. $(0, 1)^{4n+2}, n \in N$
p. $(0, 1)^{4n+3}, n \in N$

3. Show that "=" is an equivalence relation on the plane [that is, $(a, b) = (a, b)$; if $(a, b) = (c, d)$, $(c, d) = (a, b)$; and if $(a, b) = (c, d)$, $(c, d) = (e, f)$, then $(a, b) = (e, f)$].

4. Show that if $(a, b) + (c, d)$ can be given as both (e, f) and (g, h), prove that $(e, f) = (g, h)$ (that is, show that there is "only one" sum). Show the same proof for $(a, b)(c, d)$, $(a, b) - (c, d)$, and $(a, b) \div (c, d)$, $(c, d) \neq (0, 0)$.

5. If for each (a, b), $(a, b) + (c, d) = (a, b)$, what is (c, d)?

6. If (c, d) is such that $(a, b)(c, d) = (a, b)$ for all points (a, b), what is (c, d)?

7. If $(a, b)(c, d) = (-b, a)$ for all (a, b), what is (c, d)?

8. Show that $(a, b)(c, 0) = c(a, b)$ and $(a, b) \div (c, 0) = (1/c)(a, b)$ if $c \neq 0$.

9. Compute:

    a. $(0, k)(a, b)$ d. $(a, b) - (a, -b)$
    b. $(a, b) \div (0, k)$ e. $(a, b)(a, -b)$
    c. $(a, b) + (a, -b)$

10. What is the geometric relation between (a, b) and $(a, -b)$?

11. Compute:

    a. $(r \cos a, r \sin a) \cdot (s \cos b, s \sin b)$
    b. $(r \cos a, r \sin a) \div (s \cos b, s \sin b)$
    c. $(r \cos a, r \sin a) \cdot (r \cos a, r \sin a)$
    d. $(r \cos a, r \sin a)(r \cos a, r \sin a)(r \cos a, r \sin a)$

7.3 THE COMPLEX PLANE (NUMBERS)

In the preceding section we stated that the plane, with the operations defined, is an important algebraic structure. It is, in fact, a *field*. (See Exercise 7.3.4.)

Let us consider the function $i: R \to C$, C the (complex) plane with these operations, given by $i(a) = (a, 0)$. For example, i takes 6 to $(6, 0)$ and $-\frac{1}{2}$ to $(-\frac{1}{2}, 0)$. Geometrically, i is the mapping that takes the real numbers onto the x-axis in a 1:1 fashion. Thus, we say that the complex plane contains a "copy" of the real numbers. In fact, from this point we shall not differentiate between a and $(a, 0)$.

The *complex number* (a, b) can be written in the form $(a, 0) + (0, b)$. Thus, we say that a [or $(a, 0)$] is the *real part* of (a, b). The term b [or $(b, 0)$] is called the *imaginary part* of (a, b). The real and imaginary parts can be interpreted geometrically in terms of projections. (See Figure 7.3.1.)

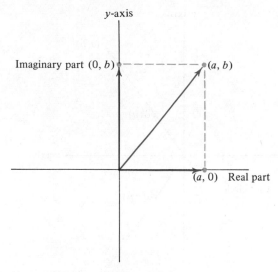

Figure 7.3.1

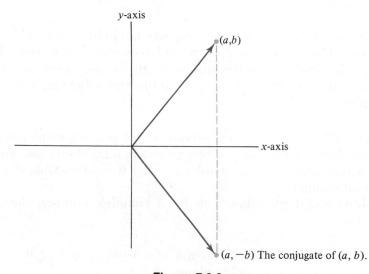

Figure 7.3.2

Figure 7.3.2 shows that (a, b) and $(a, -b)$ have the geometric relationship of being "reflections about the x-axis." The point $\alpha(p)$ is on the vector (or its extension) determined by (a, b) while $\alpha(-p)$ is on that vector (or its extension) determined by its (complex) *conjugate* $(a, -b)$. (See Figure 7.3.3.)

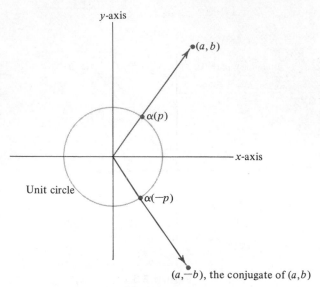

Figure 7.3.3

The *complex field* has one property not held by the field of real numbers. Each polynomial with real (complex) coefficients has a zero in the field of complex numbers. In the next sections we will see why this occurs. For now, examine the following illustrative examples.

EXAMPLE 7.3.1. The expression $x^2 + 1$ is a polynomial over the real numbers (that is, the coefficients are real numbers) but does not have a real zero. In other words, $x^2 + 1 = 0$ is impossible if x must be a real number.

However, if we allow x to be a complex number, there is a solution.

EXAMPLE 7.3.2. The equation $x^2 + 1 = 0$ or $x^2 + (1, 0) = (0, 0)$ has a solution in the complex plane.

This follows since, letting $x = (0, 1)$, we have $(0, 1)(0, 1) + (1, 0) = (-1, 0) + (1, 0) = (0, 0)$.

EXERCISES 7.3

1. Show that the mapping $i: R \to C$ by $i(a) = (a, 0)$ is $1:1$.

2. Show that i from Exercise 7.3.1 satisfies:

$$i(a + b) = i(a) + i(b)$$
$$i(ab) = a[i(b)] = i(a)i(b)$$
$$i(a) = (0, 0) \quad \text{if and only if } a = 0$$

A function i satisfying these properties is called an *isomorphism* from R into C.

3. Show (use Exercise 7.3.2) that $i(a) = ai(1)$.

4. Show that C is a field (that is, satisfies the first nine conditions from Chapter 1).

5. a. Find the real part of each number and answer in Exercise 7.2.2.
   b. Find the imaginary part of each number and answer in Exercise 7.2.2.

6. In terms of real and imaginary parts, what is $p_1(a, b)$? $p_2(a, b)$?

7. What is the length of the vector associated with (a, b)?

8. It is possible that you have seen the complex numbers in another form. Writing (a, b) as $a + bi$, we observe the classic designation of a complex number. The i is endowed with the property that $i^2 = -1$ or $i = \sqrt{-1}$. Now it is agreed that $a + 0i = a$ and $0 + bi = bi$. The student can then rewrite all the definitions of this section in terms of the $a + bi$ form. Multiplication, for instance, can be performed by the same mechanical pattern as a polynomial multiplication [that is, $(a + bx)(c + dx) = ac + adx + bxc + bdx^2 = ac + (ad + bc)x + bdx^2$]. Multiplying $(a + bi)$ $(c + di)$ in this fashion, we have:

$$(a + bi)(c + di) = ac + adi + bic + bidi$$
$$= ac + (ad + bc)i + bdi^2$$
$$= ac + (ad + bc)i - bd$$
$$= (ac - bd) + (ad + bc)i$$

The process of this manner of solution assumes all kinds of properties including commutativity and the like. It does, however, generate an answer analogous to that in the material of this section.

Division can be performed in an easy mechanical fashion:

$$\frac{a + bi}{c + di} = \frac{(a - bi)(c - di)}{(c + di)(c - di)} = \frac{(ac + bd) + (bd - ad)i}{c^2 + d^2}$$

$$= \frac{ac + bd}{c^2 + d^2} + \frac{(bd - ad)i}{c^2 + d^2}$$

Again no proof of the validity of the chain of operations is given. The process does, however, generate the analogous answer.

   a. Rewrite all definitions given in Sections 7.2 and 7.3 in terms of the $a + bi$ form.
   b. Rewrite and redo problems 1–2 of Exercise 7.2 in the $a + bi$ form.

9. Show that the complex numbers do not have an ordering similar to "$<$" as defined on R. Show, in particular, that $(0, 1)$ cannot be positive or zero in the sense of a law of trichotomy, and that $-(0, 1) = (0, -1)$ cannot be positive. [*Hint:* If $(0, 1)$ is positive, what can you say about $(0, 1)^2$?]

7.4 COMPLEX NUMBERS AND EXPONENTS

Exponential notation can be employed as a descriptive symbolism. If n is a positive integer, $(a, b)^n$ means $(a, b)(a, b)^{n-1}$ where $(a, b)^0 = (1, 0)$. The expression $(a, b)^{-n}$ means $1/(a, b)^n$, being interpreted as $(1, 0) \div (a, b)^n$.

EXAMPLE 7.4.1. $(3, 1)^3 = (3, 1)(3, 1)(3, 1) = (18, 26) = 2(9, 13)$. Since (a, b) is a point in the plane, it can be written in the form $(r \cos \theta, r \sin \theta)$ where $r = (a^2 + b^2)^{1/2}$ and θ is an appropriate real number (Section 6.1). Then $(a, b) = (r \cos \theta, r \sin \theta) = r(\cos \theta, \sin \theta)$. The latter form is called the *trigonometric form* of (a, b); r is called the absolute value of (a, b) and θ is called the *argument*. The *absolute value* of (a, b) is written $|(a, b)|$.

EXAMPLE 7.4.2. $(1, 1) = (\sqrt{2} \cos \pi/4, \sqrt{2} \sin \pi/4) = \sqrt{2}(\cos \pi/4, \sin \pi/4)$. The argument is $\pi/4$ and $|(1, 1)| = \sqrt{2}$.

Let $(a, b) = r (\cos \theta, \sin \theta)$ and $(c, d) = s (\cos \varphi, \sin \varphi)$. Then $(a, b)(c, d) = (r \cos \theta, r \sin \theta)(s \cos \varphi, s \sin \varphi) = (rs \cos \theta \cos \varphi - rs \sin \theta \sin \varphi, rs \cos \theta \sin \varphi + rs \sin \theta \cos \varphi) = rs (\cos \theta \cos \varphi - \sin \theta \sin \varphi, \cos \theta \sin \varphi + \sin \theta \cos \varphi) = rs (\cos (\theta + \varphi), \sin (\theta + \varphi))$.

That is, to multiply two pairs we first write each in trognometric form. We then form the product of the respective absolute values with the term $(\cos (\theta + \varphi), \sin (\theta + \varphi))$. *The argument of the product is the sum of the two respective arguments.*

EXAMPLE 7.4.3.   Find $(3, 3)(1, \sqrt{3})$.

$$(3, 3)(1, \sqrt{3}) = 3\sqrt{2} \ (\cos \pi/4, \sin \pi/4) \cdot 2 \ (\cos \pi/3, \sin \pi/3)$$
$$= 6\sqrt{2} \ (\cos 7\pi/12, \sin 7\pi/12)$$

(See Figure 7.4.1.)

This result, together with mathematical induction, can be used to prove *De Moivre's theorem:* if $n \in Z$ and if $(a, b) = r (\cos \theta, \sin \theta)$, $(a, b)^n = [r (\cos \theta, \sin \theta)]^n = r^n (\cos n\theta, \sin n\theta)$.

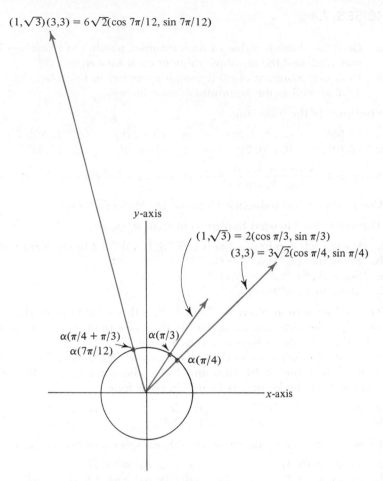

$(1,\sqrt{3})(3,3) = 6\sqrt{2}(\cos 7\pi/12, \sin 7\pi/12)$

y-axis

$(1,\sqrt{3}) = 2(\cos \pi/3, \sin \pi/3)$

$(3,3) = 3\sqrt{2}(\cos \pi/4, \sin \pi/4)$

$\alpha(\pi/4 + \pi/3)$

$\alpha(7\pi/12)$

$\alpha(\pi/3)$

$\alpha(\pi/4)$

x-axis

Figure 7.4.1

EXAMPLE 7.4.4.   To illustrate De Moivre's theorem, find $(2, 2)^4$.

$$(2, 2)^4 = [2\sqrt{2}\,(\cos \pi/4,\ \sin \pi/4)]^4$$
$$= (2\sqrt{2})^4\,(\cos \pi,\ \sin \pi) = 64(-1, 0) = (-64, 0).$$

EXAMPLE 7.4.5.   Let $z = (5, 5\sqrt{3})$. Compute z^3. Now, $z = (5, 5\sqrt{3}) = 10\,(1/2,\ \sqrt{3}/2) = 10\,(\cos \pi/3,\ \sin \pi/3)$. Thus, $z^3 = 10^3$ $(\cos 3\pi/3, \sin 3\pi/3) = 10^3\,(\cos \pi, \sin \pi) = 10^3\,(-1, 0) = (-1000, 0)$.

EXAMPLE 7.4.6.   Let $z = (-\sqrt{3}, -1)$. We find, then, that $z^5 = (-\sqrt{3}, -1)^5 = [2\,(-\sqrt{3}/2, -1/2)]^5 = 32\ (\cos 7\pi/6,\ \sin 7\pi/6)^5 = 32\,(\cos 35\pi/6, \sin 35\pi/6) = 32\,(\cos -\pi/6, \sin -\pi/6) = 32\,(\sqrt{3}/2, -1/2) = (16\sqrt{3}, -16)$.

EXERCISES 7.4

1. a. Give the absolute value of each complex number of Exercises 7.2.1 and 7.2.2, and the absolute value of each answer.
    b. Find the argument of each complex number in Exercises 7.2.1 and 7.2.2 as well as the argument of each answer.

2. Find each of the following.

    a. $(3, 3)^6$ c. $(-1, 0)^3$ e. $(0, -2)^4$ g. $(3\sqrt{3}, 3)^3$
    b. $(2, 0)^4$ d. $(0, 3)^3$ f. $(1, \sqrt{3})^6$ h. $(5, 5)^{-4}$

3. Show that $\dfrac{r\,(\cos a,\, \sin a)}{s\,(\cos b,\, \sin b)} = (r/s)\,(\cos\,(a - b),\, \sin\,(a - b))$ if $s \neq 0$.

4. Use mathematical induction to prove De Moivre's theorem.

5. The symbol $re^{i\theta}$ is used to represent $r(\cos\,\theta,\, \sin\,\theta)$.

    a. Write each number in Exercises 7.2.1 and 7.2.2 in the form $re^{i\theta}$.
    b. Show that $(re^{i\theta})(se^{i\varphi}) = rse^{i(\theta+\varphi)}$.
    c. Show that $re^{i\theta}/se^{i\varphi} = (r/s)e^{i(\theta-\varphi)}$.
    d. Show that $(re^{i\theta})^n = r^n e^{in\theta}$.

6. The $re^{i\theta}$ form from Exercise 7.4.5 is like the polar form in that the r and θ are the determining factors in the makeup of the number. Thus we call $re^{i\theta}$ the *polar form* of $r\,(\cos\,\theta,\, \sin\,\theta)$.

    Write the following in the $a + bi$ form (see Exercise 7.3.8) and then write each in the (a, b) form and locate the point in the plane. (The point $re^{i\theta}$ may be located as $(r, \theta)°$ in polar form.)

    a. $3e^{\pi i/2}$ c. $5e^{\pi i/6}$ e. $2e^{\pi i}$ g. $21e^{-3\pi i/4}$
    b. $e^{2\pi i}$ d. $6e^{5\pi i/6}$ f. $3e^{3\pi i/2}$ h. $4e^{-3\pi i/4}$

7. Find in each case at least one suitable complex number for (x, y).

    a. $(x, y)^2 = (0, 1)$ c. $(x, y)^3 = (0, 1)$
    b. $(x, y)^2 = (2, 2)$ d. $(x, y)^4 = (8, 8\sqrt{3})$

7.5 ROOTS OF COMPLEX NUMBERS

As we might expect, exponential notation can be employed in the case of roots. We must define, though, what a root of a complex number ought to be.

Suppose (c, d) is any complex number. If $n \in N$, (a, b) is an *nth root* of (c, d) means (just as we would expect) $(a, b)^n = (c, d)$. We will also show this same relationship by $(a, b) = (c, d)^{1/n}$. The fractional exponent is used to indicate any nth root. If $(c, d) \neq (0, 0)$, we will see that there are precisely n *distinct* nth roots of (c, d). This is quite different from the case for the real numbers (for example, there is only one real cube root of -27).

Let us set about finding the n different nth roots for (c, d). First, write (c, d) in its trigonometric form, say $(c, d) = r (\cos \theta, \sin \theta)$ with $0 \le \theta < 2\pi$. Recall that for any positive integer k,

$$(c, d)^k = r^k (\cos k\theta, \sin k\theta)$$

It would be a natural desire then to have $(c, d)^{1/n} = r^{1/n}(\cos \theta/n, \sin \theta/n)$ where $r^{1/n}$ is the *principal* nth root of r. This result is found by replacing k above by $1/n$. By the definition of $(c, d)^{1/n}$, $((c, d)^{1/n})^n = (c, d)$. Thus, we need to evaluate $(r^{1/n} (\cos \theta/n, \sin \theta/n))^n$. By De Moivre's theorem,

$$(r^{1/n} (\cos \theta/n, \sin \theta/n))^n = r (\cos \theta, \sin \theta)) = (c, d)$$

Necessarily, this analog of De Moivre's theorem holds for the case of roots.

However, our analysis is not complete. Let $k \in Z$ and examine the following.

$$\left[r^{1/n}\left(\cos \frac{\theta + 2k\pi}{n}, \sin \frac{\theta + 2k\pi}{n}\right)\right]^n = r(\cos (\theta + 2k\pi), \sin (\theta + 2k\pi))$$
$$= r (\cos \theta, \sin \theta) = (c, d)$$

Hence we conclude that each number of the form

$$r^{1/n} (\cos (\theta + 2k\pi)/n, \sin (\theta + 2k\pi)/n)$$

is an nth root of $(c, d) = r(\cos \theta, \sin \theta)$.

Does there exist an infinite number of nth roots of (c, d)? No! If $k \in \{0, 1, 2, \ldots, n - 1\}$, the corresponding values of $r^{1/n}(\cos (\theta + 2\pi k)/n, \sin (\theta + 2\pi k)/n)$ form a set of n distinct roots [if $(c, d) \ne (0, 0)$]. If $k = n$, $r^{1/n} (\cos (\theta + 2n\pi)/n, \sin (\theta + 2n\pi)/n) = r^{1/n} (\cos \theta/n, \sin \theta/n)$. The latter is the same root as given by $k = 0$. Similarly, if $k \in Z - \{0, 1, 2, \ldots, n - 1\}$,

$$r^{1/n} (\cos (\theta + 2k\pi)/n, \sin (\theta + 2k\pi)/n)$$

is one of the roots given when $k \in \{0, 1, 2, \ldots, n - 1\}$. (You should verify this algebraically.) Consequently there exist precisely n distinct nth roots of each $(c, d) \ne (0, 0)$.

The following worked example shows an application of the above reasoning.

EXAMPLE 7.5.1. Find the 4 fourth roots of $16 = (16, 0) = 16(\cos 0, \sin 0)$. These roots are:

$16^{1/4} (\cos (0 + 0\pi)/4, \sin (0 + 0\pi)/4) = 2 (\cos 0, \sin 0) = (2, 0)$
$16^{1/4} (\cos (0 + 2\pi)/4, \sin (0 + 2\pi)/4) = 2 (\cos \pi/2, \sin \pi/2) = (0, 2)$

$16^{1/4} \ (\cos \ (0 + 4\pi)/4, \ \sin \ (0 + 4\pi)/4) = 2 \ (\cos \ \pi, \ \sin \ \pi) = (-2, 0)$
$16^{1/4} \ (\cos \ (0 + 6\pi)/4, \ \sin \ (0 + 6\pi)/4) = 2 \ (\cos \ 3\pi/2, \ \sin \ 3\pi/2)$
$$= (0, -2)$$

It is of interest to note that the nth roots of $r \ (\cos \ \theta, \ \sin \ \theta)$ are all on the circle $x^2 + y^2 = (r^{1/n})^2$ and the difference in the argument of any two "successive" points is $2\pi/n$. The roots form the vertices of a regular n-sided polygon inscribed in the given circle. (See Figure 7.5.1.)

The following example together with Figure 7.5.2 illustrates this point.

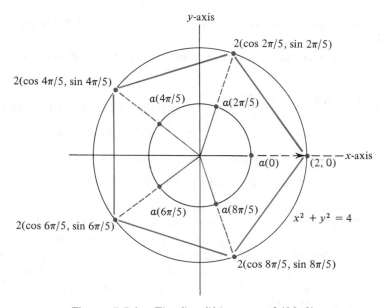

Figure 7.5.1. The five fifth roots of (32, 0).

EXAMPLE 7.5.2. Compute the three cube roots of $(4 \sqrt{2}, 4 \sqrt{2})$. Now $(4 \sqrt{2}, 4 \sqrt{2}) = 8 \ (\cos \ \pi/4, \ \sin \ \pi/4)$ and one cube root is

$$8^{1/3} \ (\cos \ \pi/12, \ \sin \ \pi/12) = 2 \ (\cos \ \pi/12, \ \sin \ \pi/12)$$

The difference in argument between successive roots is $2\pi/3$. Since $\pi/12 + 2\pi/3 = 3\pi/4$ and $3\pi/4 + 2\pi/3 = 17\pi/12$, we have for the other pair of roots:

$$2 \ (\cos \ 3\pi/4, \ \sin \ 3\pi/4)$$
$$2 \ (\cos \ 17\pi/12, \ \sin \ 17\pi/12)$$

For the geometric interpretation, see Figure 7.5.2.

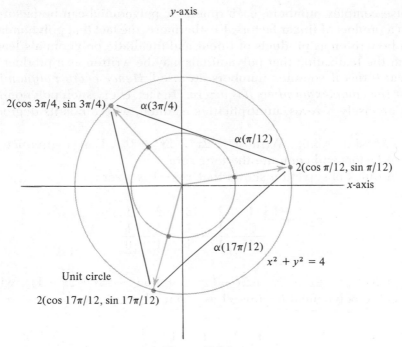

Figure 7.5.2

The notation $(a, b)^{m/n}$ means $((a, b)^{1/n})^m$ where n is any positive integer and $m \in Z$.

EXAMPLE 7.5.3. Compute $(4\sqrt{2}, 4\sqrt{2})^{2/3}$. We know that $(4\sqrt{2}, 4\sqrt{2})^{1/3}$ has three possibilities and that $(4\sqrt{2}, 4\sqrt{2})^{2/3}$ is the square of any cube root. Thus the choices for $(4\sqrt{2}, 4\sqrt{2})^{2/3}$ are

$$4 (\cos \pi/6, \sin \pi/6)$$
$$4 (\cos 3\pi/2, \sin 3\pi/2)$$
$$4 (\cos 11\pi/6, \sin 11\pi/6)$$

Let $f(x) = ax^2 + bx + c$ with $a \neq 0$ and each of a, b, and c real numbers. We saw earlier that if f has zeros, they must be of the form $(-b \pm \sqrt{b^2 - 4ac})/2a$. However, we required that $b^2 - 4ac$ be non-negative. Now, though, if $b^2 - 4ac$ is negative, we can calculate its square root as a complex number. Consequently, all quadratic polynomials (with real coefficients) have zeros in the complex numbers.

The parenthetic expression above is not really necessary since if a, b, and c are complex, $b^2 - 4ac$ is still a complex number: every complex number has square roots. We also see that if we are willing to use complex numbers, each quadratic polynomial can be factored into a product of linear factors. Furthermore, the fact that polynomials can be written as products of linear and quadratic polynomials leads us to the realization that polynomials may be written as a product of linear terms if complex numbers are used. *Hence every polynomial over the complex numbers has a zero.* (In fact, every such polynomial has precisely n zeros, multiplicities counted, where n is its degree.)

EXAMPLE 7.5.4.   Write $x^4 + 2x^3 + 2x^2 + 2x + 1$ as a product of linear factors and compute the four zeros.

A quick observation shows that $x = -1$ is a zero:

$$
\begin{array}{r|rrrrr}
-1 & 1 & 2 & 2 & 2 & 1 \\
   &   & -1 & -1 & -1 & -1 \\
\hline
   & 1 & 1 & 1 & 1 & \underline{|0} \\
\end{array}
$$

We have $x^4 + 2x^3 + 2x^2 + 2x + 1 = (x + 1)(x^3 + 2x^2 + 2x + 1)$, with the cubic polynomial having -1 as a zero also:

$$
\begin{array}{r|rrrr}
-1 & 1 & 1 & 1 & 1 \\
   &   & -1 & 0 & -1 \\
\hline
   & 1 & 0 & 1 & \underline{|0} \\
\end{array}
$$

Necessarily, $x^4 + 2x^3 + 2x^2 + 2x + 1 = (x + 1)^2(x^2 + 1)$. The zeros of $x^2 + 1$ are $i = (0, 1)$ and $-i$. We now know that $x^4 + 2x^3 + 2x^2 + 2x + 1 = (x + 1)^2(x - i)(x + i)$. That $x^2 + 1 = (x - i)(x + i)$ is also shown by the following synthetic division:

$$
\begin{array}{r|rrr}
i & 1 & 0 & 1 \\
  &   & i & -1 \\
\hline
  & 1 & i & \underline{|0} \\
\end{array}
$$

The four zeros of our quadratic polynomial are: $-1, -1, i$, and $-i$ (-1 is a double zero).

EXAMPLE 7.5.5.   Compute the zeros of $g: C \to C$ where $g(x) = x^3 - x^2 - x - 2$. Now the possible rational roots are ± 2 and ± 1:

$$
\begin{array}{r|rrrr}
2 & 1 & -1 & -1 & -2 \\
  &   & 2 & 2 & 2 \\
\hline
  & 1 & 1 & 1 & \underline{|0} \\
\end{array}
$$

That is, $g(x) = (x - 2)(x^2 + x + 1)$. If we employ the quadratic formula (even though its use has not really been justified) we find that $x^2 + x + 1 = 0$ if

$$x = (-1 \pm \sqrt{1-4})/2$$
$$= -1/2 \pm \sqrt{3}\,i/2$$

In the ordered-pair notation, $x = (-1/2, \sqrt{3}/2)$ or $x = (-1/2, -\sqrt{3}/2)$. If we check by substitution we find the solution set [for $g(x) = 0$] to be $\{2 = (2, 0), (-1/2, \sqrt{3}/2), (-1/2, -\sqrt{3}/2)\}$.

EXERCISES 7.5

Find all roots in Exercises 7.5.1–7.5.8.

1. $(16, 0)^{1/2}$ 5. $(50\sqrt{2}, 50\sqrt{2})^{1/2}$

2. $(8, 0)^{1/3}$ 6. $(1, 0)^{1/10}$

3. $(0, 27)^{1/3}$ 7. $(1, \sqrt{3})^{1/2}$

4. $(0, -27)^{1/3}$ 8. $(2\sqrt{3}, 2)^{1/4}$

Solve the equations in Exercises 7.5.9–7.5.12.

9. $(a, b)^4 = (0, -81)$ 11. $(a, b)^3 = (-125, 0)$

10. $(a, b)^2 = (16, 0)$ 12. $(a, b)^5 = (-16, 16\sqrt{3})$

13. Rewrite all the definitions of the section in terms of the $a + bi$ form. Write out the worked examples in the $a + bi$ form and observe the corresponding analysis.

14. Follow the instructions of Exercise 7.5.13 for the polar form $re^{i\theta}$.

15. Show that the quadratic formula holds even if the coefficients of $ax^2 + bx + c$ are complex numbers.

16. Using Exercise 7.5.15 if necessary, find all complex zeros for $g(x)$ where

a. $g(x) = x^4 - 1$ d. $g(x) = x^2 + 2x + 2$
b. $g(x) = x^3 - 1$ e. $g(x) = 3 + x + x^2$
c. $g(x) = x^2 + 4x + 1$ f. $g(x) = x^3 + x^2 - 2$

7.6 REVIEW EXERCISES

1. The following are in polar form. Rewrite them in rectangular coordinates.

a. $(-3, 27\pi/4)^\circ$ f. $(-8, -\sin^{-1}1/2)^\circ$
b. $(2, -21\pi/6)^\circ$ g. $(12, \pi - \cos^{-1}(8/17))^\circ$
c. $(13, 285\pi)^\circ$ h. $(-11, \cos^{-1}(-22/13))^\circ$
d. $(21, -23\pi/3)^\circ$ i. $(r, \theta)^\circ$
e. $(16, \tan^{-1}3/4)^\circ$

2. The following are in rectangular coordinates. Convert each to polar
    coordinates.

    a. $(25, 60)$ c. $(51, 68)$ e. $(30, 0)$ g. $(-51, 51\sqrt{3})$
    b. $(24, 25)$ d. $(29, 29)$ f. $(0, -18)$ h. $(13, -3/\sqrt{3})$

3. Describe each of the following curves given in polar form (that is, the
    set of points of the form $(r, \theta)^{*}$ satisfying each of the following):

    a. $\theta = \pi/4$ d. $r = 50 \sin \theta$ f. $r = 17$
    b. $\theta = \pi/6$ e. $r = 6$ g. $r = \theta$
    c. $r = 64 \cos \theta$

4. Give the absolute value and argument of each complex number. Find
    also the real and imaginary part of each.

    a. $3 + 3i$ d. $-6i$ g. $2 - i$
    b. 6 e. -16 h. $2 + i$
    c. $2 + 2\sqrt{3}i$ f. $-4 + 4i$ i. $3 - 3i$

5. Perform the following operations.

    a. $(3 - 2i) + (5 + i)$ e. $(1 + i)(1 - i)$ i. $(a + bi) \div (a + bi)$
    b. $(-2 + i) + (6 - i)$ f. $i(1 + i)$ j. $(4 - 2i) \div (-3 + i)$
    c. $(1 - i) - (-2 + i)$ g. $(3 - i)(4 + 2i)$ k. $(5 + 6i) \div (2 - 3i)$
    d. $(5 + 3i) - (3 - 3i)$ h. $(3 + 2i)(-3 - 2i)$ l. $(-i)(-3 + i)$

6. Square each number in Exercise 7.6.4.

7. Cube each number in Exercise 7.6.4.

8. Find all square roots of each number in Exercise 7.6.4.

9. Find all fourth roots of each number in Exercise 7.6.4.

Find all zeros for each of the following polynomials and write each as the
product of linear factors.

10. $f(x) = x^4 + 3x^3 + 3x^2 + 3x + 2$ 17. $x^5 + x^4 - 4x - 4$

11. $g(x) = 2x^3 - x^2 + 6x - 3$ 18. $x^2 - 3x + 4$

12. $r(x) = 3x^3 + 4x^2 + 4x + 1$ 19. $x^3 - 1$

13. $s(x) = x^3 + 4x - 5$ 20. $x^3 + 4x - 5$

14. $t(x) = 4x^3 - 7x^2 + 7x - 3$ 21. $x^4 - 7x^2 - 18$

15. $x^4 - 1$ 22. $x^5 + x^3 - x^2 - 1$

16. $z^4 - 16$ 23. $x^6 - 1$

7.7   QUIZ

Attempt to work all the problems in one hour. Check your results.

1. Write each of the following in polar form.

    a. $(3, 4)$ b. $(5, -5)$ c. $(6 \cos a, 6 \sin a)$

2. Write each in rectangular form.

   a. $(3, \tan^{-1} 5/12)^*$ b. $(-5, \sin^{-1}(-15/17))^*$ c. $(2, \pi/3)^*$

3. Write each resulting complex number in the $a + bi$ form.

   a. $(3, 2) + (5, -1)$ c. $(1, 5)(5, 1)$ e. $6(4, 0)$
   b. $(1, 7) - (-2, 3)$ d. $(9, 7) \div (1, 2)$ f. $(3, 2)^2$

4. Determine the real and imaginary parts for your solutions to parts a and d from Problem 3.

5. Determine the following powers.

   a. $(1, 1)^4$ b. $(3 \cos \pi/5, 3 \sin \pi/5)^{10}$

6. Compute all the fifth roots of $243i$.

7. Find the three zeros for $x^3 - x^2 + x - 6$.

7.8 ADVANCED EXERCISES

1. Show that $[d((r, \theta)^*, (s, \varphi)^*)]^2 = r^2 + s^2 - 2rs \cos(\theta - \varphi)$ where $(r, \theta)^*$ and $(s, \varphi)^*$ are in polar form.

2. Sketch the following curves. They are given in polar coordinates.

   a. $r = k(1 + \cos \theta)$ d. $r^2 = k^2 \cos 2\theta$
   b. $r \cos \theta = 2$ e. $r = k/2$
   c. $r^2 = 1 + \sin^2 \theta$

3. Find the point(s) of intersection of parts d and e from Exercise 7.8.2.

4. Show that $(a + bi)(c + di) = 0$ implies $a^2 + b^2 = 0$ or $c^2 + d^2 = 0$.

5. Solve the following equations, the solution set being a set of complex numbers.

   a. $z^2 + 2z + 1 = 8\sqrt{2} + 8\sqrt{2}i$. (*Hint:* The left-hand side is a perfect square.)
   b. $z^2 + 2z = 5 - 2i$
   c. $z^{1/2} = 2 + i$
   d. $z^5 = 32$

6. Let z and a be complex numbers in the ordered-pair form. Describe the set of z where

   a. $|z - a| = 2$ i. $I(z)/R(z) = 1$
   b. $|z - a| = 20$ j. $R^2(z) + 1^2(z) = 4$
   c. $|z - a| < 2$ k. $R(z) > 0$ and argument $z = \pi/3$
   d. $|z - a| > 2$
   e. $R(z) = 2$ [$R(z)$ is the real part of z]
   f. $I(z) = -2$ [$I(z)$ is the imaginary part of z]
   g. $R(z) < 0$
   h. $I(z) > 0$

CHAPTER 8

More Analytic Geometry

8.1 PARAMETRIC REPRESENTATION

We have, throughout the text, graphed sets of points given in the form $\{(x, y) : y = s(x)\}$ where s is a function in the variable x. Often $s(x)$ was given in terms of an equation, an equation generally uncomplicated and usually readily identified and examined. In this section we wish to look at point sets given by $\{(x, y) : x = f(t), y = g(t)\}$, both x and y being functions of another variable t. Frequently this *parametric representation* (with *parameter t*) affords a simpler means of describing the points of the set. The primary advantage of such a representation is brought to the fore in calculus. The following examples illustrate the concept of parametric representation.

EXAMPLE 8.1.1. Describe the set of points $\{(x, y) : x = 3 + t, y = 2 + t\}$.

The set (as it will be seen later) is a straight line passing through $(3, 2)$ and having a slope of 1. The point $(3, 2)$ is found as a result of letting $t = 0$.

We can proceed by "eliminating the parameter t." We do so by observing that $t = y - 2$ whence $x = 3 + t = y + 1$. The latter equation indeed represents a straight line through $(3, 2)$ with slope one. Thus we suspect:

Theorem 8.1.1. *If l is a line passing through* (a, b) *with slope* m, *l is given parametrically by the equations:*

$$x = a + m_1 t$$

$$y = b + m_2 t$$

$$m_1^2 + m_2^2 \neq 0,$$

$$m = m_2/m_1.$$

Conversely, such a pair of equations represents a straight line. The proof is left as an exercise.

Theorem 8.1.2. *The set l is a straight line passing through* (a, b) *and* (c, d) *if and only if l is given parametrically by:*

$$x = a + t(c - a)$$

$$y = a + t(d - b)$$

Observe that the representation is valid for vertical lines. The proof is left, as in the case of Theorem 8.1.1, as an exercise.

EXAMPLE 8.1.2. Sketch $\{(x, y) : x = t + 1, \ y = t^2\}$.

Again, we may eliminate the parameter and write $y = (x - 1)^2$. The parabola, as we now recognize the curve, is shown in Figure 8.1.1.

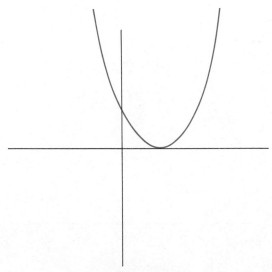

Figure 8.1.1

EXAMPLE 8.1.3. Draw a graph of $\{(x, y): x = 2^t, y = t^2 + t\}$.

We first observe that $x > 0$ and that $t = \log_2 x$. If, however, we substitute this value for t into the expression for y, we admit to $y = (\log_2 x)^2 + \log_2 x = (\log_2 x + 1) \log_2 x$, a not too nice representation.

We sketch the curve directly without eliminating the parameter by forming sample values as shown in the chart of Figure 8.1.2. These values are then transferred to $R \times R$ and the sketch formed as in Figure 8.1.3.

t	x	y
0	1	0
1	2	2
−1	1/2	0
2	4	6
−2	1/4	2
3	8	12

Figure 8.1.2

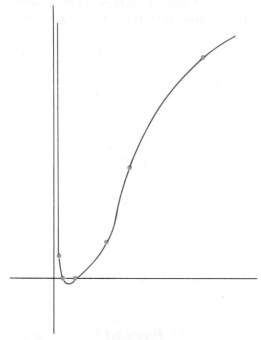

Figure 8.1.3. $\{(x, y): x = 2^t, y = t^2 + t\}$

EXERCISES 8.1

1. Prove Theorem 8.1.1.

2. Prove Theorem 8.1.2.

3. In view of Theorem 8.1.2 describe in terms of the values taken on by t the segment between (a, b) and (c, d), the ray from (a, b) through (c, d), the ray from (c, d) through (a, b), and the ray emanating from (a, b) and lying on the line through (c, d) but not passing through (c, d).

4. Describe parametrically the following lines:
   a. l passes through the origin with slope 3.
   b. l passes through $(-1, 2)$ having slope $-1/2$.
   c. l passes through $(-1, 2)$ and is perpendicular to the line in b.
   d. l passes through $(6, 3)$ and $(-4, -2)$.

5. Sketch each of the following sets.
   a. $\{(x, y): x = t - 1, y = t^2 - 2\}$ c. $\{(x, y): x = 2^t \text{ and } y = 2^t\}$
   b. $\{(x, y): x^2 = t \text{ and } y = t^2\}$ d. $\{(x, y): x = 2^t \text{ and } y = 3^t\}$

6. Sketch each of the following.
   a. $\{(x, y): x = \cos t, y = \sin t\}$
   b. $\{(x, y): x = 3t, y = \sin t\}$
   c. $\{(x, y): x = \sin t, y = t \text{ with } t \in [-\pi/2, \pi/2]\}$
   d. $\{(x, y): x = 5, y = \cos t\}$

7. Draw a graph of each set:
   a. $\{(r, a)^\circ: r = 2t, a = t\}$
   b. $\{(r, a)^\circ: r = t, a = \pi/2\}$
   c. $\{(r, a)^\circ: r = t^2, a = 2t\}$
   d. $\{(r, a)^\circ: r = 1/t, a = t, t \neq 0\}$

8.2 TRANSLATION OF AXES

Figure 8.2.1 shows a sketch of the plane with an additional set of axes formed by the vertical line $x = a$ and the horizontal line $y = b$. The point (a, b) is, in some sense, the origin of a coordinate system having x' and y' axes. That is, we may locate points in terms of horizontal and vertical displacements from (a, b) rather than from $(0, 0)$.

The point (x, y) shown in Figure 8.2.1 is located x units horizontally from $(0, 0)$ and y units vertically from that point. However, with respect to (a, b), (x, y) is displaced $x - a$ units horizontally and $y - b$ units vertically.

Alternatively if we consider (a, b) to be the origin of the "primed" coordinate system, (x, y) bears the new description $(x - a, y - b)$.

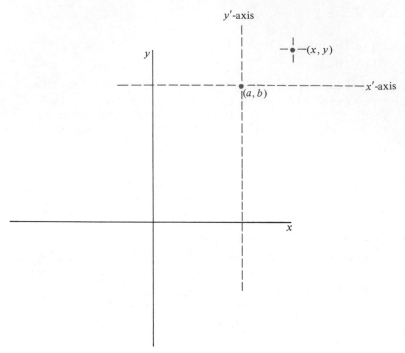

Figure 8.2.1

Thus, if the new coordinates are denoted by (x', y') we have the *translation equations*:

$$x' = x - a$$

$$y' = y - b$$

Figure 8.2.2 indicates the geometric considerations for this renaming process.

EXAMPLE 8.2.1. Figure 8.2.3 shows the plane and its usual coordinate system having x and y axes. Superimposed is a primed system having center $(2, -1)$ in the usual system. The point $(5, 4)$ is shown to have the primed designation $(3, 5) = (5 - 2, 4 - (-1))$. Indeed, if (x, y) is a point in the unprimed system having coordinates (x', y') in the alternative coordinate scheme, $x' = x - 2$ and $y' = y + 1$.

EXAMPLE 8.2.2. The circle given by $(x - 3)^2 + (y - 4)^2 = 25$ is clearly a circle of center $(3, 4)$ and radius 5. If we place a primed coordinate system with origin at $(3, 4)$, the translation equations

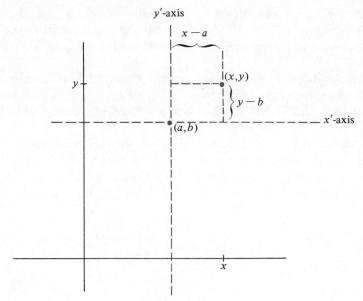

Figure 8.2.2

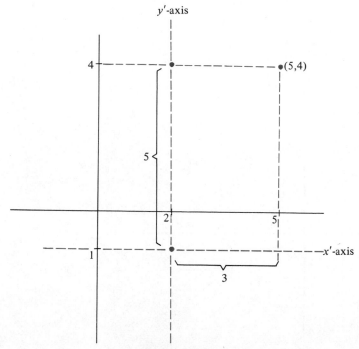

Figure 8.2.3

become $x' = x - 3$ and $y' = y - 4$ whereby the circle has the transformed equation $(x')^2 + (y')^2 = 25$. This you will recall is the form
of an equation of a circle of radius 5 and center the origin. (See Figure 8.2.4.)

This technique of introducing on a second horizontal-vertical
coordinate system is called a *translation of axes*. Use is made of this
concept in Example 8.2.4. First, however, examine the following.

EXAMPLE 8.2.3. A *parabola* having *focus P* and *directrix l* is
the set $\{(x, y) : d(P, (x, y)) = d((x, y), l)\}$ where $d((x, y), l)$ is the
distance from (x, y) along the perpendicular to l. We assume that
P does not lie on l. Form an equation depicting the parabola having

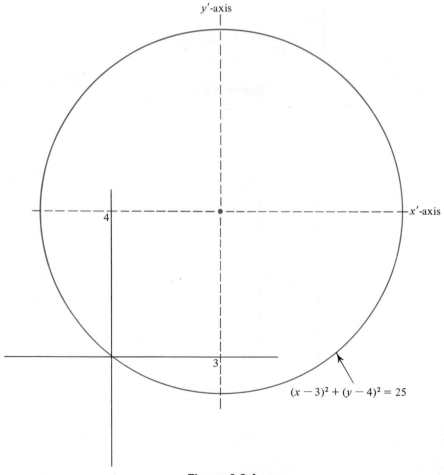

Figure 8.2.4

focus $(p, 0)$ and directrix $x = -p$. Our sketch in Figure 8.2.5 shows $p > 0$ but the algebraic results we obtain do not depend on this.

We observe quickly that the origin $(0, 0)$ lies halfway between the focus and directrix. It is therefore a point of the parabola. If we let (x, y) represent any point of our parabola, $d((x, y), (p, 0)) = d((x, y), l) = d((x, y), (-p, y))$. Alternatively, $[(x - p)^2 + y^2]^{1/2} = |x + p|$ whereby $(x - p)^2 + y^2 = (x + p)^2$. Separating the x and y terms, we find $y^2 = 4px$. This then is the desired equation.

We might note that if $p < 0$, the whole sketch of Figure 8.2.5 would be reflected about the y-axis.

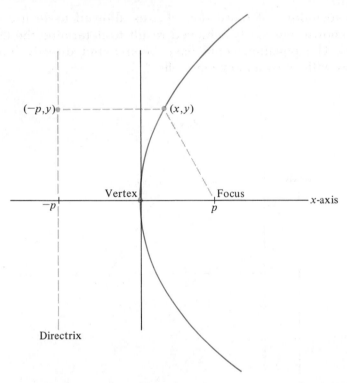

Figure 8.2.5

EXAMPLE 8.2.4. We are now ready to use the idea of the translation of axes to determine an equation of the parabola having focus $(a + p, b)$ and directrix $x = a - p$.

The parabola has, as in Example 8.2.3, a vertical directrix separated from the focal point $2p$ units. Alternatively, if $(a, b) = (0, 0)$ we are right back to Example 8.2.3. Since the direction of the directrix and its separation from the focus determine the "shape" of the parabola, this parabola is like that of Example 8.2.3 except that the *vertex* is at (a, b) rather than $(0, 0)$.

Figure 8.2.6 shows a coordinate system with origin at (a, b). In the primed system the equation of the parabola is, from Example 8.2.3, $(y')^2 = 4px'$. The translation equations $x' = x - a$ and $y' = y - b$ transform the equation to the standard form $(y - b)^2 = 4p(x - a)$.

In this form the equation gives the center (vertex, here) and the separation between directrix and focus (given in terms of p). Since it is the x term that is linear (not squared) the directrix is vertical. If $p > 0$, the parabola opens to the right and if $p < 0$, it opens to the left.

The technique of translation of axes allowed us to use a previously known and easily achieved result to determine the desired equation. The equation could have been found directly from the definition with considerably more effort.

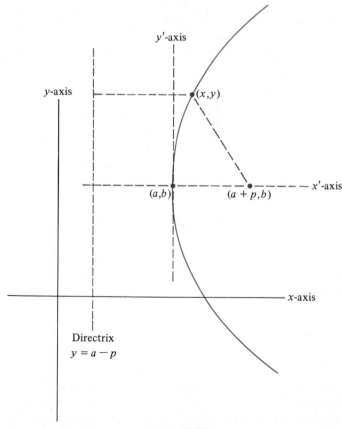

Figure 8.2.6

The parabola we have found is not a function having domain on the x-axis (why?). It is, though, a relation on $R \times R$.

EXAMPLE 8.2.5.   An *ellipse* having focal points P_1 and P_2 and *major axis length* $2a > d(P_1, P_2) > 0$ is the set $\{(x, y) : d((x, y), P_1) + d((x, y), P_2) = 2a\}$. Find an equation for the ellipse having focal points $(c, 0)$ and $(-c, 0)$, $c > 0$, with major axis length $2a$.

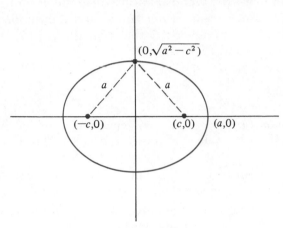

Figure 8.2.7

Figure 8.2.7 indicates that we have again, as in Example 8.2.3, chosen the origin as the "center" of our figure. The point $(0, \sqrt{a^2 - c^2})$ lies on the ellipse and it will be advantageous to call this *semiminor axis length* $b(b = \sqrt{a^2 - c^2})$. The following analysis brings about the standard form of the desired equation.

$$d((x, y), (-c, 0)) + d((x, y), (c, 0)) = 2a$$
$$[(x + c)^2 + y^2]^{1/2} + [(x - c)^2 + y^2]^{1/2} = 2a$$
$$[(x + c)^2 + y^2]^{1/2} = 2a - [(x - c)^2 + y^2]^{1/2}$$

Squaring, we find

$$(x + c)^2 + y^2 = 4a^2 - 4a[(x - c)^2 + y^2]^{1/2} + (x - c)^2 + y^2$$

$$\frac{cx - a^2}{a} = [(x - c)^2 + y^2]^{1/2}$$

Squaring again, we notice

$$\frac{c^2x^2 - 2a^2cx + a^4}{a^2} = x^2 - 2cx + c^2 + y^2$$

or

$$a^2 - c^2 = \left(\frac{a^2 - c^2}{a^2}\right) x^2 + y^2.$$

Substituting b^2 for $a^2 - c^2$, we derive

$$b^2 = (b^2/a^2)x^2 + y^2$$

or finally

$$1 = x^2/a^2 + y^2/b^2$$

In this form we have expressed the lengths of the semimajor axis and semiminor axis. We also know (since the a^2 term divides the x^2 term) that the major axis is horizontal and the minor axis is vertical. Note also from Figure 8.2.7 that $(-a, 0)$ and $(a, 0)$ lie on the ellipse as do $(0, b)$ and $(0, -b)$.

EXERCISES 8.2

1. The equation $(y - b)^2 = 4p(x - a)$ represents a parabola with vertical directrix. For what values p does the parabola "open" to the right? to the left?

2. Derive an equation of the parabola with directrix $y = -p$ and focus $(0, p)$. What direction does the parabola open if $p < 0$? if $p > 0$?

3. Use Exercise 8.2.2 to find an equation of a parabola having directrix $y = b - p$ and focus $(a, b + p)$. What is the vertex of such a parabola?

4. Describe an axis of symmetry for a parabola, telling what an axis of symmetry ought to be.

5. Write equations for each of the following parabolas. Sketch each curve.
    a. directrix $x = 5$, focus $(9, 4)$
    b. directrix $y = -2$, vertex $(5, 0)$
    c. focus $(5, 1)$, $p = 1$, opens upward
    d. $p = -1$, vertex at $(1, 2)$, opens to the left
    e. opens downward and passes through $(5, 3)$ with vertex $(1, 5)$.

6. a. Does an ellipse represent a relation? a function? Explain.
    b. Which axis (major or minor) of an ellipse contains the focal points?

7. Derive an equation for an ellipse with foci $(0, c)$ and $(0, -c)$ and having major axis of length $2a > 2c > 0$. (*Hint:* Exercise 8.2.6b might be helpful.)

8. a. Write an equation of an ellipse having focal points $(h + c, k)$ and $(h - c, k)$ with major axis of length $2a > 2c > 0$. What is the point (h, k)?
    b. Use Exercise 8.2.7 to derive an equation for an ellipse of focal points $(h, k + c)$, and $(h, k - c)$, and major axis length $2a > 2c > 0$.

9. Using Exercises 8.2.7 and 8.2.8, write equations for each of the following ellipses. Sketch each curve:

    a.   foci $(1, 6)$ and $(3, 6)$ with $a = 3$
    b.   foci $(1, 4)$ and $(1, 8)$ with $b = 3$
    c.   foci $(-1, 2)$ $(-1, 6)$ and passing through $(0, 4)$
    d.   intersection of major and minor axes is $(-1, -1)$ and passes through $(-1, 5)$ and $(4, -1)$

10. How many axes of symmetry does an ellipse have? What are they?

11. The set $\{(x, y): |d((x, y), P_1) - d((x, y), P_2| = 2a > 0\}$ is called a *hyperbola* with focal points P_1 and P_2.

    a.   Write an equation of the hyperbola with foci $(c, 0)$ and $(-c, 0)$, $0 < a < c$. (*Hint:* Let $b = \sqrt{c^2 - a^2}$ and the equation will become $x^2/a^2 - y^2/b^2 = 1$.) (See Figure 8.2.8.)
    b.   Write an equation of the hyperbola having foci $(0, c)$ and $(0, -c)$, $0 < a < c$. Compare with your result from part a.
    c.   Using part a, write an equation of the hyperbola with foci $(h + c, k)$ and $(h - c, k)$, $0 < a < c$.
    d.   Using b, write an equation of the hyperbola with foci $(h, k + c)$ and $(h, k - c)$, $0 < a < c$.
    e.   How many axes of symmetry does a hyperbola have? What are they?

12. Derive an equation for each of the following hyperbolas. Sketch each curve.

    a.   foci $(1, 2)$ and $(1, 8)$ with $a = 2$
    b.   vertices $(3, -2)$ and $(9, -2)$ with $c = 5$
    c.   foci $(2, 1)$ and $(2, 7)$ with one vertex $(2, 3)$
    d.   foci $(5, -6)$ and $(15, -6)$ passing through $(10 + \sqrt{7}, 6)$

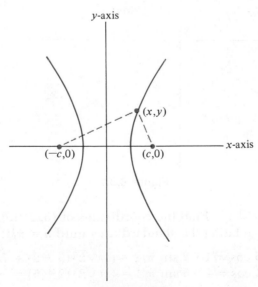

Figure 8.2.8

8.3   ROTATION OF AXES

Figure 8.3.1 shows a copy of the plane with a secondary set of axes that appear to have been positioned by rotating the x and y axes. The x' axis is seen to be the line $y = (\tan a)x$, the point $\alpha(a)$ lying on the x' axis. The point having coordinates (x, y) has, as in the case of a translation, another designation in the primed system.

The point (x, y) can be written as $(r \cos \theta, r \sin \theta)$ for some $\theta \in R$ and $r = \sqrt{x^2 + y^2}$. If the new coordinates are given as (x', y'), it is clear that $r = \sqrt{x'^2 + y'^2}$ as well. Moreover, $(x', y') = (r \cos(\theta - a), r \sin(\theta - a))$, whereby:

$$x' = r \cos(\theta - a) = r \cos \theta \cos a + r \sin \theta \sin a$$
$$= x \cos a + y \sin a$$
$$y' = r \sin(\theta - a) = r \sin \theta \cos a - r \cos \theta \sin a$$
$$= y \cos a - x \sin a$$

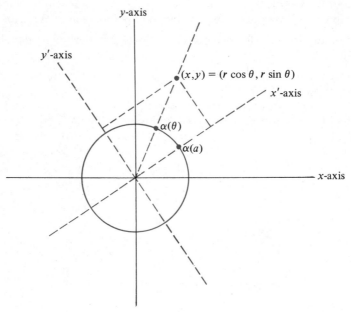

Figure 8.3.1

EXAMPLE 8.3.1. Find the coordinates of $(5, 2)$ relative to a system formed by rotating the standard axes until $a = \pi/4$.

$$x' = 5 \cos \pi/4 + 2 \sin \pi/4 = (1/\sqrt{2})(5 + 2) = 7/\sqrt{2}$$
$$y' = 2 \cos \pi/4 - 5 \sin \pi/4 = (1/\sqrt{2})(2 - 5) = -3/\sqrt{2}$$

In the primed system, our point is $(7/\sqrt{2}, -3/\sqrt{2})$.

EXAMPLE 8.3.2.   From the *rotation equations*,

$$x' = x \cos a + y \sin a$$
$$y' = y \cos a - x \sin a$$

we find

$$x \cos^2 a = x' \cos a - y \sin a \cos a$$
$$x \sin^2 a = y \cos a \sin a - y' \sin a$$

Adding, we see that

$$x = x(\cos^2 a + \sin^2 a) = x' \cos a - y' \sin a$$

Similarly,

$$y = x' \sin a + y' \cos a$$

We now have x and y given in terms of x' and y'. We will use this idea shortly.

If A, B, C, D, E, and F are real numbers, $Ax^2 + Bxy + Cy^2 + Dx + Ey + F = 0$ is called the *general quadratic* equation in two variables. The equation gives rise to a set of points forming a *conic section* (circle, ellipse, parabola, hyperbola, two intersecting lines, a single line, or a single point). We have investigated quadratics generating all such curves. What we have not done is to examine a quadratic involving an xy term (Bxy in this case). If $B = 0$, we can handle the equation just as we have in the past. (If $B = 0$, the x and y terms can be separated and the resultant conic sections, other than lines and single points, have horizontal or vertical axes of symmetry.) If $B \neq 0$, the axes are not horizontal or vertical. A *rotation* of axes, then, can bring the axes to "apparently" horizontal or vertical positions.

EXAMPLE 8.3.3.   Suppose $Ax^2 + Bxy + Cy^2 + Dx + Ey + F = 0$ and $B \neq 0$. Find a rotation such that the transformed equation has the form $A'(x')^2 + C'(y')^2 + D'x' + E'y' + F' = 0$. That is, find a rotation to eliminate the cross-product (xy) term.

Suppose the rotation is such that the x' axis passes through $\alpha(a)$. Then $x = x' \cos a - y' \sin a$ and $y = x' \sin a + y' \cos a$. Thus

$$Ax^2 + Bxy + Cy^2 + Dx + Ey + F =$$
$$A(x' \cos a - y' \sin a)^2 + B(x' \cos a - y' \sin a)(x' \sin a + y' \cos a)$$
$$+ C(x' \sin a + y' \cos a)^2 + Dx + Ey + F$$

The substitutions were made only in the second-degree terms Ax^2, Bxy, and Cy^2 since only these will generate a term of the form $Kx'y'$. The terms involving $x'y'$ are $-2Ax'y' \cos a \sin a$, $B(x'y' \cos^2 a - y'x' \sin^2 a) = Bx'y' \cos 2a$, and $2Cx'y' \cos a \sin a$.

Adding these, we find the term involving $x'y'$ to be

$$[2(C - A) \cos a \sin a + B \cos 2a]\, x'y'$$
$$= [(C - A) \sin 2a + B \cos 2a]\, x'y'.$$

Our goal is to have the $x'y'$ term disappear. That is, we want

$$(C - A) \sin 2a + B \cos 2a = 0.$$

Therefore

$$(A - C) \sin 2a = B \cos 2a$$

$$\frac{A - C}{B} = \frac{\cos 2a}{\sin 2a} = \cot 2a$$

Thus a may be chosen in $(0, \pi/2)$ (why?) and a is determined by this equation $[a \in (0, \pi/2) \cap \mathbf{arccot}\, 2a]$.

EXAMPLE 8.3.4. Describe the conic section given by the equation $x^2 + xy + y^2 = 2/3$. In the form $Ax^2 + Bxy + Cy^2 + Dx + Ey + F = 0$ the equation is written $1 \cdot x^2 + 1 \cdot xy + 1 \cdot y^2 + 0 \cdot x + 0 \cdot y - 2/3 = 0$. Consequently, $\cot 2a = (1 - 1)/1 = 0$. Thus $2a = \pi/2$ and $a = \pi/4$. Our rotation formulas become: $x = x' \cos \pi/4 - y' \sin \pi/4 = (1/\sqrt{2})(x' - y')$, while $y = (1/\sqrt{2})(x' + y')$. Our original equation therefore becomes

$$2/3 = [1/\sqrt{2})(x' - y')]^2 + [(1/\sqrt{2})(x' - y') + (1/\sqrt{2})(x' + y')]$$
$$+ [(1/\sqrt{2})(x' + y')]^2$$
$$= (1/2)[(x')^2 - 2x'y' + (y')^2 + (x')^2 - (y')^2 + (x')^2 + 2x'y' + (y')^2]$$
$$= (1/2)[3(x')^2 + (y')^2]$$

$$= \frac{(x')^2}{2/3} + \frac{(y')^2}{2}$$

Alternatively,

$$1 = (x')^2/(2/3)^2 + (y')^2/(2/\sqrt{3})^2$$

The result is seen to be an ellipse with major axis along the y'-axis (the line $y = -x$). The semimajor axis length is $2/\sqrt{3}$ and the semiminor or axis has length $2/3$. The origin is the "center" of the ellipse and $C = [4/3 - 4/9]^{1/2} = (8/9)^{1/2} = 2\sqrt{2}/3$. The focal points *in the primed system* are $(0, 2\sqrt{2}/3)$ and $(0, -2\sqrt{2}/3)$. (See Figure 8.3.2.)

EXERCISES 8.3

Use a rotation to remove the "cross-product" term xy. Then describe the resulting curve and draw a sketch.

1. $9x^2 - 2xy + 9y^2 = 40$
2. $12x^2 + 8\sqrt{3}\,xy\,;\, + 4y^2 - x + \sqrt{3}y = 0$
3. $x^2 + 14\sqrt{3}\,xy - 2y^2 = 4$
4. $2x^2 + 4\sqrt{3}\,xy - 2y^2 - (4\sqrt{3} + 8)x + (8\sqrt{3} - 4)y = 16$
5. $x^2 - 2xy + y^2 + 8\sqrt{2}\,x + 48 = 0$
6. $7x^2 + 6\sqrt{3}\,xy + 5y^2 + (16 + 4\sqrt{3})x + (16\sqrt{3} - 4)\,y + 16 = 0$

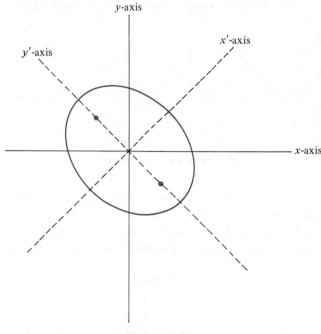

Figure 8.3.2

8.4 *R*³ AND PLANES

We remarked earlier that $R^3 = R \times R \times R$ is a natural setting in which to graph real functions of two real variables. A model of R^3 may be sketched in several ways; we choose to use that found in Figure 8.4.1.

The y-axis is horizontal; the z-axis, vertical; and the x-axis is drawn at a 45° angle with the y and z axes (that is, the x-axis traces the line $y = z$). The unit length of the x-axis is approximately 7/10 the unit lengths on the y and z axes. This gives us a three-dimensional appearance.

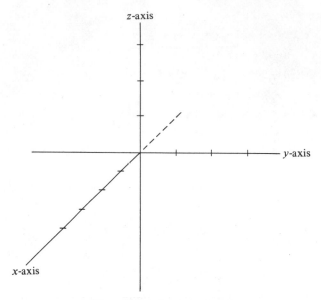

Figure 8.4.1. $R \times R \times R$.

Of principal interest are the xy, xz, and yz planes (Figures 8.4.2–8.4.4, respectively), called the *coordinate planes*. These planes are to prove useful in sketching graphs of functions of two variables. We will have need also of planes "parallel" to these coordinate planes.

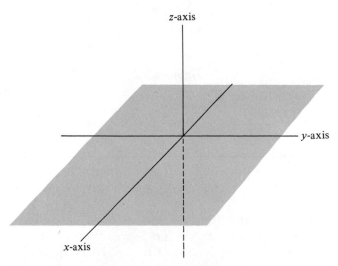

Figure 8.4.2. xy-plane.

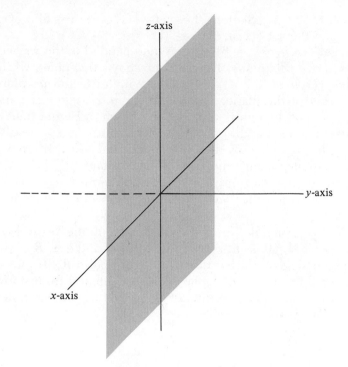

Figure 8.4.3. *xz*-plane.

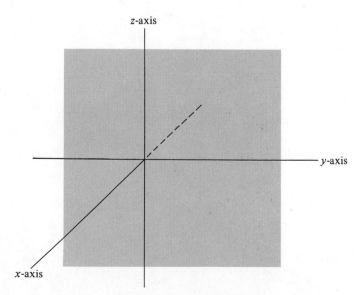

Figure 8.4.4. *yz*-plane.

EXAMPLE 8.4.1. Sketch the sets $\{(x, y, z): z = 6\}$, $\{(x, y, z): y = -2\}$, and the set $\{(x, y, z): x = 4\}$.

The set $\{(x, y, z): z = 6\}$ is the plane parallel to the xy plane and six units above this plane. Figure 8.4.5 shows this plane while 8.4.6 illustrates $\{(x, y, z): x = 4\}$, the plane parallel to the yz plane four units in front of the plane. The plane $\{(x, y, z): y = -2\}$ parallel to the xz plane and two units to its left is shown in Figure 8.4.7.

These planes parallel to the coordinate planes clearly have the role played in R^2 by the horizontal and vertical lines. Their equations are analogous: one variable is constant.

As a more general representative: $z = k$ represents a plane parallel to the xy plane (the "missing" variables) and k units from that plane.

Planes in general are defined as sets of the form $\{(x, y, z): ax + by + cz + d = 0; a, b, c \in R$, not all zero and $d \in R\}$. The form is entirely analogous to that of the straight line in R^2. If one of a, b, and c is zero the result is a plane easily sketched from its *trace* in (intersection with) the coordinate plane of the two variables with nonzero coefficients.

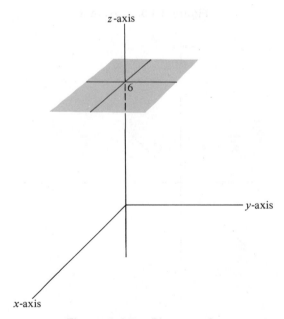

Figure 8.4.5. Plane $z = 6$.

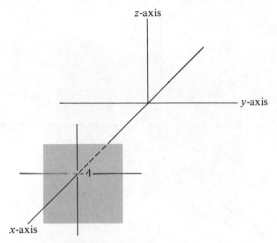

Figure 8.4.6. Plane $x = 4$.

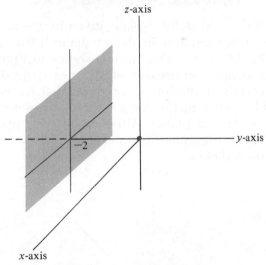

Figure 8.4.7. Plane $y = -2$.

EXAMPLE 8.4.2. Sketch the plane given by $3x + 2y = 1$.

We see that the plane is the set $\{(x, y, z) : 3x + 2y - 1 = 0\}$ and the determination is independent of z. Letting $z = 0$, we find that the trace in the xy plane (the plane $z = 0$) is the straight line given by $3x + 2y = 1$. In fact, for every plane $z = k$, the intersection of this plane with the desired plane is a "copy" of that same line and lies directly above (or below). Figure 8.4.8 shows the plane, its trace, and some *slices* (intersections with planes of the form $z = k$).

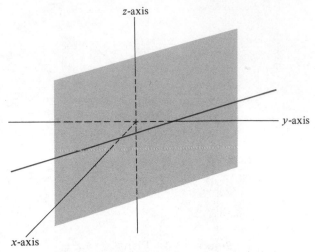

Figure 8.4.8.   Plane $3x + 2y - 1 = 0$.

EXAMPLE 8.4.3.   Sketch the plane given by $x = z$.

The trace of interest lies in the xz plane (plane $y = 0$) and is the line described by $x = z$. This trace is shown in Figure 8.4.9. The plane under consideration consists of all lines lying directly to the right of and directly to the left of the trace (that is, those lines that can be formed by sweeping the line $x = z$ in the xz plane in a direction perpendicular to the xz plane). Alternatively, the plane is the set of all points having the same first- and third-coordinate values. See Figure 8.4.10 for a sketch.

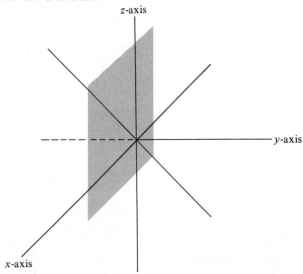

Figure 8.4.9.   Line $x = z$ lying in the xz-plane.

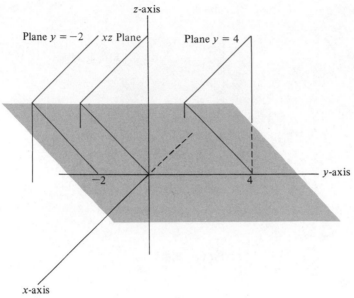

Figure 8.4.10. Plane $x = 2$.

EXAMPLE 8.4.4. Sketch the rectangular parallelepiped formed by the coordinate planes, and the planes $x = 5$, $y = 4$, and $z = 3$.

Figure 8.4.11 shows the geometric entity. We might observe that the solid figure enclosed is given by the set of points (x, y, z) satisfying:

$$0 \leqslant x \leqslant 5$$
$$0 \leqslant y \leqslant 4$$
$$0 \leqslant z \leqslant 3$$

and that it is therefore the product $[0, 5] \times [0, 4] \times [0, 3]$.

What is the length of a diagonal of the parallelepiped from above? Figure 8.4.12 shows that a double application of the Pythagorean theorem yields the length $(5^2 + 4^2 + 3^2)^{1/2} = 5\sqrt{2}$. That is, we feel $d((0, 0, 0), (5, 4, 3)) = [(5 - 0)^2 + (4 - 0)^2 + (3 - 0)^2]^{1/2} = 5\sqrt{2}$. For this reason let us define:

$$d((x, y, z), (u, v, w)) = [(x - u)^2 + (y - v)^2 + (z - w)^2]^{1/2}.$$

Our definition of *distance* is completely in line with our work in R and $R \times R$.

How do we find an equation describing a plane? (A plane is determined by three noncollinear points.)

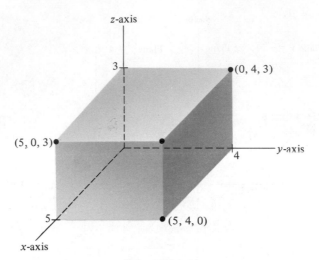

Figure 8.4.11

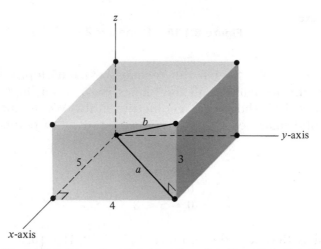

Figure 8.4.12. $a^2 = 5^2 + 4^2$; $b^2 = a^2 + 3^2 = 5^2 + 4^2 + 3^2$.

EXAMPLE 8.4.5.   Find an equation of the plane passing through $(2, -1, 3)$, $(1, 2, 3)$, and $(-1, 2, 1)$.

Write the equation of the plane as $ax + by + cz + d = 0$ and evaluate this equation in terms of the three given points:

$$2a - b + 3c + d = 0$$
$$a + 2b + 3c + d = 0$$
$$-a + 2b + c + d = 0$$

Since we have only three equations and four unknown values, we will find a simultaneous solution for the coefficients a, b, and c in terms of d. We find the solution to be: $a = 3d/4$, $b = d/4$, and $c = -3d/4$.

Since $d \neq 0$ (if $d = 0$, $a = b = c = 0$) we may let $d = 4$ (or any other nonzero real number), whence $a = 3$, $b = 1$, $c = -3$ and the plane has the equation $3x + y - 3z + 4 = 0$.

Note that if we had let d be some other number, say 16, then $a = 12$, $b = 4$, and $c = -12$ would have resulted along with the equation $12x + 4y - 12z + 16 = 0$. However, dividing by 4, we have our original solution. Thus any choice for $d \neq 0$ will give an equation yielding the same set of points. The motivation for our choice $d = 4$ is clear: the calculations were simplest.

EXAMPLE 8.4.6. Find the equation of the plane whose trace in the xy plane is the line $y = 3x$ and that passes through the point $(0, 5, 3)$.

Again, let the equation be $ax + by + cz + d = 0$. From $y = 3x$ we determine that $(0, 0, 0)$ and $(1, 3, 0)$ are in the plane [points of the trace are of the form $(x, 3x, 0)$]. The three points give us the equations:

$$0 \cdot a + 0 \cdot b + 0 \cdot c + d = 0$$
$$a + 3b + 0 \cdot c + d = 0$$
$$0 \cdot a + 5b + 3c + d = 0$$

Thus $d = 0$, $a = -3b$ and $c = -5b/3$. Letting $b = 3$, we have $a = -1$ and $c = -5$, whence the equation of the plane must be $-x + 3y - 5z = 0$.

EXERCISES 8.4

1. Sketch each plane and describe each in terms of parallel and so many units above, below, etc.

   a. $x = -2$ b. $y = 3$ c. $z = 4$

2. Sketch the parallelepiped formed by the planes $x = 3$, $x = -2$, $z = 1$, $z = 3$, $y = -6$, $y = -3$. Describe the solid figure (enclosed by the parallelepiped) in terms of inequalities. In terms of a Cartesian product.

3. Find equations for the plane passing through the triple of points in each case. Sketch each plane.

   a. $(1, 1, 1)$, $(0, 1, 5)$, $(-2, 4, 1)$ d. $(4, 0, 7)$, $(0, 0, 3)$, $(0, 2, 2)$
   b. $(6, 2, 1)$, $(2, 2, 1)$, $(0, 6, 1)$ e. $(5, 1, 1)$, $(-5, 1, 0)$, $(0, 1, 0)$
   c. $(-3, 1, 6)$, $(3, 1, 0)$, $(0, 1, 1)$

4. Find equations for the planes described as:

    a. having xy trace $x + y = 3$ and passing through $(1, 1, 1)$.
    b. having xz trace $z = 2x - 1$ and passing through $(2, 1, 1)$.
    c. having yz trace $y = 2z + 5$ and passing through $(1, 0, 0)$.

5. Theorem 8.1.2 gave a parametric representation for a line in $R \times R$ passing through two points. In R^3 the concept is similar. The line l passing through (a, b, c) and (d, e, f) is given by the equations

    $$x = a + t(d - a) \qquad y = b + t(e - b) \qquad z = c + t(f - c)$$

    Write parametric equations describing the line passing through each pair of points.

    a. $(0, 0, 0)$, (a, b, c) d. $(1, 1, 1)$, $(2, 2, 2)$
    b. $(1, -1, 2)$, $(6, 0, 7)$ e. $(-3, -2, 1)$, $(1, 3, -5)$
    c. $(5, 3, -3)$, $(2, -3, 1)$

6. Using Exercise 8.4.5, write parametric equations describing the line that is the intersection of the planes from

    a. Exercises 8.4.3, parts a and b.
    b. Exercises 8.4.3, parts d and a.
    c. Exercises 8.4.1a and 8.4.3a.
    d. Exercises 8.4.1b and 8.4.3b.
    e. Exercises 8.4.1c and 8.4.3c.

    (*Hint:* Find two points from the intersection.)

8.5 CYLINDERS

A *cylinder* is generally taken to be a set (in R^3) that can be written as the union of parallel lines (our concept of parallelism in R^3 is strictly intuitive). In this section we are concerned with what might be termed *right cylinders,* cylinders given in equation form involving no more than two variables. Some of the planes we investigated fit the category of right cylinders, whereas all planes are cylinders.

EXAMPLE 8.5.1. Graph the cylinder $x^2 + y^2 - 1 = 0$.

The set of points $\{(x, y, z) : x^2 + y^2 = 1\}$ is a cylinder since the describing equation is independent of z. The xy trace is the unit circle. Moreover, every point of the cylinder lies directly above or below the unit circle.

For example, the set $\{(x, y, 5) : x^2 + y^2 = 1\}$ is a copy of the unit circle five units above the xy plane. We call this set the *slice* formed by the plane $z = 5$. For each real number k, the slice generated by $z = k$ is a copy of the unit circle located k units from the xy plane. (See Figure 8.5.1 for a sketch of this right circular cylinder.)

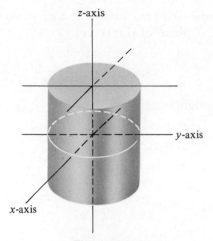

Figure 8.5.1

EXAMPLE 8.5.2. Sketch the parabolic cylinder $z = y^2$.

The yz trace is a parabola we have seen on several previous occasions. The cylinder, then, is formed intuitively by moving this parabolic trace in a direction perpendicular to the yz plane. (See Figure 8.5.2.)

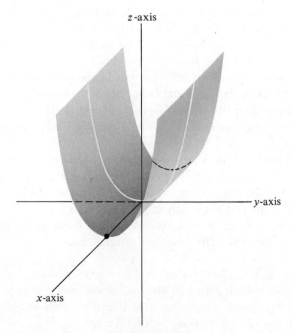

Figure 8.5.2

All right cylinders can be envisioned as a "sweeping" of a trace perpendicular to the plane of the trace.

EXERCISES 8.5

Sketch the following right cylinders.

1. $\{(x, y, z): (x - 2)^2 + (y - 3)^2 = 4\}$
2. $\{(x, y, z): z = x^2 + 2x + 1\}$
3. $zy = 1$ (There are two sections.)
4. $z^2 + y^2 = 25$
5. $x^2 + (z - 3)^2 = 9$
6. $y = x^3$
7. $y^2 = x^2$
8. $\{(x, y, z): y = x^2\} \cup \{(x, y, z): y = 4\}$
9. The "oval" cylinder formed by the cylinders $y = x^2 - 4$ and $y = 4 - x^2$

8.6 REAL-VALUED FUNCTIONS OF TWO REAL VARIABLES

We know from previous remarks that the domain for a real-valued function of two real variables is a subset of R^2.

Thus if f is such a function, f consists of ordered pairs of the form $((x, y), f(x, y))$, or, if we write $z = f(x, y)$, $((x, y), z)$. By removing the inner parentheses, we could denote these elements of f in the form $(x, y, f(x, y))$ or (x, y, z), remembering that the first two coordinates give the domain element, while the third represents the corresponding functional value. Consequently we can graph f as a subset of R^3.

The functions of two variables of interest are most often given in equation form. For example, $f(x, y) = \Sigma_{i=0}^n a_i x^i y^{n-i}$ is a *polynomial* in x and y. Our technique for examining cylinders by slices is useful in sketching graphs of such functions.

By setting $z = f(x, y) = 5$, we simply examine $5 = \Sigma_{i=0}^n a_i x^i y^{n-i}$ to see what appearance this particular slice has.

EXAMPLE 8.6.1. Let $f(x, y) = (x^2 + y^2)^{1/2}$. Sketch a graph of f.

If we square both sides of the equation we find it to be of form $f^2(x, y) = x^2 + y^2$ and we recognize that for each value $f(x, y)$ the equation represents a circle of radius $f(x, y)$.

If $f(x, y) = 0$, $x^2 + y^2 = 0$ gives the point $(0, 0, 0)$. If $f(x, y) = 1$, $x^2 + y^2 = 1$ is the equation describing that slice. The slice given by $z = f(x, y) = 1$, then, is a copy of the unit circle one unit above the xy plane. A slice $z = f(x, y) = 2$ shows a copy of the circle of center $(0, 0, 2)$ and radius 2.

Figure 8.6.1 shows the resulting graph of f to be a right circular cone.

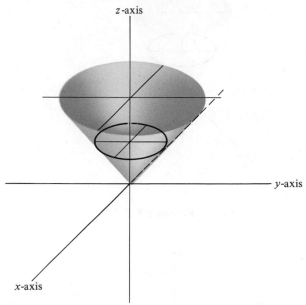

Figure 8.6.1

EXAMPLE 8.6.2. Let $f(x, y) = x^2 + y^2$. Graph f as a subset of $R \times R \times R$.

Again, $f(x, y) \geqslant 0$ and for each plane $z = f(x, y) = k$, the corresponding slice is a circle of center $(0, 0, k)$ and radius $\sqrt{k}$. The sketch is shown in Figure 8.6.2.

EXAMPLE 8.6.3. Sketch $f(x, y) = |x| - |y|$.

The trace $f(x, y) = 0$ shows $|x| = |y|$ and is the sketch in R^2 shown in Figure 8.6.3. The slice $f(x, y) = k > 0$ is shown in Figure 8.6.4 and the slice $f(x, y) = k < 0$ is found in 8.6.5. These bits of information are pieced together in Figure 8.6.6.

In general our sketches will be formed by setting $f(x) = k$ and sketching that particular slice. The relationship might be given implicitly rather than explicitly. However, the same general technique applies.

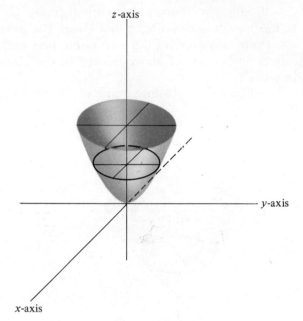

Figure 8.6.2

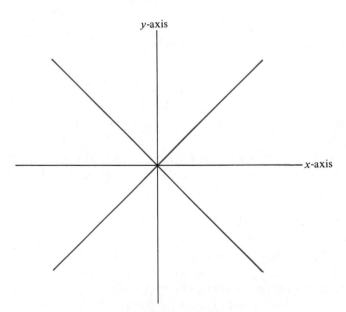

Figure 8.6.3. $|x| = |y|$.

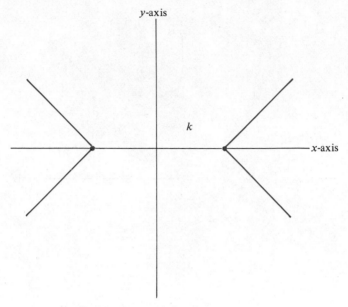

Figure 8.6.4. $|x| = |y| + k, k > 0$.

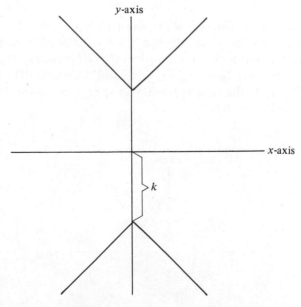

Figure 8.6.5. $|x| = |y| + k, k < 0$.

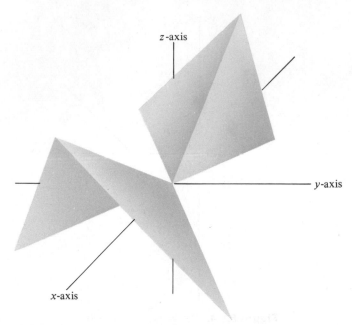

Figure 8.6.6. That part of $\{(x, y, z): z = |x| - |y|\}$ above the xy-plane.

EXAMPLE 8.6.4. Sketch a graph of $x^2 + y^2 + z^2 = 1$.

This graph can be done by the slice method, but is done much more easily by noting that $x^2 + y^2 + z^2 = d((x, y, z), (0, 0, 0))^2$, the square of the distance between (x, y, z) and the origin. Thus $\{(x, y, z): x^2 + y^2 + z^2 = 1\}$ is the unit sphere, the sphere of center $(0, 0, 0)$ and radius 1. (See Figure 8.6.7.)

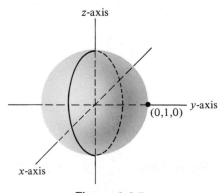

Figure 8.6.7

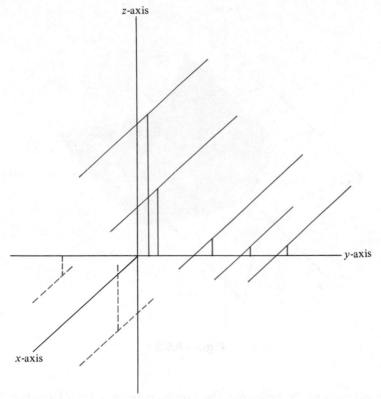

Figure 8.6.8

EXAMPLE 8.6.5. Sketch $\{(x, y, z) : y$ and z are reciprocals$\}$.

We can write the equation in the form $yz = 1$. Fixing $y = k$, we find the slice represented by the equation $z = 1/k$ ($k \neq 0$ obviously) to be a straight line. Figure 8.6.8 shows several slices, while Figure 8.6.9 shows a sketch of the function.

EXERCISES 8.6

Sketch each of the functions given below.

1. $f(x, y) = 1$
2. $g(x, y) = x$
3. $h(x, y) = y^2$
4. $\eta(x, y) = x^2 - y$
5. $\mu^2(x, y) = x^2 + y^2$
6. $\theta(x, y) = xy$
7. $x \, \gamma(x, y) = y$
8. $\zeta(x, y) = x^2 + y^2 - 1$
9. $\xi(x, y) = x + y^2$
10. $s(x, y) = |x|$

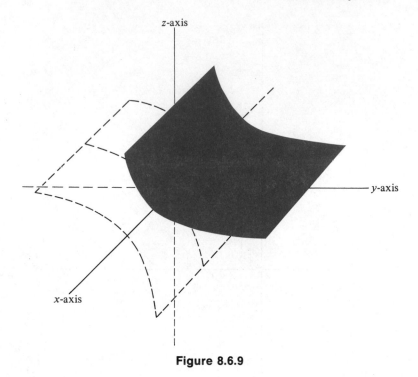

Figure 8.6.9

Evaluate each of the following, the functions coming from Exercises 8.6.1–8.6.10.

11. $f(0, 0)$ 16. $\theta(-1, -1)$

12. $g(5, -2)$ 17. $\gamma(0, 1)$

13. $h(0, 1)$ 18. $\zeta(1, 0)$

14. $\eta(1, 1)$ 19. $\xi(1, -1)$

15. $\mu(1, -1)$ 20. $s(-11, 4)$

8.7 REVIEW EXERCISES

1. Sketch each plane.

    a. $z = -3$ d. $z - y + x = 3$
    b. $y = -x$ e. $z - y = 3$
    c. $y + 3x = 2$

2. Sketch the figure enclosed between the planes $y = x$, $y = 0$, $y = 5 - x$, $z = 0$, and $z = 5$. Can this figure be given as a Cartesian product of intervals? Why or why not?

3. Derive an equation for the plane

    a. passing through $(1, 1, 2)$, $(-1, 1, 2)$, and $(0, 1, 0)$
    b. passing through $(0, -1, -2)$, $(-2, -1, 0)$ and $(0, 0, 0)$
    c. having xy trace $x + y = 1$ and passing through $(1, 0, 1)$
    d. having yz trace $y - z = 6$ and containing $(-1, -1, -2)$

4. Find parametric equations for the straight line that

    a. contains $(1, 1, 1)$ and $(0, 0, 0)$
    b. contains $(1, 1, 0)$ and $(0, 0, 0)$
    c. contains $(2, 4, -1)$ and $(-1, 4, 2)$
    d. is the intersection of the planes in Exercises 8.7.1b and 8.7.3a
    e. is the intersection of the planes in Exercises 8.7.3b and 8.7.3d

5. Sketch the subsets of R^3 described in each case:

    a. $y^2 + z^2 = 1$ d. $x^2 + 3x - 4 = y$
    b. $y^2 - x^2 = 1$ e. $x^4 = z^4$
    c. $x^2 + 4z^2 = 1$ f. $x^2 + y^2 + z^2 = 9$

6. Sketch the following functions.

    a. $f(x, y) = 3y$ d. $f(x, y) = \sqrt{x^2 + y^2 - 9}$
    b. $f(x, y) = x + y$ e. $f(x, y) = 2^x$
    c. $f(x, y) = x^2 + y$

7. Sketch each of the curves described by the given equation.

    a. $16x^2 + 32x + y + 13 = 0$
    b. $8\sqrt{2}\,x^2 + 16\sqrt{2}\,xy + 8\sqrt{2}\,y^2 + 31x + 33y + 13\sqrt{2} = 0$
    c. $144x^2 + 25y^2 + 288x - 50y = 3431$
    d. $2529x^2 + 2856xy + 1696y^2 + 6510x + 3320y = 85775$
    e. $x^2 + 4x - y^2 + 2y + 2 = 0$
    f. $119x^2 + 240xy + 119y^2 - 494x + 572y + 338 = 0$

8. Write an equation for each of the following curves.

    a. The parabola with vertex at $(-1, 7)$ and focus at $(5, 7)$.
    b. The parabola with vertex $(1, 1)$ and focus $(3, 3)$.
    c. The ellipse with minor axis length 6 and foci $(1, 6)$ and $(1, 10)$.
    d. The ellipse with major axis length 20 and foci $(3, 4)$ and $(6, 8)$.
    e. The hyperbola with focal points $(5, 2)$ and $(25, 2)$ and vertices $(8, 2)$ and $(22, 2)$.
    f. The hyperbola with vertices $(-2, 2)$ and $(2, -2)$ with foci separated by $8\sqrt{2}$ units.

8.8 QUIZ

Give yourself no more than one and a half hours to complete this test. Then carefully check your work.

1. Derive an equation for the following plane and then sketch it. The plane has xz trace described by $z + x = 5$ and contains the point $(8, 0, 5)$.

2. Derive a parametric representation of the intersection of the plane in 1 with the yz plane.

3. Sketch the union of the following three sets:

$$\{(x, y, z): x = y, 0 \leqslant x \leqslant 1/\sqrt{2}, 0 \leqslant z \leqslant 1\}$$
$$\{(x, y, z): y = |x|, -1/\sqrt{2} \leqslant x \leqslant 0, 0 \leqslant z \leqslant 1\}$$
$$\{(x, y, z): y = \sqrt{1 - x^2}\}$$

4. Sketch the function f given by the equation $f(x, y) = 1/(x^2 + y^2)^{1/2}$. What is the "largest acceptable" domain for f?

5. Describe in detail the curve given by $x + 5 = 24(y - 3)^2$.

6. Find an equation for the ellipse having focal points $(7, 10)$ and $(7, 16)$ with major axis length 10.

7. Sketch the curve given by $12y^2 + 250x = 37x^2 + 168xy + 325$.

8.9 ADVANCED EXERCISES

1. Sketch the following point sets, using the slice method.

   a. $x^2 + y^2 - z^2 = 1$ d. $25(x^2 + y^2) = z^2$
   b. $z^2 = x^2 + y^2 + 1$ e. $y - 3 = 4(x - 1)^2$
   c. $z = x^2/4 + y^2/9$

2. Sketch graphs of the following functions of two variables.

   a. $f(x, y) = x^2/9 + y^2/25$
   b. $g(x, y) = x^2/9 - y^2/25$
   c. $\mu(x, y) = 3x + 2y$

3. Sketch the figures bounded by the surfaces described.

   a. $y^2 + z^2 = 4x$, $y^2 = x$, $x = 3$
   b. $x = 0$, $y = 0$, $z = 0$, $x/3 + y/4 + z/5 = 1$
   c. That part of $x^2 + y^2 + z^2 = 25$ above $z^2 = x^2 + y^2$
   d. That part of $x^2 + y^2 + z^2 = 4z$ above $x^2 + y^2 = 2z$

4. Sketch graphs of the following subsets of $R \times R \times R$ by eliminating the parameter:

   a. $x = 3 \sin t$, $y = 2 \cos t$ (*Hint*: Examine $x^2/9 + y2/4$.)
   b. $x = 3 \sec t$, $y = 2 \tan t$
   c. $x = \sec t$, $y = \cos t$
   d. $x = 5 + \sin t$, $y = 5 \cos t - 2$
   e. $x = \sec t - 1$, $y = 2 - 3 \tan t$
   f. $x = 1 + \csc t$, $y = 2 - \sin t$

5. If we transform the equation $Ax^2 + Bxy + Cy^2 + Dx + Ey + F = 0$ by rotating the axes and have the resulting equation $A'x'^2 + B'x'y' + C'y'^2 + D'x' + E'y' + F' = 0$, then $B^2 - 4AC = B^2 - 4A'C'$. The expression $B^2 - 4AC$ is called the *discriminant*.

6. Show that the following hold for the quadratic equation $Ax^2 + Bxy + Cy^2 + Dx + Ey + F = 0$. (Use Exercise 8.9.5.) If the resulting figure does not consist of straight lines or individual points, and

   a. if $B^2 - 4AC < 0$, the curve is an ellipse.
   b. if $B^2 - 4AC = 0$, the curve is a parabola.
   c. if $B^2 - 4AC > 0$, the curve is a hyperbola.

7. Using the notation of Exercise 8.9.5, show that $A + C = A' + C'$.

APPENDIX A

Logic

A.1 PROPOSITIONS

Mathematics makes use of—in fact is said by some to consist solely of—communication via symbols and the written word. In our case we are making use of the English language for the purpose of exchanging ideas. Even our symbols can be verbalized. It is therefore not surprising that we find ourselves confronted with logical and semantical problems. Our brief efforts here are given in the hope of establishing agreements and understanding relative to the use of the language as it pertains to the material of the basic text. Our discussions will be of a quite naïve nature, but nevertheless should suffice.

Two types of (declarative) statements can occur. There are those that make sense in the context of the discussion (those that actually communicate an idea or image) and those that do not transmit any intelligible information. We are clearly interested only in statements of the former type (these are those that can be determined to be either true or false within the context of our discussion) and choose to give to such statements the title *proposition*. Let us assume that all statements of our future discussions are propositions.

Our use of the language is not confined to such simple propositions as "A dog is a canine." We make use of sentences of a more compound nature by introducing connectives such as the conjunctive *and* and the disjunctive *or*. For purposes of discussion let us denote by p the proposition "A dog is a canine" and introduce q, the proposition "A cat is a canine."

The (compound) proposition "*p* and *q*" (A dog is a canine and a cat is a canine) is indeed a proposition whose truth value (that is, its characteristic of being correct or false within the context of the discussion) is as yet undetermined. Such a value is to be determined by the truth value of *p*, the truth value of *q*, and our agreement as to how these values are to dictate the truth value of the conjunctive statement.

Axiom (Agreement) **A.1.1.** *The* conjunction "p *and* q" *is a true statement if* p *and* q *are true statements. Otherwise,* "p *and* q" *is a false statement. The truth table of Figure A.1.1 diagrams the truth value relationships inherent in this agreement.*

p	*q*	*p* and *q*
T	T	T
T	F	F
F	T	F
F	F	F

Figure A.1.1

Accordingly, our statement "A dog is a canine and a cat is a canine" is false. This follows from the fact that "A cat is a canine" is false.

Axiom A.1.2. *The* disjunction "p *or* q" *is a true statement if at least one of* p *and* q *is true. If both* p *and* q *have a false value, the disjunction is also false. Observe the diagram of Figure A.1.2.*

p	*q*	*p* or *q*
T	T	T
T	F	T
F	T	T
F	F	F

Figure A.1.2

The disjunctive statement "A dog is a canine or a cat is a canine" is, because of our agreement, a true statement.

The statement "A cat is not a canine" has a very definite relationship to the proposition "A cat is a canine." In simplest terms, the new proposition is a *negation* of the idea of the original sentence. Its truth value is opposite that of the original.

Axiom A.1.3. If p *is a proposition, "not* p*" is any statement having at all times a truth value opposite that of* p. *Figure A.1.3 illustrates.*

p	not p
T	F
F	T

Figure A.1.3.

One feature of our excursion here is to indicate how different statements that might arise in any given study are to be compared. Crudely speaking, we are interested in how the same thing can be said in different ways. How do we determine whether two statements really say the same thing? What does it mean for two propositions to say the same thing?

Definition A.1.1. Two propositions are equivalent *if they have the same truth table (values).*

This will be our accepted concept for determining when two propositions convey the same information. The examples to follow illustrate such determinations and give insight into relationships existent between conjunction, disjunction, and negation.

EXAMPLE A.1.1. The truth tables in Figure A.1.4 show that the negation of "p and q" is equivalently given by "(not p) or (not q)." The truth tables are formed by calculating truth values in each column through the use of appropriate columns to the left.

p	q	not p	not q	p and q	not (p and q)	(not p) or (not q)
T	T	F	F	T	F	F
T	F	F	T	F	T	T
F	T	T	F	F	T	T
F	F	T	T	F	T	T

Figure A.1.4

EXAMPLE A.1.2. In a manner similar to that of Example A.1.1, we have constructed the truth tables of Figure A.1.5 to show the equivalence of the negation of "p or q" and the proposition "(not p) and (not q)."

p	q	not p	not q	p or q	not $(p$ or $q)$	(not p) and (not q)
T	T	F	F	T	F	F
T	F	F	T	T	F	F
F	T	T	F	T	F	F
F	F	T	T	F	T	T

Figure A.1.5

EXAMPLE A.1.3. Let us write a negation for each of the following propositions:

1. A dog is a canine and a cat is a canine.
2. A dog is a canine or a cat is a canine.

We will make use of the results of Examples A.1.1 and A.1.2. A negation of 1 reads: A dog is not a canine or a cat is not a canine. The procedure is straightforward: we negate each individual proposition and exchange the conjunctive *and* for the disjunctive *or*. The equivalent statement is more than the word equivalent might connote. The latter statement is actually an interpretation or clarification of the message of the desired negation.

The negation for 2 is contrived in an analogous manner. This time, however, the disjunction becomes a conjunction: A dog is not a canine and a cat is not a canine.

Further examples are given in the exercises at the end of this section.

There are further general types of propositions that are prevalent in mathematics (even everyday life). Among these is found the class of *implications*. We are herewith concerned with a proposition, say q, whose truth value depends in some way on the truth value of another proposition p. If the truth of p guarantees the truth of q, we say that "p implies q." Observe that we say nothing about the truth value of q under the condition that p is false. Accordingly we set forth a pattern for determining the truth values of a proposition.

Axiom A.1.4. The statement "p implies q" is to be equivalent to "(not p) or q." The truth tables of Figure A.1.6 illustrate.

We might well note that "p implies q" is true except for the single case when p is true and q is false. This agrees with our earlier remarks. Note also the two ways by which an implication can be proved true. First, we may assume that the *hypothesis p* is true and verify (in the context of the discussion) that the *conclusion q* is likewise true. Secondly, we may assume that the hypothesis is true but

p	q	not p	(not p) or q	p implies q
T	T	F	T	T
T	F	F	F	F
F	T	T	T	T
F	F	T	T	T

Figure A.1.6

that the conclusion q is false. According to Figure A.1.6, the resulting implication must be false. If in reality the implication is valid (true), a contradiction exists. The demonstration of such a contradiction proves that q must be true, not false. This is a proof by contradiction or an indirect proof (as opposed to the direct-proof technique first discussed).

There is yet a third alternative available. The truth value for "p implies q" is given by the values for "(not p) or q." It should be clear, though, that the truth values for "q or (not p)" agree also. Observing that q and *not (not q)* have the same truth values, we find that "[not (not q)] or (not p)" and "p implies q" have the same truth tables. Furthermore, "[not (not q)] or (not p)" is by Axiom A.1.4 equivalent to "(not q) implies (not p)."

In summary, we now know that "p implies q" is equivalent to "(not q) implies (not p)." The latter implication is called a *contrapositive* of the former. Figure A.1.7 illustrates the argument of this paragraph.

p	q	not p	p implies q	q or (not p)	(not q) implies (not p)
T	T	F	T	T	T
T	F	F	F	F	F
F	T	T	T	T	T
F	F	T	T	T	T

Figure A.1.7

EXAMPLE A.1.4. Consider the implication: "If Max is a dog, Max is a canine." The contrapositive of this statement is: "If Max is not a canine, Max is not a dog."

The contrapositive is easily formed: we merely negate the hypothesis and conclusion and then interchange the order (roles). This reversal of propositions brings up an interesting, important, and yet often confusing point.

If the implication "p implies q" is true, does it follow that its *converse* "q implies p" is also true? The answer is negative. It is not generally the case that the truth of an implication dictates the truth of its converse. *In fact, this happens in only those cases where* p *and* q *are equivalent propositions.*

Figure A.1.8 shows that both "*p* implies *q*" and its converse are true whenever *p* and *q* have the same truth value while one of the implications is false if the truth values of *p* and *q* differ.

p	q	not p	not q	p implies q	q implies p
T	T	F	F	T	T
T	F	F	T	F	T
F	T	T	F	T	F
F	F	T	T	T	T

Figure A.1.8

EXAMPLE A.1.5. Consider the implication: "If Max is a dog, Max is a canine." A converse reads: "If Max is a canine, Max is a dog." Clearly, one might feel that the converse might not be correct (depending on one's definition of dog and canine) even though the original proposition is true.

We close the exposition of this section by examination of the transitive property of the implication and one of the resulting consequences.

Theorem A.1.1. If p *implies* q *and* q *implies* r, *then* p *implies* r.

Proof. We have four implications involved. They are:

1. *p* implies *q*   (Call this proposition *A*),
2. *q* implies *r*   (Call this proposition *B*),
3. *p* implies *r*   (Call this proposition *C*),   and
4. (*A* and *B*) implies *C*.

It is 4 that we need to show to be correct. Figure A.1.9 shows (perhaps surprisingly) that the statement of our theorem (implication 4) has only *T* values. The theorem must be correct.

EXAMPLE A.1.6. Let *p* be "$x \in A$", let *q* be "$x \in B$" and let *r* be "$x \in C$." The statement of Theorem A.1.1 becomes: If $x \in A$ implies $x \in B$ and if $x \in B$ implies $x \in C$, then $x \in A$ implies $x \in C$. Restated: If $A \subset B$ and $B \subset C$, then $A \subset C$. This establishes the transitive property for "$\subset$." (See Theorem 0.3.1.)

p	q	r	A p implies q	B q implies r	C p implies r	A and B	(A and B) imply C
T	T	T	T	T	T	T	T
T	T	F	T	F	F	F	T
T	F	T	F	T	T	F	T
T	F	F	F	T	F	F	T
F	T	T	T	T	T	T	T
F	T	F	T	F	T	F	T
F	F	T	T	T	T	T	T
F	F	F	T	T	T	T	T

Figure A.1.9

EXERCISES A.1

1. Let p be "$x \in A$," let q be "$x \in B$," and let r be "$x \in C$." Write out each of the propositions diagrammed below.

    a. p or q.
    b. p and q.
    c. p and (not q).
    d. p or (not q).
    e. p implies q.
    f. (not q) implies (not q).
    g. q implies p.
    h. (p implies q) and (q implies p). (*Note:* This statement is often written: p if and only if q.)
    i. (p and q) and (not r).
    j. (p and q) or (not r).
    k. A negation of a.
    l. A negation of b.
    m. A negation of e. (*Hint:* Use the equivalent form for "p implies q.")
    n. A contrapositive for e.
    o. A contrapositive for g.
    p. A contrapositive for f.
    q. A converse for e.
    r. A converse for g.
    s. A converse for f.

2. A proposition is a *tautology* if it can have only the truth T and it is a *self-contradiction* if F is its only attainable truth value. Form truth tables to determine which of the following are tautologies and which are self-contradictions.

    a. p and (not p).
    b. p or (not p).
    c. p and [not (not p)].
    d. {not [not (not p)]} and (not p).
    e. [(not p) or q] and [q or (not p)].
    f. p if and only if q; where p and q are equivalent (See Exercise A.1.1h).
    g. p implies [p or (not p)].

A.2 QUANTIFIERS

Another matter of logic that arises in our study is the concept of the quantifier, the phrase that asks "How many?" We break the ideas into three categories described, albeit somewhat cryptically, as all, some, and none.

For each *propositional form* (a proposition involving a variable) there is associated a defining class or truth set. For example, our proposition "Max is a dog" becomes a propositional form by introducing a variable x and writing: "x is a dog." Now, the associated *truth set* for the propositional form p is $\{x : x$ is a dog$\}$. That is, it is the set of all elements that, when replacing x in p, yield a true statement.

The truth set for (not p) must be the "false set" for p. That is, the truth set for (not p) must be the complement of the truth set for p.

EXAMPLE A.2.1. The truth set for "p or q" is the set of all elements x such that p is true for x or q is true for x. If P and Q are the truth sets for p and q respectively, the truth set for the disjunction is seen to be $P \cup Q$.

These concepts allow us to relate set theory and logic. In particular, it is now advantageous to examine the question, "How many?"

Suppose that we have a universal set X and a set $S \subset X$. The following statements connote the same idea, that of the *universal quantifier*.

1. If $x \in S$, x is a dog.   4. Any $x \in S$ is a dog.
2. Each $x \in S$ is a dog.   5. All x's in S are dogs.
3. Every $x \in S$ is a dog.

Statement 1 can be reinterpreted in view of our set theory: $S \subset D$, D being the set of all dogs in X. The adjective *universal* is used because, in a sense, "all of S has the described property."

How can the choice of x be given so that perhaps not all of S has the given property? The following all relate the concept of the *existential quantifier*.

6. There is an $x \in S$ with x a dog.
7. For some $x \in S$, x is a dog.

Setwise we might give 6 and 7 by: $S \cap D \neq \phi$.

Finally, the disjoint condition $S \cap D = \phi$ is given by

8. If $x \in S$, x is not a dog.

9. No x in S is a dog.

A major question that occurs frequently is that of the negation of any of the quantifiers (rather, statements 1–9). The negation of 1 is formulated as follows.

Statements 1–5 can be given by $S \subset D$ and the negation is then $S \not\subset D$. Intuitively, a Venn diagram shows us the existence of an $x \in S$ with x outside D ($x \notin D$). That is, a negation of 1–5 can be given in terms of the existential quantifier: "There is an $x \in S$ with x not a dog."

Similarly, a negation of 6 and 7 reads: "There is no $x \in S$ with with x a dog." A negation of 8 and 9 is: "There is an $x \in S$ such that x is a dog."

These results lead us to believe that only two quantifiers really exist, those being the universal and existential. A quantifier indicating "none" is really a universal quantifier saying: "All do not have a given property." In this context, the universal and existential quantifiers are used to yield negations of each other.

EXERCISES A.2

1. What is the truth set for a tautology? a self-contradiction?

2. Determine truth sets for:

    a. p and q.   d. q implies p.
    b. p implies q.   e. p if and only if q.
    c. (not q) implies (not p).

3. Use truth sets to give set statements describing the logical (verbal) statements below.

    a. "p implies q" is a self-contradiction.
    b. "p implies q" is a tautology.
    c. "(p and q) implies r" is a tautology.
    d. "p if and only if q" is a self-contradiction.
    e. "p if and only if q" is a tautology.

4. Let P and Q be the truth sets for p and q respectively. Find the desired truth sets given the described conditions.

    a. The truth set for "p and q" where P and Q are disjoint. Where $P \subset Q$; $P = Q$; $P \subset Q'$.
    b. The truth set for "p or q" where $P \subset Q$; $Q \subset P$; $Q = P$; $P \subset Q'$.

5. Let P and Q be the truth sets for p and q respectively.

    a. If $P \subset Q$, what is the truth set for "p implies q"?

    b. If $P = Q$, what is the truth set for "p if and only if q"?

    c. If "p implies q" is a tautology, show that $P \subset Q$.

    d. If "p if and only if q" is a tautology, show that $P = Q$.

Real-Number System

INTRODUCTION

This appendix is intended primarily to serve as a refresher study aimed at bringing into focus certain facts and ideas used in the basic text. Certain portions of the material presented here are viewed in Chapter 2. The attitude in that chapter is such that the material warrants the different (pre)viewing found herein. We will find the appendix to be a good reference source for our basic study of the elementary functions.

B.1   THE FIELD STRUCTURE

The basic text does not make a great effort to discuss and display the number system underlying our study. However, it may be that we need to examine some of the elementary behavior patterns of this system that determine the mechanical or manipulative techniques governing our every computation. Some of these are so well known as to be of second nature; some are discussed in Sections 2.1 and 2.3, and others, while no more important or subtle, are not so well known.

These patterns of behavior give rise to what is called an *algebraic structure*, the particular structure at hand being a *complete ordered field*. The prototype of this structure is the system of *real numbers*. It is this system we wish to discuss presently.

We use various types of numbers in our study and they include the set of *natural numbers* or positive integers ($\{1, 2, 3, 4, 5, \ldots\}$)

denoted by N, the set Z of *integers* given by $Z = \{\ldots, -3, -2, -1,$
$0, 1, 2, 3, \ldots\}$, and the set $Q = \{m/n: m, n \in Z, n \neq 0\}$ of *rational*
forms (*numbers*). Other numbers, such as $\sqrt{2}$ (a number x such that
$x \cdot x = 2$), in the set R of real numbers constitute the set I of *irrational numbers*. If we agree to let n and $n/1$ represent the same
element in R, we have the following set relationships:

$$N \subset Z \subset Q \subset R$$
$$R = Q \cup I$$
$$Q \cap I = \phi$$

The difference between rational and irrational numbers may
seem clouded here. Let us only remark that rational numbers have
a decimal form that is terminating (.25, for example) or nonterminating
and periodic (.62515151 ..., where the concatenation 51 is indicated as being repeated indefinitely). On the other hand, irrational
numbers have nonterminating and nonperiodic decimal expansions
(thus, $I \cap Q = \phi$).

Given the set R of real numbers, we wish to focus our attention
on the operations of addition $(+)$ and multiplication $(\cdot)$. We say that
R together with $(+)$, $(\cdot)$, and the equivalence relation $(=)$ forms
a field because this system $(\{R, +, \cdot, =\})$ behaves according to the
following rules:

1. If $a, b \in R, a + b \in R$ and $a \cdot b = ab \in R$.

2. If $a, b \in R, a + b = b + a$ and $ab = ba$.

3. If $a, b, c \in R, (a + b) + c = a + (b + c)$ and
    $(ab)c = a(bc)$.

4. If $a, b, c \in R, a(b + c) = ab + ac$.

5. R has an element 0 satisfying $a + 0 = a$ for all $a \in R$. More-
    over, for each $a \in R, R$ has a term $(-a)$ where $a + (-a) = 0$.

6. R has an element 1 satisfying $a \cdot 1 = a$ for all $a \in R$. More-
    over, if $a \neq 0, R$ has a term a^{-1} where $aa^{-1} = 1$.

The six rules emphasize well-known facts and are viewed in
Section 2.3. Rule 1 tells that the sum (or product) of one real number
with another is itself a real number. We say that R is *closed* with
respect to $(+)$ [or $(\cdot)$]. Alternatively, the rule shows that $+: R \times R \to R$
and $\cdot: R \times R \to R$ are functions.

Statement 2 illustrates the long-accepted fact that the order of
addition (multiplication) is immaterial so far as the outcome is concerned. This is called the *commutative* property.

The *associative* property (property 3) indicates that the grouping of terms for addition (or multiplication) does not affect the sum. In all actuality, rule 3 gives the means whereby more than two numbers may be added (multiplied).

The *distributive* property 4 describes a combining of the operations of addition and multiplication. Roughly speaking, it shows that the operation of multiplication distributes itself over the elements of the sum.

Behavior patterns 5 and 6 show the special roles of 0 and 1. For rather obvious reasons, 0 and 1 are called respectively the *additive* and *multiplicative identities* (the functions taking $a \to a + 0$ and $a \to a \cdot 1$ are identity functions). The inverse elements $(-a)$ and a^{-1} for a are also displayed [$(-a)$ being the *additive inverse* for a and a^{-1} the *multiplicative inverse*]. We might say that the inverse elements "neutralize" a relative to the particular operation involved.

The following examples illustrate how the six rules can be used to draw further well-known results.

EXAMPLE B.1.1.   Suppose that $a = b$ and $c = d$ belong to R. Since $(a, c) = (b, d)$ in $R \times R$, and since $+$ and $\cdot$ are functions from $R \times R$ into R, $+(a, c) = +(b, c)$ while $\cdot (a, c) = \cdot (b, c)$. However, $+(a, c) = a + c$ and $\cdot (a, c) = ac$, whereby $a + c = b + d$ and $ac = bd$. In very unsophisticated terms we may say that equals added to (or multiplied by) equals result in equal numbers.

EXAMPLE B.1.2.   If $a = b$, $a + [(-a) + (-b)] = b + [(-a) + (-b)]$. The left-hand expression is, however, $-b$, while the right-hand expression is $b + [(-a) + \cdot (-b)] = b + [(-b) + (-a)] = [b + (-b)] + (-a) = 0 + (-a) = -a$. That is, if $a = b$, $-a = -b$ as well. Similarly, if $a = b \neq 0$, $a^{-1} = b^{-1}$.

EXAMPLE B.1.3.   We can calculate $a \cdot 0$ by using the field properties. That is, $a \cdot 0 = a(0 + 0) = a \cdot 0 + a \cdot 0$. By adding $-(a \cdot 0)$ to both sides (we are using Example A.1.1), we have $a \cdot 0 + [-(a \cdot 0)] = [a \cdot 0 + a \cdot 0] + [-(a \cdot 0)] = a \cdot 0 + a \cdot 0 + [-(a \cdot 0)]$. Since $a \cdot 0 + [-(a \cdot 0)] = 0$, we have, from the string of equalities, $0 = a \cdot 0$.

EXAMPLE B.1.4.   We can compute $a(-1)$ by: $0 = a[1 + (-1)] = a \cdot 1 + a(-1) = a + a(-1)$. By adding $(-a)$ to both sides, we show $-a = 0 + (-a) = [a + a(-1)] + (-a) = a + [(-a) + a(-1)] = [a + (-a)] + a(-1) = 0 + a(-1) = a(-1)$. That is, $a(-1) = -a$.

EXAMPLE B.1.5. Consider the equality $ab = 0$. We know now that if $a = 0$ or $b = 0$, the statement $ab = 0$ is true. Is the converse true? That is, if $ab = 0$ can we conclude that $a = 0$ or $b = 0$?

Suppose that $ab = 0$ but that $a \neq 0$. Then a has a multiplicative inverse a^{-1} and $a^{-1}(ab) = a^{-1} \cdot 0 = 0$. But $a^{-1}(ab) = (a^{-1}a)b = 1 \cdot b = b \cdot 1 = b$. That is, if $ab = 0$ and $a \neq 0$, then $b = 0$. Likewise, we can show that if $b \neq 0$ while $ab = 0$, then $a = 0$.

The exercises below give a means of reviewing your mechanical operations through practice and also afford you the opportunity of proving basic truths through the use of the field properties (usually called the field axioms) and Examples B.1.1–B.1.5. Note that the demonstrations given in the examples were simply collections of mechanical changes, each of which was justified on the basis of a field axiom or previous example.

EXERCISES B.1

1. Without referring to the material of this section, indicate whether true or false for each of the following.

   a. $N \subset Z$ h. $Q \subset I$ o. $R \cap I = \phi$
   b. $N \subset Q$ i. $Q \subset R$ p. $Q \cup I = R$
   c. $N \subset I$ j. $I \subset Q$ q. $0 \in N$
   d. $N \subset R$ k. $I \subset R$ r. $0 \in Z$
   e. $Z \subset Q$ l. $N \cap I = \phi$ s. $0 \in Q$
   f. $Z \subset I$ m. $Z \cap I = \phi$ t. $0 \in R$
   g. $Z \subset R$ n. $Q \cap I = \phi$

2. The additive inverse $(-a)$ for a is sometimes called the *negative* of a. Find $(-a)$ where a is given as

   a. 2   b. 31   c. 0   d. -3 e. -21

3. What real number is its own additive inverse? What two real numbers are their own multiplicative inverses?

4. Let $n \in N$ with $n = ab$, $a \in N$ and $b \in N$. We say that a and b are *factors* of n. If $n \neq 1$ has no factors other than itself and 1, it is *prime* (nonprime natural numbers are *composite*). Find all possible factors for the following:

   a. 1   c. 15   e. 390   g. 828
   b. 2   d. 95   f. 663   h. 3604

5. List all the *prime factors* (factors that are prime numbers) of each number from Exercise B.1.6.

6. A natural number is *even* if it has a factor of 2. (See Exercise B.1.6 for terminology.) List all even prime numbers.

7. Justify, using the six field properties and Examples B.1.1–B.1.5, the following (a, b, and c represent real numbers).

    a. $a = a$
    b. $a = b$ implies that $b = a$
    c. $a = b$ and $b = c$ yields $a = c$

8. If $a + p = a$ for all $a \in R$, $p = 0$.

9. If $ap = a$ for all $a \in R$, $p = 1$.

10. If $a + p = 0$, $p = (-a)$.

11. If $ap = 1$, $p = a^{-1}$.

12. If $a = b \neq 0$, $a^{-1} = b^{-1}$.

13. If $a + c = b + c$, $a = b$.

14. If $ac = bc$, $c \neq 0$, then $a = b$.

15. If $ab = 0$, $b \neq 0$, then $a = 0$.

16. $(a + b)c = ac + bc$.

17. $a = -(-a)$.   (*Hint:* You may use Exercise B.1.10.)

18. $(b^{-1})^{-1} = b$ if $b \neq 0$. (*Hint:* You may use Exercise B.1.11.)

19. $(-a)b = -(ab) = a(-b)$.

20. $(-a)(-b) = ab$.

21. If $a \neq 0$, $b \neq 0$, $(ab)^{-1} = a^{-1}b^{-1}$.

B.2 SUBTRACTION AND DIVISION

In Section B.1 we saw the field structure held by $\{R, +, \cdot, =\}$. The operations of subtraction and division did not enter implicitly. However, the inverse elements $[(-a)$ and $a^{-1}]$ hint at these operations. Let us merely define the operations, discuss them briefly, and then present more exercises.

Definition B.2.1. If a *and* b *are real numbers,* a − b = c, *if* c ∈ R *and* b + c = a.

The obvious connection one expects to find existing is $a - b = a + (-b)$. Since $b + [a + (-b)] = b + [(-b) + a] = [b + (-b)] + a = 0 + a = a$, $a + (-b)$ is indeed the value c specified in Definition B.2.1. Because of this, addition and subtraction are said to be *inverse operations*.

EXAMPLE B.2.1. $a - (-b) = a + (-(-b)) = a + b$.

EXAMPLE B.2.2. $(-a) - b = (-a) + (-b) = (-1)a + (-1)b = (-1)(a + b) = -(a + b)$.

Definition B.2.2. If a, b $\in$ R, b $\neq$ 0, a $\div$ b $=$ a/b $=$ c *where* c $\in$ R *and* bc $=$ a.

Similar to the case for subtraction, $a \div b = a/b$ ought to be ab^{-1}. Clearly, $b(ab^{-1}) = b(b^{-1}a) = (bb^{-1})a = 1 \cdot a = a$. Consequently, division and multiplication are inverse operations. (*Note: Division by zero is not allowed.* Roughly speaking, it is because zero has no multiplicative inverse.)

EXERCISES B.2

1. Subtraction is a function with domain $R \times R$. Is this assertion true or false?

2. Is the following assertion true? R is closed relative to division.

3. Find a multiplicative inverse for each of the following.

   a. 6   b. 1/2   c. 2/3   d. 2   e. .9   f. 0

4. Justify (using definitions, field properties, and Exercises B.1.19 and B.1.20) the following rules of signs:

   a. $(-a)^{-1} = -(a^{-1})$ if $a \neq 0$.
   b. $(-a) \div b = -(a \div b) = a \div (-b)$ if $b \neq 0$.
   c. $(-a) \div (-b) = a \div b$, if $b \neq 0$.

5. Compute (or simplify) the following.

   a. $7 + (-8)$ f. $-9 - (-2)$
   b. $-7 + 8$ g. $21 \div (-3)$
   c. $6 - 8$ h. $(-15) \div 5$
   d. $-6 - 4$ i. $(-15) \div (-3)$
   e. $6 - (-5)$ j. $(3 + 2/3) \div [1 - 1/(3 + 1/4)]$

6. Show that if $a = b$ and $c = d$, $a - c = b - d$. If, in addition, $c = d \neq 0$, $a/c = b/d$.

7. As converses to Exercise B.2.4, show that if $a - c = b - c$, $a = b$ and if $d = c \neq 0$, $a/c = b/d$ implies that $a = b$.

8. Show that $a/b = c/d$ if and only if $ad = bc$.

9. Show that if $b \neq 0$ and $x \neq 0$, $a/b = ax/bx$.

10. The rational number $x = .12121212\ldots$, can be converted to fractional form by the following process. Multiplying x by 100, we have $100x = 12.121212\ldots$. Then

$$100x = 12.121212\ldots$$
$$\underline{-\quad x = -.121212\ldots}$$
$$99x = 12$$
$$x = 12/99 = 4/33$$

Using some variation of the illustrated technique, find a fractional form for each of the following:

a. 0.125
b. 1.0625
c. 1.666 . . .

d. 0.141414 . . .
e. 0.230230230 . . .
f. 3.1111 . . .

g. 0.0121212 . . .
h. 0.9999 . . .
i. 0.249999 . . .

B.3   THE ORDERING "$<$"

At the beginning of section B.1 we mentioned a system called a *complete ordered field*. The field properties have been discussed, and thus we are to take a look at an ordering and eventually the concept of completeness (given in Section 2.1).

The ordering discussed above is a result of the relation induced by "$<$", the relation "is less than." The real numbers are split into three disjoint subsets called the positive reals, the negative reals, and zero. This statement is given in the form of the *law of trichotomy*:

7. If $a \in R$, then precisely one of the following holds:
    a. a is zero.
    b. a is positive.
    c. $(-a)$ is positive.

Any field satisfying a law of trichotomy and statement 8 below is an *ordered field*.

8. If a and b are positive, $a + b$ and ab are positive.

The real number field with "$<$" $(R, +, \cdot, <)$ is an ordered field where $<$ is defined below.

Definition B.3.1.   If a, b $\in$ R, a $<$ b *means that* a $+$ c $=$ b *for some positive number* c. *To write* b $>$ a *is to mean* a $<$ b.

Theorem B.3.1.   The law of trichotomy becomes: if a, b $\in$ R *precisely one of the following holds.*

$$a. \quad a = b$$
$$b. \quad a < b$$
$$c. \quad b < a$$

Proof.   To say that $a < b$ is equivalent to saying that $b - a$ is positive. Thus, if $a = b$, $a - b = b - a = 0$ and both $a < b$ and $b < a$ are false conclusions.

If $a \neq b$, $b - a \neq 0$. Either $b - a$ is positive or it is negative [$b - a$ negative means $-(b - a) = a - b$ is positive]. Thus either $b > a$ or $a < b$ but not both. The theorem therefore holds.

The exercises ask you to establish the fact that "<" gives a transitive relation that is not symmetric and is not reflexive.

EXAMPLE B.3.1. If a is positive, $0 + a = a$, implying that $0 < a$. If b is negative, $-b$ is positive and $b + (-b) = 0$, whence, $b < 0$. That is, positive real numbers are those greater than zero, while negative reals are those less than zero.

EXAMPLE B.3.2. If $a < b$ then there is a positive real number c ($c > 0$) such that $a + c = b$. Then $a + c + d = b + d$ or rewriting, we have $(a + d) + c = b + d$. The last equality shows that $a + d < b + d$.

EXAMPLE B.3.3. Suppose $a < b$ and $c < d$. Then there are positive numbers x and y with $a + x = b$ and $c + y = d$. Now, $(a + x) + (c + y) = b + d$. Rewriting, we find $(a + c) + (x + y) = b + d$. Since $x + y > 0$ (property 8), $a + c < b + d$.

These two examples give rules for the behavior of "<" as regards addition. These are similar to several rules for "adding equals to equals," and so forth.

In terms of the ordering "<" we may define the absolute value of a real number.

Definition B.3.2. *The absolute value* $|x|$ *of* x *is given by*

$$|x| = x \text{ if } x \text{ is positive}$$
$$|x| = 0 \text{ if } x = 0, \text{ or}$$
$$|x| = -x \text{ if } x \text{ is negative}$$

Thus $|x| = x$ if $x \geq 0$ ($x > 0$ or $x = 0$) while $|x| = -x$ if $x < 0$.

EXAMPLE B.3.4. The values $|x|$ and $|-x|$ are equal. This follows since: (1) if $x > 0$, $(-x) < 0$ (law of trichotomy), $|x| = x$ and $|-x| = x$, (2) if $x < 0$, $(-x) > 0$ whence $|x| = -x$ and $|-x| = -x$, and (3) if $x = 0$, $-x = 0$ and $|x| = |-x| = 0$.

EXAMPLE B.3.5. The expression $-|x| \leq x \leq |x|$ holds. This follows from the argument below.

If $x > 0$, $|x| = x$, $-|x| = -x < 0$ so that $-|x| < 0 < x = |x|$. If $x < 0$, $|x| = (-x) > 0$ and $-|x| = -(-x) = x < 0$. Thus $-|x| = x < |x|$. If $x = 0$, $-x = 0 = |x| = -|x|$. Thus in any case $-|x| \leq x \leq |x|$.

EXAMPLE B.3.6. That expression $|x| < y$ holds if and only if $-y < x < y$ is seen below.

The expression $|x| < y$ means either $0 \leqslant x < y$ or $0 < -x < y$. However, $0 < -x < y$ can be interpreted as $-y < x < 0$. In either case, $-y < x < y$ ($y > 0$ is obviously necessary).

For the converse, suppose $-y < x < y$. If $0 < x$, $x = |x|$ whence $|x| < y$ since $x < y$. On the other hand, if $x < 0$, $|x| = -x$. But $-y < x$ can be given as $-x < y$ whereby $|x| < y$.

EXAMPLE B.3.7. As a corollary to previous examples, we find the following to hold: $|x + y| < |x| + |y|$.

Since $-|x| \leqslant x \leqslant |x|$ and $-|y| \leqslant y \leqslant |y|$, we find $-|x| + (-|y|) \leqslant x + y \leqslant |x| + |y|$ which may be restated in the form $-(|x| + |y|) \leqslant x + y \leqslant (|x| + |y|)$. Example B.3.5 dictates the result $|x + y| \leqslant |x| + |y|$.

A number of other "inequalities" involving absolute values are given in the exercises.

EXERCISES B.3

1. Why do each of the following hold?

    a. $7 < 10$ b. $8 < 19$ c. $-20 < 2$ d. $-35 < -20$

2. Find $|x|$ for x given as

    a. 5   c. 0   e. $-6 - 4$
    b. -5 d. $5 - 6$ f. $(-3)(2)$

Give justification to each in Exercises B.3.3–B.3.10.

3. $a < a$ is not true.

4. If $a < b$ then $b < a$ is false.

5. If $a < b$ and $b < c$, then $a < c$.

6. If $a < b$ and $0 < c$, $ac < bc$.

7. If $0 < c$, $1/c = c^{-1} > 0$.

8. If $0 < c$ and $a < b$, $a/c < b/c$.

9. If $c < 0$, $1/c = c^{-1} < 0$.

10. If $a < b$ and $c < 0$, $ac > bc$ and $a/c > b/c$.

11. Using Exercises B.3.3–B.3.10 and Examples B.3.1–B.3.3, determine the numbers x where

    a. $x + 2 < 5$ c. $3x - 1 < x + 5$ e. $x - 1 < 2x - 1$
    b. $3 - x < 6$ d. $3 - 2x < 15 + 4x$ f. $3x + 6 > 3 - x$

Using Examples B.3.4–B.3.7 and similar techniques, give justification to the following Exercises B.3.12–B.3.15.

12. $|x| \geqslant 0$

13. $|xy| = |x| \, |y|$

14. $|x| \geqslant a \geqslant 0$ if and only if $x \leqslant -a$ or $x \geqslant a$.

15. $||x| - |y|| \leqslant |x - y|$ [*Hint:* Write $x = (x - y) + y < x - y(+ y)$.]

16. Find the values of x for which the following hold.

   a. $|x - 1| \leqslant 3$ c. $|1 - x| < 5$
   b. $|x - 1| \geqslant 3$ d. $|1 - x| > 5$

17. Define x^+ to be the larger of x and zero while defining x^- to be the lesser of x and zero. Calculate $x^+ + x^-$ and $x^+ - x^-$.

B.4 THE POSITIVE INTEGERS AND INDUCTION

The most basic system of numbers, the system of natural numbers, finds many uses both practical and theoretical. The following axiom (Peano's postulates) generates, in a very abstract way, this system. Rather than being appalled at its abstractness, we ought to rejoice in being able to use our concept of a function in this novel way. Our knowledge will allow us to unravel the mystery and make clear the meaning.

Axiom (Peano) B.4.1. The set N *of natural numbers contains an element 1 and there is a 1:1 function* s:N $\rightarrow$ N *such that* $1 \notin$ s[N]. *Furthermore, if* S $\subset$ N, S = N *if and only if* $1 \in$ S *and whenever* n $\in$ S, s(n) $\in$ S.

We first observe that $N \neq \phi$ since $1 \in N$. The set N contains more than one element $[1 \notin s[N] \subset N$, whence $1 \neq s(1)]$. A chain of elements is established: $1, s(1), s(s(1)), s(s(s(1))), \dots$. It can be shown that none of the four elements are equal (the same). The natural number 1 is singled out in the sense that 1 is not $s(n)$ for any $n (1 \notin s[N])$.

Is there another element $n \in N$ where $n \notin s[N]$? The answer is negative. The following theorem shows that

$$s[N] = N - \{1\}$$

Theorem B.4.1. N = s[N] $\cup$ {1}.

Proof. Let $S = s[N] \cup \{1\}$. Now $1 \in S$ and $s(n) \in S$ for every

$n \in N$. Thus if $n \in S$, $s(n) \in S$ as well. According to the final statement of our axiom, $S = N$.

The statement "Furthermore, if $S \subset N$, $S = N$ if and only if $1 \in S$ and whenever $n \in S$, $s(n) \in S$" is known as the *axiom of mathematical induction*. We make use of this axiom again in the theorem below.

Theorem B.4.2.   *If* n ∈ N, s(n) ≠ n.

Proof.   Let $S = \{n \in N : s(n) \neq n\}$. The proof will be complete when we show that $S = N$. The set S is seen to be the "truth set" for Theorem B.4.2.

Now $1 \notin s[N]$, whereby $1 \neq s(1)$. That is, $1 \in S$. Now suppose that $n \in S$ (alternatively, $s(n) \neq n$). Does it follow that $s(n) \in S$? To say that $s(n) \in S$ means that $s(s(n)) \neq s(n)$.

We know that $s(n) \neq n$ and that s is a $1:1$ function. Consequently $s(s(n)) \neq s(n)$. No question about it, $s(n) \in S$. By the axiom of mathematical induction, $S = N$.

The technique of proof in Theorem B.4.2 involved two steps:

1.   A demonstration that $1 \in S$.

2.   A demonstration that $s(n) \in S$ *on the assumption that* $n \in S$.

The assumption in 2 is often called the *inductive hypothesis*. That is, $n \in S$ is assumed, not proved (except for $n = 1$).

What is this mysterious function s? It is generally the case that $s(n) = n + 1$. That is, $s(1) = 2$, $s(2) = 3$, $s(3) = 4$, ... This gives a natural order to N: $1 < s(1) = 2 < s(s(1)) = s(2) = 3 < s(s(s(s(1))))$ $= s(3) = 4 < \ldots$. The Peano axiom may be restated as follows.

(Peano.) The set N of natural numbers contains the element 1. Moreover, for each $n \in N$ there is a unique element $n + 1 \in N$ and if $n + 1 = m + 1$, $n = m$. The number 1 is not $n + 1$ for any $n \in N$. Furthermore, if $S \subset N$, $S = N$ if and only if $1 \in S$, and whenever $n \in S$, $n + 1 \in S$ also.

We can derive (from the induction axiom) the *Well-Ordering Principle* for N: if $S \neq \phi$ and $S \subset N$, S has a least element x. (That is, there is an element $x \in S$ such that $x \leqslant y$ for all $y \in S$.)

Theorem B.4.3.   *If* A ⊂ N, *then* A = φ *or* A *has a least element.*

Proof.   We can prove the theorem by showing that if A has no least element, $A = \phi$. Consequently let us assume that A has no least element.

Does $1 \in A$? Clearly, the answer is negative since 1 must be a least element of any set of natural numbers (remember that A

has no least element). Suppose, then, that for $1 \leqslant k \leqslant n$, $k \notin A$. The element $k + 1 \notin A$ since, if it were, it would be the least such element ($x \in A$ implies that $x \neq 1$, $x \neq 2$, ..., $x \neq k$, whence $x \geqslant k + 1$). By the induction axiom, no element belongs to A (we have shown that $N - A = N$). Thus $A = \phi$.

Our inductive hypothesis (for $1 \leqslant k \leqslant n$, $k \notin A$) is different in form from the previous ones. Our set S in this case must be described as $S = \{n \in N : k \notin A \text{ for all } k, \ 1 \leqslant k \leqslant n\}$. This form of the inductive hypothesis is seen quite often.

EXERCISES B.4

1. Define, for $x \in R$, $x^1 = x$ and $x^{n+1} = x^n x$ for each $n \in N$. Use induction to show that:

    a. $x^m x^n = x^{m+n}$ for $m, n \in N$.
    b. $x^m / x^n = x^{m-n}$ for $x \neq 0$, $m, n \in N$, $m - n \in N$.
    c. $(x^m)^n = x^{mn}$ for $m, n \in N$.
    d. $(xy)^n = x^n y^n$, $x, y \in R$, $n \in N$.
    e. $(x/y)^n = x^n / y^n$, $y \neq 0$, $x, y \in R$, $n \in N$.

    These results (a–e) constitute the general mechanical behavior of exponents (even if the exponents are real but not positive integers). They should be known, understood, and remembered.

    f. Show that $(1)^n = 1$ for each $n \in N$.
    g. Using $(-1)^2 = 1$, show that $(-1)^{2n} = 1$ for $n \in N$.
    h. Show that $n < 2^n$ for each $n \in N$.
    i. Show that if $0 < a < 1$, $a \in R$ then $0 < a^n < 1$ for $n \in N$. (*Hint*: $0 < a < 1$ implies $0 < a^2 < a$.)

2. Show that $n(n + 1)$ is even [that is, $n(n + 1) = 2x$ for some $x \in N$] for each $n \in N$.

3. Suppose that $x_1, x_2, \ldots,$ and x_n satisfy $a < x_i < b$. Show then that the *arithmetic mean* $(x_1 + x_2 + x_3 + \cdots + x_n)/n$ of the x's satisfies: $a < (x_1 + x_2 + \cdots + x_n)/n < b$.

4. An *arithmetic progression* is a function $f : N \to R$ such that for some $d \in R$, $f(n + 1) = f(n) + d$. We say that f is defined *inductively* or *recursively* [$f(2) = f(1) + d$, $f(3) = f(2) + d = f(1) + 2d$, $f(4) = f(3) + d = f(1) + 3d, \ldots$]. Each value $f(n)$ is defined in terms of $f(n - 1)$ ($n > 1$). Write out the values $f(1)$ through $f(5)$ for the following arithmetic progressions.

    a. $f(1) = 5$, $d = 2$
    b. $f(1) = 3$, $f(2) = 7$
    c. $f(6) = 19$, $d = -1$
    d. $f(1) = 30$, $f(5) = 60$

Show that the following are true.

e. If f is an arithmetic progression with $f(2) = f(1) = d$, then $f(n) = f(1) + (n-1)\,d$.

f. The progression in part e satisfies $f(1) + f(2) + \cdots + f(n) = na_1 + n(n-1)d/2$.

5. Another recursively defined (see Exercise B.4.4) function-type is the *geometric progression*. A function $f: N \to R$ is a geometric progression if there is an $r \in R$ such that $f(n+1) = rf(n)$. Calculate the values $f(1)$ through $f(5)$ for the following geometric progressions.

a. $f(1) = 6$, $r = 2$
b. $f(1) = 2$, $f(2) = 0$
c. $f(1) = 32$, $f(6) = 1$
d. $f(3) = 27$, $r = 3$

Show that the following hold [f is a geometric progression with $f(n+1) = rf(n)$].

e. $f(n) = r^{n-1}f(1)$.
f. $f(1) + f(2) + \cdots + f(n) = f(1)(1 - r^n)/(1 - r)$

6. Functions $g: N \to R$ are usually called *(real) sequences*. That is, a sequence is a function having domain N. The arithmetic and geometric progressions afford examples. The function s in Peano's Postulates is another example.

a. Give an example of two sequences that are both arithmetic and geometric progressions.
b. Give an example of a sequence whose values approach zero as the function samples arbitrarily large domain elements.
c. Give an example of a sequence whose functional values approach 6 as f samples arbitrarily large domain elements.
d. Give an example of a sequence $f: N \to R$ where $f(n)$ gets arbitrarily large as n becomes large.

7. The following discussion purports to show that everyone has the same color hair. Give a critical analysis of the argument.

We will consider S to be the set of all natural numbers n such that if we form any group of n persons (with hair) everyone in that particular group has the same color hair. We will attempt to show by mathematical induction that $S = N$. Since the number of persons in the world is a positive integer k, $k \in S$. That is, the group of k persons (all people) is such that they all have the same color hair. That $S = N$ is the objective of the following.

Let $n = 1$. It is clear that if G is any group of people having only one person in the group, everyone in G has the same color hair. Thus $1 \in S$.

Now, the *inductive hypothesis* becomes: let n be an integer such that if G is any group consisting of exactly n persons, everyone in G has the same color hair. Examine, then, any set G^* having precisely $(n + 1)$ people.

Imagine if you will that the members of G^* are in a single room. Label the persons in G^* by $1, 2, 3, \ldots, n + 1$. We ask person $(n + 1)$ to leave the room. The room now holds just n persons. By the inductive hypothesis, these n persons (forming a group of n people) all have the same color hair. If there is a person in G^* with a "different color hair" he must be number $n + 1$.

Therefore, we ask $n + 1$ to reenter the room and we escort person 1 out. Again, there are n persons in the room and they all have the same color hair (by the induction hypothesis). That is, the hair color of $n + 1$ agrees with the hair color of the other persons in the room. However, we already know that person 1 has the hair color of those same people. Thus everyone in G^* has the same color hair. That is, $n + 1 \in S$. The induction axiom claims that $S = N$.

B.5 MISCELLANEOUS EXERCISES

1. The field postulates show that the sum $a_1 + a_2 + a_3$ of three real numbers is a real number. Show that if n is any positive integer and $a_1, a_2, \ldots, a_n \in R$, $a_1 + a_2 + \cdots + a_n$ is in R also.

2. If $a \in R$, $a \neq 0$, call $1/a = 1 \div a$ the *reciprocal* of a. What is another name for $1/a$?

3. Suppose $(x - 1)(x - 2) = 0$. What must be true of x?

4. Can you give an example of a real number that is both rational and irrational?

5. Show that if $0 < x < y$, then $0 < x^n < y^n$ for $n \in N$.

6. How is it possible that $ab > 0$? $ab < 0$? Under what conditions on x then is each of the following true?

    a. $(x - 1)(x + 1) > 0$ b. $(x - 1)(x + 1) < 0$

7. Define $n!$ (*n factorial*) in the following recursive manner: $0! = 1$, $1! = 1$, $(n + 1)! = (n + 1)n!$ for $n \in N$. Calculate:

    a. $2!$ d. $5!$
    b. $3!$ e. $(n + 1)!/n!$
    c. $4!$ f. $(n + j)!/n!$ where $j \in N$.

8. For $n, r \in N$, $0 \leqslant r \leqslant n$, define $C(n, r) = n!/r!(n - r)!$ Calculate:

    a. $C(5, 0)$ c. $C(5, 1)$ e. $C(5, 3)$ g. $C(n\,0)$
    b. $C(5, 5)$ d. $C(5, 4)$ f. $C(5, 2)$ h. $C(n, n)$

    Show that: (*i*) $C(n + 1, r) = C(n, r) + C(n, r - 1)$ if $n > r > 1$.

9. Define a to be an *nth root* ($n \in N$) of x if $a^n = x$. Calculate all nth roots indicated.

    a. Second (square) roots of 16   d. Fourth roots of 16
    b. Third (cube) roots of 27   e. Square roots of $1/4$
    c. Cube roots of (-27) f. Square roots of (-1)

10. From Exercise B.5.9 we might observe that each real number has exactly one nth root if n is odd. We can then denote the nth root of a by $a^{1/n}$ if n is odd. However, if n is even, three possibilities exist. If $a > 0$, a has two nth roots, while if $a < 0$, a has no nth root. If $a = 0$, a has exactly one nth root. We therefore have a decision to make if we are to assign to $a^{1/n}$ a unique value. Thus if n is even and $a > 0$, we define: $a^{1/n}$ is the nonnegative nth root of a. Calculate the following:

a. $9^{1/2}$ b. $(-27)^{1/3}$ c. $8^{1/3}$ d. $16^{1/4}$

11. Let n be a negative integer and define x^n to be $1/x^{(-n)}$ for $x \neq 0$. Calculate:

a. 2^{-4} b. 3^{-2} c. $(1/5)^{-3}$ d. $(1/8)^{-1}$

12. Let $x^0 = 1$ if $x \neq 0$ and define $x^{m/n} = (x^{1/n})^m$ for $n \in N$, $m \in Z$. (We assume x is such that $x^{1/n}$ is defined.) Calculate:

a. $8^{2/3}$ b. $27^{4/3}$ c. $16^{-1/2}$ d. 27^0

13. Show that if $a > 1$, $a^{m/n} > 1$ where $m, n \in N$.

14. We may introduce the sigma Σ notation inductively. We define

$$\sum_{i=k}^{i=k} a_i = a_k, \quad \sum_{i=k}^{k+1} a_i = a_k + a_{k+1}$$

and

$$\sum_{i=k}^{k+p+1} a_i = \sum_{i=k}^{k+p} a_i + a_{k+p+1}$$

For example,

$$\sum_{i=0}^{5} i^2 = \sum_{i=0}^{4} i^2 + 5^2 = \sum_{i=0}^{3} i^2 + (4^2 + 5^2)$$

$$= \cdots = 0^2 + 1^2 + 2^2 + 3^2 + 4^2 + 5^2 = 55$$

Calculate:

a. $\displaystyle\sum_{i=0}^{5} (i + 1)$

b. $\displaystyle\sum_{i=1}^{6} (i^2 + i)$. State how this compares with

$$\sum_{i=1}^{6} i^2 + \sum_{i=1}^{6} i$$

c. $\displaystyle\sum_{i=3}^{9} 6i$. State how this compares with $6 \left(\displaystyle\sum_{i=3}^{9} i \right)$.

d. $\displaystyle\sum_{i=1}^{n} a_i$ where $a_1 = 1$ for each indicated i.

e. $\displaystyle\sum_{i=1}^{3} i^3$

Show that the following hold.

f. $\sum_{i=1}^{n} i = n(n+1)/2$

g. $\sum_{i=1}^{n} i^2 = n(2n^2 + 3n + 1)/6$

h. $\sum_{i=1}^{n} i^3 = [n(n+1)/2]^2 = \left[\sum_{i=1}^{n} i\right]^2$

APPENDIX C

Answers to Exercises

EXERCISES 0.1

1. a. True   c. False   e. False   g. True   i. False
   k. False   m. True   o. True   q. True   s. False
   u. False   w. False   y. False   a'. False   c'. True
   e'. False   g'. False   i'. False

2. $P \subset P$ is true.

3. a. A c. F e. F g. A h. $\{0, 1, 3, 5\}$
   k. B m. E o. C q. F s. F

5. a. B c. $\{1, 3\}$ e. A g. None   i. None
   k. B m. C o. C q. D s. E

7. No

8. a. C c. $\{2, 4, 5\}$ e. 0   g. B k. All elements deleted.
   m. All elements deleted.   p. $\{1, 3\}$ r. All elements deleted.

9. If $P \subset Q$, all elements are deleted. If P and Q have no common elements, P remains.

10. a. 0   c. 0, 1, 3   e. B

EXERCISES 0.2

1. a. A c. A e. F g. F i. A
   k. $\{0, 1, 3, 5\}$ m. C o. $\{0, 1, 2, 3, 4\}$ q. E s. D
   u. F w. E y. F

2. $P \cup P = P$

3. a. A c. B e. $\{1,3\}$ g. A i. ϕ
    k. $\{1,3\}$ m. B o. ϕ q. C s. D
    u. D w. E y. E

4. $P \cap P = P$

5. a. ϕ c. ϕ e. $\{2,4,5\}$ g. ϕ i. B
    k. $\{5\}$ m. ϕ p. D r. ϕ t. $\{1,3\}$
    v. C x. ϕ z. ϕ

6. $P - P = \phi$

7. a. $\{0\}$ c. $\{0,1,3,5\}$ e. B g. F

8. b. See Figure C.1.   i. See Figure C.2.   q. See Figure C.3.
    m. See Figure C.4.

9. Yes

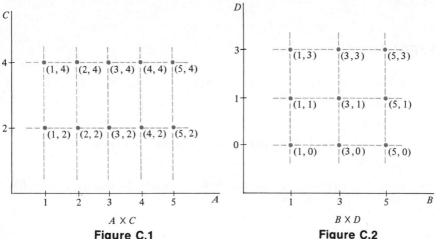

A × C
Figure C.1

B × D
Figure C.2

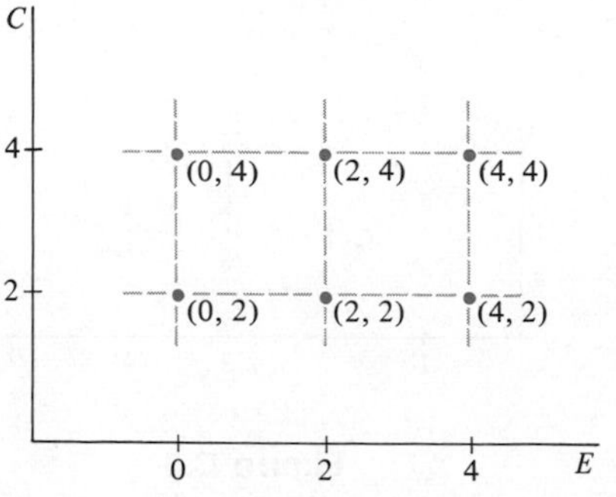

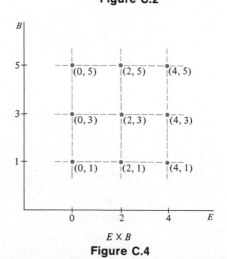

E × C
Figure C.3

E × B
Figure C.4

EXERCISES 0.3

1. a. The horizontally shaded portion.
   d. The portion shaded horizontally along with that shaded vertically.
   g. The portion having any shading.
   h. The portion having vertical and horizontal cross hatching.
   k. The part having all three shadings.
   l. The portion having horizontal but not vertical shading.
   r. That part having either horizontal or vertical shading but not diagonal shading.
   u. That part not having horizontal shading.
   x. That part not having either horizontal or vertical shading.
   a′. That part not having horizontal and vertical cross hatching.

3. 33

EXERCISES 0.4

1. a. See Figure C.5.   f. See Figure C.6.   k. See Figure C.7.
2. See Figure C.8.
3. No, $A \times A$
4. ϕ, $\{(a,a)\}$, $\{(b,b)\}$, $\{(a,a),(b,b)\}$, $\{(a,b),(b,a)\}$, $K \times K$
5. δ_K, $K \times K$
6. ϕ, δ_K, $\{(a,a)\}$, $\{(b,b)\}$, $\{(a,a),(a,b)\}$, $\{(a,a),(b,a)\}$, $\{(b,b),(a,b)\}$, $\{(b,b),(b,a)\}$, $K \times K$

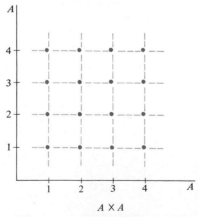

$A \times A$

Figure C.5

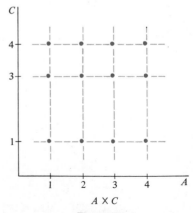

$A \times C$

Figure C.6

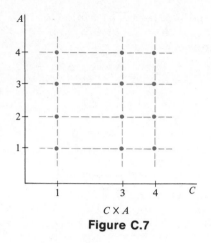

$C \times A$

Figure C.7

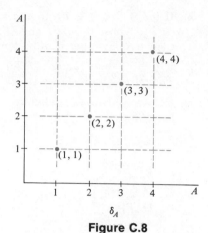

δ_A

Figure C.8

EXERCISES 0.5

1. L is not reflexive since a cannot have different color hair than a. Symmetry is clear.

2. Reflexivity fails since a cannot be taller than himself. If b is taller than a, a *cannot* be taller than b and therefore symmetry fails. Let c be six feet tall with black hair, let b be five feet nine inches tall with brown hair, and let c be five feet six inches tall with black hair. Now, $(a, b) \in M$ and $(b, c) \in M$ but $(a, c) \notin M$ since a and c have the same color hair.

3. $(a, a) \in M \cup F$ since a is as tall as a. Reflexivity follows; nonsymmetry is clear. Let a be six feet tall, let b be five feet nine inches tall, and let c be five feet seven inches tall, have a different color hair than b, and the same color hair as a. Now, $(a, b) \in M \cup F$, $(b, c) \in M \cup F$, but $(a, c) \notin M \cup F$.

5. a. $A \times B$; transitive
    c. $A \times A$, $A \times C$, $C \times A$, $A \times A$; reflexive as a subset of $C \times C$; transitive
    e. $E \times F$ where E is A or C and F is B or D; transitive
    g. $B \times D$; transitive
    i. $E \times F$ where E and F are chosen from $\{A, C, D\}$; symmetric; transitive.
    k. None.

7. b. $A = B$ means $A \subset B$ and $B \subset A$ and $A = A$ implies $A \subset A$. If $A = B$ and $B = C$, we have $A \subset B$, $B \subset A$, $B \subset C$, and $C \subset B$. Using part a. of this problem, $A \subset C$ and $C \subset A$ implies $A = C$. Moreover if $A = B$, $A \subset B$ and $B \subset A$, therefore $B \subset A$ and $A \subset B$ implies $B = A$.

8. If (a, b), $(b, c) \in \delta_A$, then $a = b = c$ and $(a, b) = (b, b)$ while $(b, c) = (b, b)$. Thus, if (a, b) and $(b, c) \in \delta_A$, $(b, b) = (a, c) \in \delta_A$ and δ_A is transitive. Symmetry and reflexivity are clear.

9. If $a \in A$, then $(a, c) \in F$ for some $c \in A$. By symmetry, $(c, a) \in F$. Transitivity then shows $(a, a) \in F$ whence $\delta_A \subset F$.

10. F is an equivalence relation.

REVIEW EXERCISES 0.6

1. a. $B \subset A, B \subset C, D \subset C$ b. No   c. A and D; B and D

2. a. $\{1, 2, 3, a, b, c, d, e\}$ c. Universal set   g. B
    j. $\{b, d\}$ w. D y. $\{b, d\}$

3. $\delta_B = \{(a, a), (c, c), (e, e)\}$

4. c. The set C is the set of all students taking trigonometry or algebra.
    f. F is the set of all algebra students not taking trigonometry.

6. a. False   c. False   e. True   g. True
    i. True   k. False   m. True

7. The Cartesian product is the set of playing cards.

QUIZ 0.7

1. a. x is an element of A.   b. x is not an element of A.
    c. A is the set of all carrots.

2. a. $X \subset Y$ means each element of X is also an element of Y.
    b. $X = Y$ means $X \subset Y$ and $Y \subset X$.

3. a. Yes   b. Yes   c. No   d. No

4. a. A b. $\{a, p, q, 1, 3, \frac{2}{3}\}$ c. C d. C
    e. $\{p, q, \frac{2}{3}\}$ f. $\{p, q, \frac{2}{3}\}$ g. ϕ
    h. $\{(a, 1), (a, 3), (1, 1), (1, 3)(3, 1), (3, 3)\}$
    i. $\{(1, 1), (3, 3), (1, 3), (3, 1)\}$
    j. $\{(p, p), (q, q), (1, 1), (3, 3), (\frac{2}{3}, \frac{2}{3})\}$

5. a. $\delta_C, \{(1, 1), (3, 3), (1, 3)\}, \{(1, 1), (3, 3), (3, 1)\}, C \times C$
    b. $\phi, \delta_C, C \times C, \{(1, 3), (3, 1)\}$
    c. $\phi, \delta_C, \{(1, 1)\}, \{(1, 3)\}, \{(3, 1)\}, \{(3, 3)\}$
    d. $\delta_C, C \times C$

ADVANCED EXERCISES 0.8

1. If $x \in A$, then it is true that $x \in A$ or $x \in B$. Thus, $x \in A \cup B$ and $A \subset A \cup B$.

3. If $x \in A - B$, then $x \notin B$. If $x \in B$, then $x \notin A - B$. Thus, $x \in (A - B) \cap B$ is not possible and $(A - B) \cap B = \phi$.

6. If $\phi \not\subset A$, there must be an $x \in \phi$ with $x \notin A$; this is not possible.

10. If $x \in A \cup (B \cap C)$, then $x \in A$ or x belongs to both B and C. If $x \in A$, then $x \in A \cup B$ and $x \in A \cup C$. If $x \in B$ and C, then $x \in A \cup B$ and $x \in A \cup C$. Thus, if $x \in A \cup (B \cap C)$, then $x \in (A \cup B) \cap (A \cup C)$ $[A \cup (B \cap C) \subset (A \cup B) \cap (A \cup C)]$.

Conversely, if $x \in (A \cup B) \cap (A \cup C)$, then $x \in A$ or $x \in B$ and $x \in A$ or $x \in C$. If $x \in A$, then $x \in A \cup (B \cap C)$. If $x \notin A$, then $x \in B$ and $x \in C$ and therefore $x \in B \cap C$ and $x \in A \cup (B \cap C)$. In either case, $x \in A \cup (B \cap C)$ and $(A \cup B) \cap (A \cup C) \subset A \cup (B \cap C)$.

EXERCISES 1.1

1. a. Every element of X is mated to just one element of Y.
    c. For each $x \in X$ there is exactly one choice for $f(x) \in Y$.
    e. Yes, yes.

2. a. No. b. g, h is not an even function since $-1, -2, \ldots$, are not in A.

3. a. a c. d e. a g. c

4. a. a c. i e. b g. i i. a k. i

EXERCISES 1.2

1. f, h, t, and k have inverses.

2. Only $h: A \to B$ of Exercise 1.1.1 has an inverse.

EXERCISES 1.3

1. a. $0 \to \frac{1}{2}$
        $1 \to \frac{1}{3}$
        $2 \to \frac{1}{4}$
        $3 \to \frac{1}{2}$
        $4 \to \frac{1}{5}$
    b. $f \circ g$ does not make sense. [What is $f(\frac{1}{2})$, for example?]

2. a. a c. a e. $\frac{1}{2}$ g. $\frac{1}{4}$ i. $\frac{1}{2}$ k. $\frac{1}{2}$

3. $h \circ f, h \circ g, h \circ k, k \circ h$

5. From 1.3.1, f has domain X and range Y while g has domain Y and range Z.

EXERCISES 1.4

1. a. See Figure C.9.   c. See Figure C.10.   f. See Figure C.11.
2. f has domain $\{a, b, c, d, e, f, g\}$ and range $\{\alpha, \beta, \gamma, \delta\}$ while H has domain $\{a, b, c, d, e, f, g\}$ and range $\{\alpha\}$.
3. g, T, u.
4. All.
5. g, T, u. See Figure C.12 for a graph of g^{-1}.

EXERCISES 1.5

1. Yes
3. In Exercise 1.1.1, $g|A = \{(0, a), (1, a), (2, a)\}$
4. b. Yes   d. $\{(0, 0), (4, 3), (7, \frac{1}{2})\}$
5. a. 6   c. $\frac{1}{2}$ e. 1   g. 3   i. 3   k. -1
6. a. a c. b e. c g. $*$ i. 1
7. Let $F: A \times (A \times B) \to A \times A \times B$ be given by $F(a, (b, c)) = (a, b, c)$.

REVIEW EXERCISES 1.6

1. a. No   b. Yes   c. Yes   d. No   h. Yes, in each case
    i. $u[X] = \{a, b, c, e\}$, $u[A] = B$, $u^{-1}[Y] = X$, $u^{-1}[B] = \{0, 2, 3, 5, 6\}$, $u^{-1}[u[X]] = X$, $u^{-1}[u[A]] = \{0, 2, 3, 5, 6\}$, $u[u^{-1}[B]] = B$
    k. No.   m. Onto B
    p. $u|A = \{(3, b), (5, a), (6, e)\}$
    q. $(u|A)^{-1} = \{(b, 3), (a, 5), (e, 6)\}$
    s. c u. a w. c y. e
2. b. $x = 0$ e. -3 g. $\frac{1}{3}$ i. 1
    m. 0   o. -1 q. 1

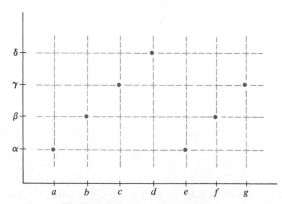

Figure C.9 A graph of f.

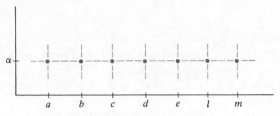

Figure C.10 A graph of *H*.

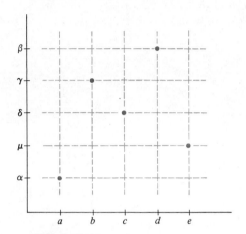

Figure C.11 A graph of *u*.

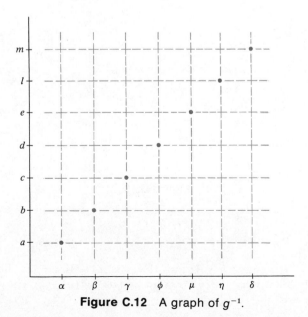

Figure C.12 A graph of g^{-1}.

QUIZ 1.7

1. a. See Figure C.13   b. $f: A \to B$, $g: B \to C$
    c. The domains for f and g are A and B, respectively, while the corresponding ranges are B and $\{1, 2\}$.
    d. If we define $E = \{1, 2, 3\}$, then $k: E \to A$ is a function.
    e. g and k are $1:1$.   f. f and k are onto.
    g. No, any two do not have the same number of elements.
    h. $g \circ f$ and $f \circ k$ are defined.   See Figure C.14.
    i. See Figure C.15, δ_A.
    j. $f | D = \{(\text{Oscar}, \text{Rolls}), (\text{James}, \text{Cad})\}$ k. Rolls, 2

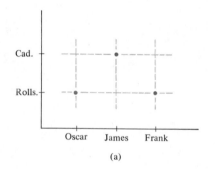

(a)

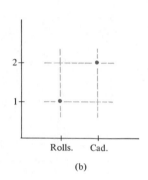

(b)

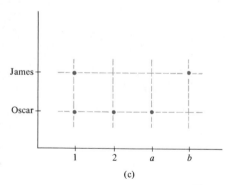

(c)

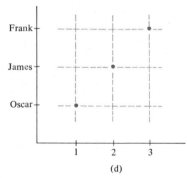

(d)

Figure C.13

ADVANCED EXERCISES 1.8

1. If $x \ne y$, $i(x) = x \ne y = i(y)$ so that i is $1:1$. If $x \in X$, then $x = i(x)$ and i is onto. If $x = i^{-1}(y)$, $x = i(x) = [i(i^{-1})](y) = y$. Thus $i^{-1}(y) = y$ and $i^{-1} = i$.

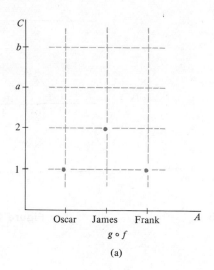

$g \circ f$

(a)

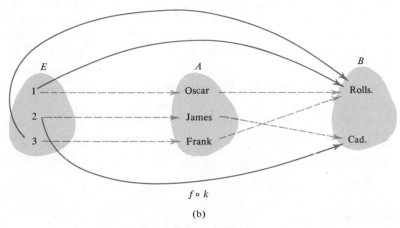

$f \circ k$

(b)

Figure C.14

3. If $f: X \to Y$ has an inverse f^{-1} and $(x, y) \in f$ and $(z, y) \in f$, then $(y, x) \in f^{-1}$ and $(y, z) \in f^{-1}$. But, by the definition of a function, $x = z$. Thus, f is $1:1$. If $y \in Y$, then $f^{-1}(y) = x$ for some $x \in X$. Thus $f(x) = y$ and f is an onto function.

6. a. Yes.
    b. $(a, b) \in F$ implies that a and b have the same remainder, say x, upon division by 7; x can be chosen to be among $\{0, 1, 2, 3, 4, 5, 6\}$. Then $(a, b) \in F$ implies (a, b) is among $\boxed{0} \cup \boxed{1} \cup \boxed{2} \cup \boxed{3} \cup \boxed{4} \cup \boxed{5} \cup \boxed{6}$.
    c. If $\boxed{a} = \boxed{b}$, then $a - b$ has remainder 0 upon division by 7 [$a = 7x + y$ and $b = 7z + y$ for some x, y, and z. Thus, $a - b = 7(x - z)$].
    g. $\boxed{6}$ i. $\boxed{4}$

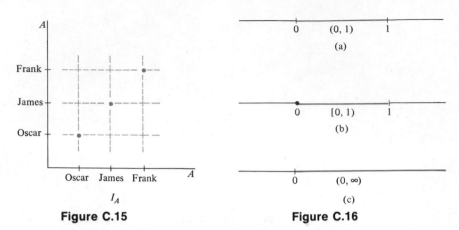

Figure C.15   **Figure C.16**

EXERCISES 2.1

1. Since $0 < 1/n$ for each positive integer n, 0 is a lower bound for A. If $x > 0$ there is an n so that $0 < 1/n < x$. Thus, x is not a lower bound for A. 0, therefore, is the largest lower bound. Similarly, $1 \geqslant 1/n$ for any positive integer n implies 1 is an upper bound for A, if $x < 1$, x is not an upper bound for $A (1 \in A$ and $1 > x)$. Thus, 1 is the smallest upper bound for A.

2. See Figure C.16 for graphs for a., c., and e.

3. a. g.l.b. $= 0$, l.u.b. $= 1$ c. Same as a.
   e. g.l.b. $= 0$, no upperbound.   g. Same as e.

4. g.l.b. $A = 1$; l.u.b. $A = 2$

5. g.l.b. $A = 0$; l.u.b. $A = 2$

6. a. $(0, 2)$ c. $(0, \frac{1}{2}]$

EXERCISES 2.2

1. a. -1 c. 0 e. 1, [ ] is neither even nor odd.

2. $[x] = n - 1$ for $x \in [n - 1, n)$; [ ] does not have an inverse.

3. R

4. See Figure C.17 for a., c., e., g., and i.

5. See Figure C.18 for a. and c.

6. See Figure C.19.

7. No.

EXERCISES 2.3

1. a. 0 c. -2 e. 1 g. 3

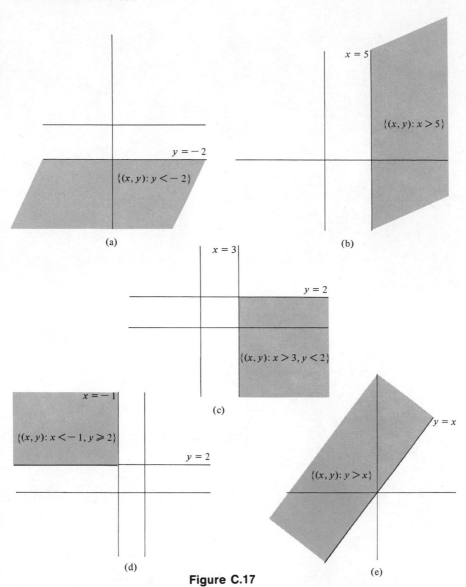

Figure C.17

2. See Figure C.20.

3. See Figure C.21 for b. and c.

4. a. 1   c. 1/3   e. 3   g. 5   i. −3   k. −5

5. $x + n(x) = 0$, when $n(x) + x = 0$ and since n is $1:1$, $n(n(x)) = x$. Similarly, $1 = xr(x) = r(x)x$ and since r is $1:1$, $x = r(r(x))$.

6. Since addition is unique, to every pair (a, b) of real numbers m associates a unique number $m(a, b) = a + (-b)$. Let $c \in R$, then $m(a, a - c) = a + [-(a - c)] = a + [-(a + (-c))] = a + [-a + c] = c$.

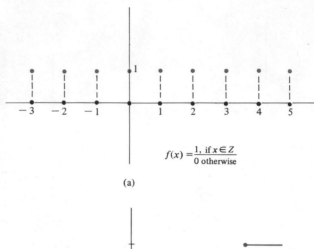

$$f(x) = \frac{1, \text{ if } x \in Z}{0 \text{ otherwise}}$$

(a)

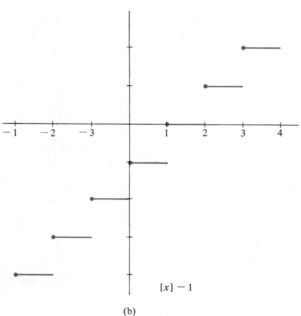

$[x] - 1$

(b)

Figure C.18

8. a. See Figure C.22.   c. i_P e. Even

9. a. 0   c. −1   e. 1   g. 2   i. 0
    k. 0   m. −4   p. 1   r. 9   t. −1/3
    v. 6   x. Not defined   z. 20   b′. −2   d′. 1

10. They do not have the same domains.

11. See Figure C.23 for e. and f.

12. a. 54   c. $-\frac{5}{8}$

13. See Figure C.24 for a. and c.

EXERCISES 2.4

1. a. 5   c. 10
2. a. 5   c. $\sqrt{2}$ e. $6\sqrt{2}$ g. $3\sqrt{2}$
3. Right triangle only.   c. Right isosceles triangle.
4. a. $(x-1)^2 + (y-1)^2 = 100$ c. $(x-1)^2 + (y+1)^2 = 9$
    e. $(x-4)^2 + (y+2)^2 = 4$
    g. $(x-2)^2 + (y-3)^2 = [d((2,3),(6,6))]^2 = 25$

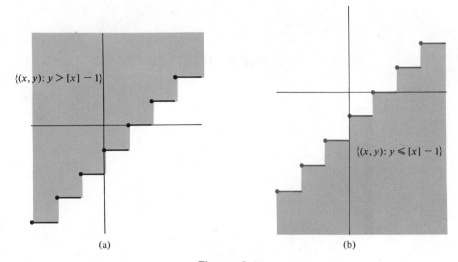

$\{(x,y): y > [x] - 1\}$

$\{(x,y): y \leqslant [x] - 1\}$

(a)   (b)

Figure C.19

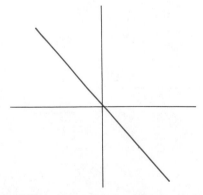

Figure C.20   A graph of n where $n(x) = -x$.

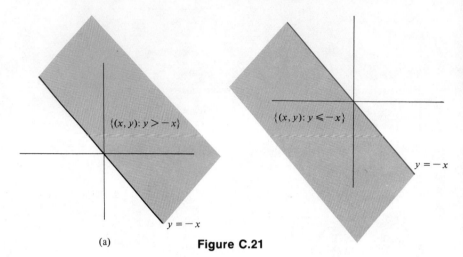

(a)   **Figure C.21**

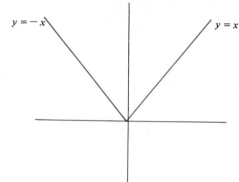

Figure C.22   A graph of $y = |x|$.

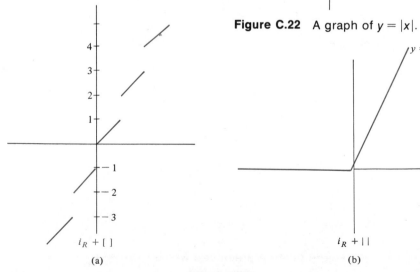

(a)   (b)

Figure C.23

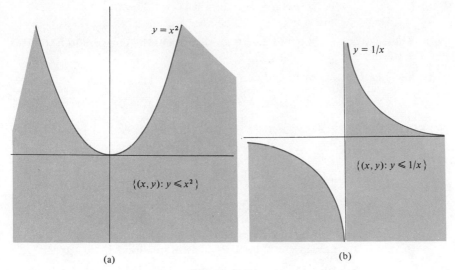

(a)   (b)

Figure C.24

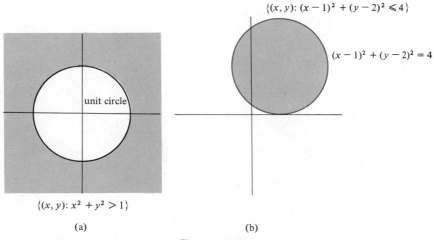

(a)   (b)

Figure C.25

5. Begin with $(x-1)^2 + (y-3)^2 = (x+1)^2 + (y-2)^2$ and remove the squared terms.

6. Begin as in 5.

7. a. $\pi/2$ c. $3\pi/2$ e. $\pi/4$ g. $2\pi/3$ i. $5\pi/6$
    k. $5\pi/4$ m. $5\pi/3$ o. $11\pi/6$

8. a. $-3\pi/2$ c. $-\pi/2$ e. $-7\pi/4$ g. $-4\pi/3$ i. $-7\pi/6$
    k. $-3\pi/4$ m. $-\pi/3$ o. $-\pi/6$

9. a. $5\pi/2$ c. $15\pi/2$ e. $5\pi/4$ g. $10\pi/3$ i. $25\pi/6$
   k. $25\pi/4$ m. $25\pi/3$ o. $55\pi/6$

11. The answers in 2.4.9 and 2.4.10 are five times the answers in Exercises 2.4.7 and 2.4.8.

12. See Figure C.25 for b. and c.

EXERCISES 2.5

1. a. $(1, 0)$ c. $(1, 0)$ e. $(1, 0)$

2. a. If $b \in [0, 2\pi)$, then $b = a$.

3. a. Since $0 < 5\pi/24 < \pi/2$, $\alpha(5\pi/24)$ is in quadrant I.
   c. Since $\pi < 17\pi/16 < 3\pi/2$, $\alpha(17\pi/16)$ is in quadrant III.
   e. Since $3\pi/2 < 23\pi/12 < 2\pi$, $\alpha(23\pi/12$ is in quadrant III.
   g. Since $0 < 24\pi/53 < \pi/2$, $\alpha(24\pi/53)$ is in quadrant I.

4. a. Since $\alpha(\pi/6) = (\sqrt{3}/2, 1/2)$, $\alpha(-\pi/6) = (\sqrt{3}/2, -1/2)$.
   c. Since $\alpha(\pi) = (-1, 0)$, $\alpha(-\pi) = (-1, 0)$ also.
   e. Since $\alpha(3\pi/2) = (0, -1)$, $\alpha(-3\pi/2) = (0, 1)$.
   g. Since $\alpha(11\pi/6) = (\sqrt{3}/2, -1/2)$, $\alpha(-11\pi/6) = (\sqrt{3}/2, 1/2)$.

5. a. IV   c. II

6. a. $(1, 0)$ c. $(1/2, \sqrt{3}/2)$ e. $(-1/2, \sqrt{3}/2)$
   g. $(-\sqrt{3}/2, 1/2)$ i. $(-1/\sqrt{2}, -1/\sqrt{2})$ k. $(0, -1)$
   m. $(1/\sqrt{2}, -1/\sqrt{2})$ o. $(\sqrt{3}/2, -1/2)$

7. a. That arc in quadrant I together with $(1, 0)$ and $(0, 1)$.
   c. All except that arc in quadrant IV.
   e. That arc in quadrants II and III with $(0, 1)$ and $(0, -1)$.
   g. That arc in quadrant III with $(-1, 0)$ and $(0, -1)$.
   i. That arc in quadrant IV together with $(0, -1)$ and $(1, 0)$.

8. a. $\alpha[[9\pi/4, 11\pi/4]]$, $\alpha[[-7\pi/4, -5\pi/4]]$ and $\alpha[[3\pi/4, 9\pi/4]]$ are examples.

9. a. $\alpha(2\pi - a) = \alpha((2\pi - a) + 2\pi) = \alpha(-a)$
   c. $\alpha(a + \pi) = \alpha((a + \pi) + 2n\pi) = \alpha(a + (2n + 1)\pi)$

10. a. $p_1 \circ \alpha(0) = p_1(1, 0) = 1$ b. $p_2 \circ \alpha(0) = p_2(1, 0) = 0$
    c. $(1/p_1 \circ \alpha)(0) = 1/p_1(1, 0) = 1$ d. $(1/p_2 \circ \alpha)(0)$ is not defined.
    e. $(p_1 \circ \alpha/p_2 \circ \alpha)(0)$ is not defined.
    f. $(p_2 \circ \alpha/p_1 \circ \alpha)(0) = 0/1 = 0$
    g. $[(p_1 \circ \alpha)^2 + (p_2 \circ \alpha)^2](0) = 1^2 + 0^2 = 1$

12. a. If $\alpha(a) = (3/5, 4/5) = (p_1 \circ \alpha)(a) = 3/5$, $(p_2 \circ \alpha)(a) = 4/5$, then $(1/p_1 \circ \alpha)(a) = 5/3$, $(1/p_2 \circ \alpha)(a) = 5/4$, $(p_1 \circ \alpha/p_2 \circ \alpha)(a) = 3/4$, $(p_2 \circ \alpha/p_1 \circ \alpha)(a) = 4/3$, and $[(p_1 \circ \alpha)^2 + (p_2 \circ \alpha)^2](a) = (3/5)^2 + (4/5)^2 = 1$.

14. A period for $p_1 \circ \alpha$ is 2π, while a period for $p_1 \circ \alpha / p_2 \circ \alpha$ is π (2π is also a period in this case). See Exercise 2.5.15 also.

17. a. $(1/\sqrt{2}, 1/\sqrt{2})$ c. $(5/13, 12/13)$, $(12/13, 5/13)$
     e. $(7/25, 24/25)$, $(24/25, 7/25)$

18. a. $d(\alpha(7\pi/12), (0, 1)) = d(\alpha(\pi/12, (1, 0))$,   hence
         $\alpha(7\pi/12) = ((1 - \sqrt{3})/2\sqrt{2}, (1 + \sqrt{3})/2\sqrt{2})$
     c. $\alpha(143\pi/12) = \alpha(12\pi - \pi/12) = \alpha(-\pi/12) = ((\sqrt{3} + 1)/2\sqrt{2}$,
         $(1 - \sqrt{3})/2\sqrt{2})$

19. See Figure C.26 for a sketch of $p_2 \circ \alpha$.

EXERCISES 2.6

1. See Figure C.27 for d.

2. a. 21   c. 9/2   e. $-8/3$ g. $-3\sqrt[3]{9}$

3. $f(x) = -f(-x)$ since $f(x) = x^{2n+1}$ is an odd function. Let $y = f(x)$, then
    $f^{-1}(y) = f^{-1}[f(x)] = x = -(-x) = -f^{-1}[f(-x)] = -f^{-1}(-y)$.

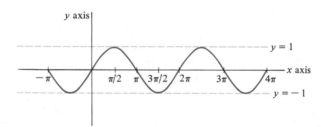

Figure C.26

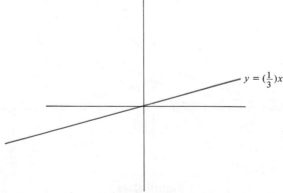

Figure C.27 A graph of f^{-1} where $f(x) = 3x$.

EXERCISES 2.7

1. See Figure C.28 for a. and b.

2. a. $2^1 = 2$, $(1/2)^1 = 1/2$, $5^1 = 5$ c. $2^2 = 4$, $(1/2)^2 = 1/4$, $5^2 = 25$
    e. $2^{1/2} = \sqrt{2}$, $(1/2)^{1/2} = 1/\sqrt{2}$, $5^{1/2} = \sqrt{5}$

3. See Figure C.29.

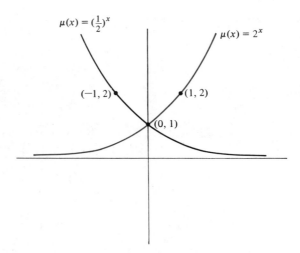

Figure C.28

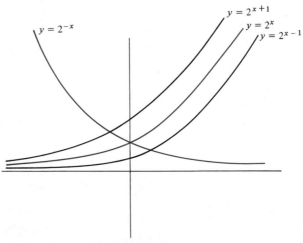

Figure C.29

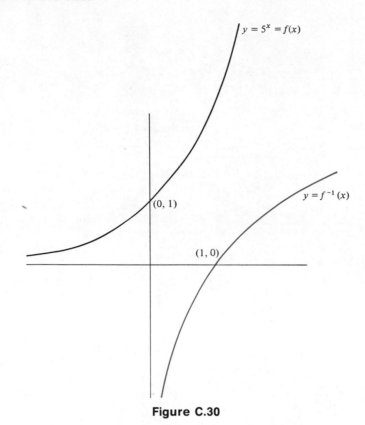

$y = 5^x = f(x)$

$(0, 1)$

$y = f^{-1}(x)$

$(1, 0)$

Figure C.30

4. See Figure C.30.
5. a. 1   c. 5   e. 25   g. 125   i. 625
6. a. 0   c. 1   e. 2   g. 3   i. 4
7. $a = f(x + 1)/f(x)$ if $a \neq 0$
8. a. $x = \frac{1}{2}$ c. $x = 2$ e. $x = 1$
9. a. $x < 0$ c. $x < 0$
10. See Figure C.31 for a. and c.

EXERCISES 2.8

1. a. 0   c. 4   e. −10   g. 32
2. a. 5   c. 25

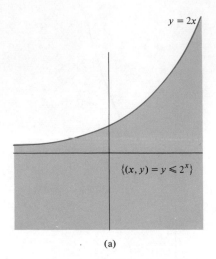

$$\{(x, y) = y \leqslant 2^x\}$$

(a)

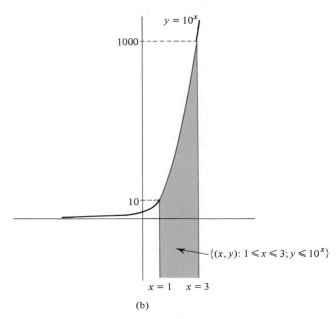

$$\{(x, y): 1 \leqslant x \leqslant 3; y \leqslant 10^x\}$$

$x = 1 \quad x = 3$

(b)

Figure C.31

4. a. .602   c. .125   e. 301/477   g. 301/954
    i. 2   k. 10

5. a. 3.845   c. $-2 + .845 = 8.845 - 10$ e. $n + .845$

6. a. $x = 64$ c. $x = 16$ e. $x = 2$

7. a. $x \leqslant 3x - 1$ or $x \geqslant \frac{1}{2}$ c. $x + 1 \geqslant 2x - 2$ or $3 \geqslant x$

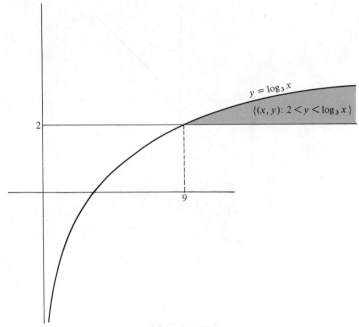

$y = \log_3 x$

$\{(x, y): 2 < y < \log_3 x\}$

2

9

Figure C.32

8. See Figure C.32 for a.

EXERCISES 2.9

1. A is bounded above means A has a least upper bound.

3. a. $b \geqslant c$ c. Always unless $a \leqslant c \leqslant d \leqslant b$ with $a \neq c$ or $b \neq d$.

4. a. ϕ

6. None are bounded below.

7. a. 13   c. 25

8. See Figure C.33 for b.

10. The functions in a. and c. are $1:1$ while each is onto.

11. If $f(a, b) = a^2 - b^2$, write $h(a, b) = a + b$ and $k(a, b) = a - b$. Therefore $f(a, b) = h(a, b) k(a, b)$.

13. The functions in a., b., and c. are bounded and even. The function in d. is not bounded and is odd. None are $1:1$ and none are onto.

15. a. $(x - 2)^2 + (y + 3)^2 = 4$

16. a. 2π

17. a. $(-1/\sqrt{2}, -1/\sqrt{2})$ c. $(-1, 0)$ e. $(1/\sqrt{2}, 1/\sqrt{2})$

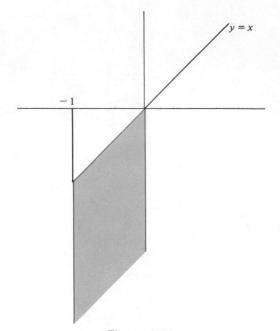

Figure C.33

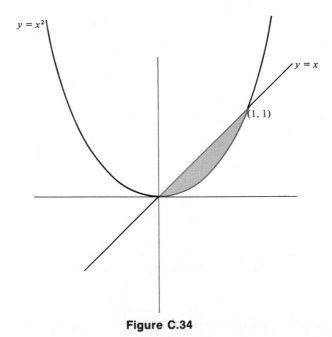

Figure C.34

18. a. $\{x: x = \pi/4 + 2n\pi,\ n \in Z\}$ c. $\{x: x = 5\pi/4 + 2n\pi,\ n \in Z\}$
     e. $\{x: x = (2n + 1)\pi/4,\ n \in Z\}$ g. $\{x: x = n\pi,\ n \in Z\}$

19. For a.: $f^{-1}[\{0\}] = Z$

20. a. $[0, 1)$ c. $\{0\}$

21. a. 0   c. 2   e. 2   g. 0   i. 8
     k. 2   m. x^2 o. 2   q. 4

22. a. $\{-2\}$ c. $\{0, -1\}$ e. $\{x^{1/4}, -x^{1/4}\}$ if $x \geqslant 0$.

23. a. 6   c. -4 e. $\frac{3}{2}$ g. a

24. See Figure C.34 for b.

QUIZ 2.10

1. Yes, No, Yes, Yes; Not unless the domain has only one element.

2. a. $(a, b) = \{x \in R: a < x < b\}$ b. $(-\infty, b) = \{x \in R: x < b\}$
    c. $|x| = x$ if $x \geqslant 0$ and $|x| = -x$ if $x < 0$.
    d. $[x] = n$ where $n \in Z$ and $n + 1 > x$, $n \leqslant x$.
    e. $f(x + p) = f(x)$ for each x in the domain of f.
    f. There are numbers m and M so that if $x \in A$, $m \leqslant f(x) \leqslant M$.

3. a. $x, x + h, 1$ (if $h \neq 0$)   b. $x^2, (x + h)^2, 2x + h$ (if $h \neq 0$)
    c. $k, k, 0$ (if $h \neq 0$)   d. $\sqrt{3}/2$ e. -1 f. 4   g. 12
    h. $-k$ i. k j. $25/k$ if $k \neq 0$

4. See Figure C.35.

5. a. $x = 2$ b. $x = 1$

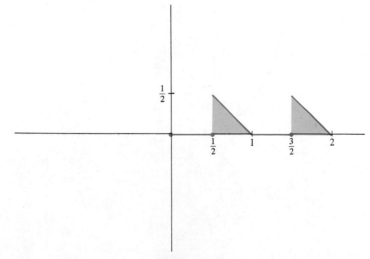

Figure C.35

EXERCISES 2.11

4. a. 55   c. 100   e. 8
6. a. The proof follows from the fact that $\sqrt{x} = 0$ if and only if $x = 0$.
    c. $\sqrt{26}$ e. $\sqrt{91}$
7. a. 3   d. 12

EXERCISES 3.1

1. The expressions in a., b., d., and e. represent monomials.
2. All except c. and i. represent polynomials. All except c. represent rational functions.
3. Each of the following is for f as in b. and g., respectively:
    a. $0, 0$ c. $-2, -6$ e. $16, 48$
5. a. 5   d. 0   f. 2   g. 3   h. 1
6. The coefficients of $6(\ )^5$, $2x^3$, $(\ )^0$, 0, and $2x$ are, respectively: 6, 2, 1, 0, and 2.
7. a. $2(\ )^2 + 2(\ )^1 + 9$
    c. $-2(\ )^3 + 11(\ )^2 + 19(\ )^1 + 14$
    e $-2(\ )^2 - 3(\ )^1 + 5$
    g. 7   i. 6   k. 8   m. 13/2   o. 15/2

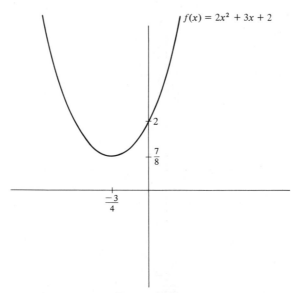

$f(x) = 2x^2 + 3x + 2$

Figure C.36

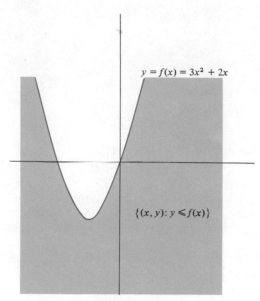

$$y = f(x) = 3x^2 + 2x$$

$$\{(x, y): y \leqslant f(x)\}$$

Figure C.37

8. See Figure C.36 for a graph of f.

9. See Figure C.37 for $f(x) = 3x^2 + 2x$.

11. Zero is a zero for $15x^5 + 2x^2 + 7x$. One is a zero for $3x^3 - 3x^2 + 2x + 2 - 4$ and $x^7 + 3x^6 - 4x^5 - 2x^3 + 2$.

12. a. Not greater than the larger of m and n.   c. $n + m$
    e. $n \cdot m$

13. a. $x = -2$ c. $x = 0$ e. $x = 1/3, \ x = -1/3$
    g. $x = 3, \ x = 4, \ x = 1$

EXERCISES 3.2

1. a. $(x - 5)(x^2 + x + 1) + 0$ c. $(x - 1/2)(2x^2 - 2x - 12) + 0$
    e. $(x - 2)(x^3 + 2x^2 + 4x + 8)$ g. $(x + 1)(x^2 - x - 1) + 2$
    h. $\beta^2 - 4\alpha\gamma \geqslant 0$

2. h. $\alpha x^2 + \beta x + \gamma = \left(x + \dfrac{\beta - \sqrt{\beta^2 - 4\gamma\alpha}}{2\alpha}\right)\left(x + \dfrac{\beta + \sqrt{\beta^2 - 4\gamma\alpha}}{2\alpha}\right)$

3. The following are answered for $x^4 + (1/2)x^3 - x - 1/2$:
    a. $1/81 + 1/54 - 1/3 - 1/2 = -130/162 = -65/81$
    c. $455/2$ e. 126910

4. Synthetic division becomes easier as the powers increase and the value of a is not zero or ± 1.

5. a. $k = -10$ c. $k = 756$ e. $k = -8$

EXERCISES 3.3

1. a. $-7/3$ c. $3/2$ e. -1
3. a. 7 c. 1 e. $1/3$
4. a. 3 c. $-2/3$ e. $1/3$
5. The following are for $f(x) = (x + 1)/3$:
    a. 5   b. -4 c. 2   d. 0
6. a. $y - 4 = (-1/2)(x + 3)$ c. $y = 1$ e. $y - 2 = 4(x - 1)$
    g. $y + 4 = 2(x + 1)$ i. $y - 1 = 2(x - 1)$
    k. $y = -2$ m. $y = 1 - x$
7. a. $(-8, -23)$ c. $(0, 1)$
10. $y - 2 = (3/4)(x - 8)$
12. See Figure C.38 for a., c., and e.
13. m

EXERCISES 3.4

1. a. 0   c. ± 6 e. 4, 3   g. -1 i. $-2 \pm \sqrt{5}$
    k. $(1/2)(1 \pm \sqrt{29})$ m. No real zeros   o. No real zeros

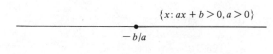

$\{x: 3x - 1 < 2\}$

1

(a)

$\{x: -3 - 2x \geqslant -5 - x\}$

2

(b)

$\{x: ax + b > 0, a > 0\}$

$-b/a$

(c)

Figure C.38

2. a. $5xx$ c. $(x+6)(x-6)$ e. $-(x-3)(x-4)$
   g. $(x+1)^2$ i. $(x+2+\sqrt{5})(x+2-\sqrt{5})$

3. a. $-6, 1$ c. None   e. None

4. a. $(-42/5, 864/25)$ c. $(4, 42)$ and $(-1, 2)$
   e. $[(1/4)(-1+\sqrt{17}), \ 15/8-3\sqrt{17}/8]$ and
      $[(1/4)(-1-\sqrt{17}), \ 7/4+3\sqrt{17}/8]$

5. Let $r_1 = (-b + \sqrt{b^2-4ac})/2a$ and $r_2 = (-b - \sqrt{b^2-4ac})/2a$ and compute $r_1 + r_2$ and $r_1 r_2$.

6. For $P(x) = 5x^2$, a minimum of zero occurs at $x = 0$. For $P(x) = x^2 + 5x + 6$, a minimum of $-1/4$ occurs at $x = -5/2$. For $P(x) = 3x^2 - x - 2$, a minimum of $-75/36$ occurs at $x = 1/6$. For $f(x) = -x^2 + 4x - 5$, a maximum of -1 occurs at $x = 2$.

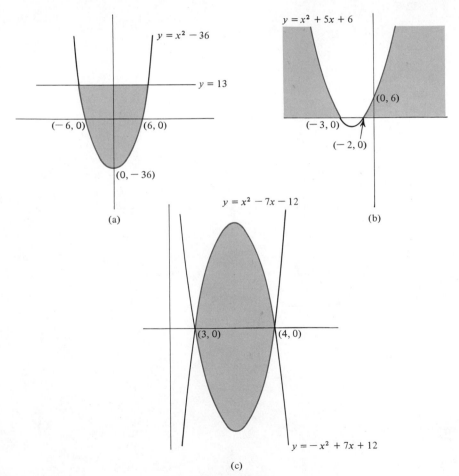

Figure C.39

7. a. ϕ, R c. $(-3,-2), (-\infty,-3) \cup (-2, \infty)$ e. ϕ, R
    q. ϕ, R

8. a. $2, -7/3, 1/3$ c. $2, 5, -8$ e. $2, 5, 3$

10. b. ϕ c. $(-1, 0)$ e. $(-6, 1)$ q. $R - \{0\}$
     i. $R - [-5, 5] = (-\infty, -5) \cup (5, \infty)$

11. See Figure C.39 for b., d., and f.

12. a. $(-3, -3), (-2, -2)$ c. None

13. a. See Figure C.40. $p_1[A] = [-3, -2]$ $p_2[A] = [-3, -2]$

14. $84x^2 + 100y^2 = 525$

15. a. argcosh $x = \ln [x + \sqrt{x^2 - 1}]$
     c. argcoth $x = (1/2) \ln [(x + 1)/(x - 1)]$

16. 22 feet

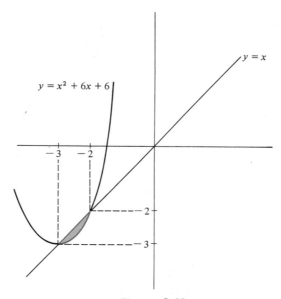

Figure C.40

EXERCISES 3.5

1. a. 1   c. 3 or 1   e. 2 or 0   g. 3 or 1

2. a. 2 or 0   c. 3 or 1   e. 1   g. 1

3. a. None   c. 0   e. None   g. 2

6. a. Root between 0 and 1   c. Roots 0, $\sqrt{2}$, $-\sqrt{2}$
    e. Roots between 2 and 3, 0 and 1, -2 and -1.

8. For $-x$ replacing x, we have all coefficients preceded by a "$-$" sign and there is no change of signs.

10. If the coefficient of the highest power of x is positive, the value of the polynomial grows large positively as x gets larger, and the value grows large negatively as x becomes increasingly negative. That is, the polynomial must be positive for some x and negative for some other x. The polynomial must be zero between any such values of x. If the "leading coefficient" is negative, the value of the polynomial grows large negatively as x becomes large, and positive while it grows large positively as x takes on large negative values.

11. $3, \frac{1}{2}, -2$

13. $1, -1, -2 \pm \sqrt{3}$

15. $2, -2$

17. Locate zeros between -1 and 0 and 1 and 2.

EXERCISES 3.6

1. a. $\frac{3}{2}$ c. None   e. $5, \frac{1}{3}, -\frac{1}{2}$ g. None

2. a. 5   c. 0

4. a. 1; 3 or 1   c. 4, 2, or 0; 4, 2, or 0   e. 4, 2, or 0; 1

5. a. $6 = x$ c. $y - 3 = 2(x - 2)$ e. $y + 2 = (-\frac{2}{3})(x - 3)$
    g. $y = (\frac{1}{2})x + 2$

6. $l_1: y - x_2 = \dfrac{z_2 - x_2}{z_1 - x_1}(x - x_1)$ $l_2: y - \dfrac{z_2 + x_2}{2} = m\left(x - \dfrac{x_1 + z_1}{2}\right)$
    Lines l_2, l_3, and l_4 will be parallel if and only if they are all horizontal.
    No.

7. a. m and m' are different.   d. $3y + x = 6$ e. $2y + x = 0$
    g. $y = 2$ and $y = -8$

8. a. $y^2 = 20x$

9. a. $\{(0, -2),\ (6/5, 8/5)\}$

15. a. $r_1 \cap r_2$ where $r_1 = \{(x, y): x_1 = 0, 1 \leqslant y\}$ and
        $r_2 = \{(x, y): x = 0, y \leqslant 5\}$.
    c. $r_1 \cap r_2$ where $r_1 = \{(-5, y): y \geqslant 4\}$ and $r_2 = \{(-5, y): y \leqslant 12\}$.
    e. $r_1 \cap r_2$, where r_1 emanates from $(2, 4)$ and passes through $(1, 6)$
        while r_2 emanates from $(1, 6)$ and passes through $(2, 4)$.

QUIZ 3.7

1. $(x - 2)(x^4 + 2x^3 + 6x^2 + 13x + 27) + 51$
2. 2 or 0; 3 or 1
3. $\pm 1, \pm 3$; No
4. No
5. Yes
6. Try 4/10
7. $\left(\dfrac{1 + \sqrt{6}}{5}, \dfrac{3\sqrt{6} - 32}{5}\right)$, $\left(\dfrac{1 - \sqrt{6}}{5}, \dfrac{-3\sqrt{6} - 32}{5}\right)$
8. See Figure C.41.
9. $2 < x < 8$

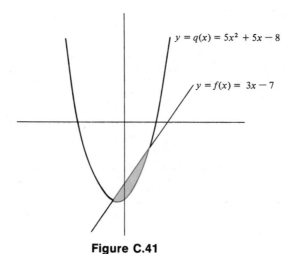

$y = q(x) = 5x^2 + 5x - 8$

$y = f(x) = 3x - 7$

Figure C.41

ADVANCED EXERCISES 3.8

2. $f(x) = c + \left(\dfrac{d - c}{b - a}\right)(x - a)$
3. Use $g(x) = f(x)$ for $x \in (a, b)$ and f from Exercise 3.8.2 and $g(a) = c$.
5. Use $f(x) = x$ for $x \in (a, b)$ with $f(a) = b$.
8. $(a, b) \sim [a, b) \sim [a, b]$

EXERCISES 4.1

1. Vertical asymptote $x = 3$; horizontal asymptote $y = 0$; No.

3. Zero at $x = 2$; vertical asymptotes $x = 3$ and $x = -3$; horizontal asymptote $y = 0$; Yes.

5. Zero at $x = 2$; vertical asymptote $x = 1$; oblique asymptote $y = x - 4$; Yes.

7. Vertical asymptotes $x = \frac{1}{2}$ and $x = -3$; horizontal asymptote $y = 3$; Yes.

9. Zeros at $x = 2$, $x = 1$; vertical asymptotes $x = 4$ and $x = -1$; horizontal asymptote $y = 1$; No.

11. If a rational function is bounded, it has no vertical asymptotes although it may have poles; it must have a horizontal asymptote.

12. a. $(1, 1)$ c. None   e. $(\frac{1}{2}, \frac{5}{2})$

13. See Figure C.42 for a., c., and e.

14. For $f(x) = 1/(x - 3)$; $-1/(x - 3 + h)(x - 3)$

EXERCISES 4.2

1. $1/3(x - 2) - 1/3(x + 1)$

3. $1/4(x - 1) + 3/4(x + 3)$

5. $1/(x - 1) - (x + 1)/(x^2 + x + 1)$

7. $3 + 1/(x + 1) + 12/(x - 3)$

9. $1/8(x - 1) - 1/8(x + 1) - 1/4(x + 1)^2 - 1/2(x + 1)^3$

EXERCISES 4.3

1. $x \geq 3$; $x = 7$ is a zero.

3. $x \geq 0$; No zeros.

5. $-6 \leq x \leq 7$; $x = 3$ and $x = -2$ are zeros.

7. $-4 \leq x < 21$, $x \neq 0$; No zeros.

9. Domain of r is ϕ.

11. $y = \sqrt{r^2 - x^2}$, $y = -\sqrt{r^2 - x^2}$

13. $1/\sqrt{(x + h)x}$

15. $(x^{1/3} - a^{1/3})(x^{2/3} + x^{1/3}a^{1/3} + a^{2/3})$

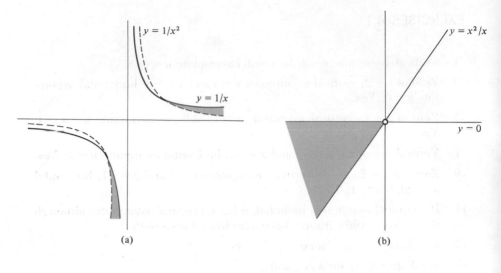

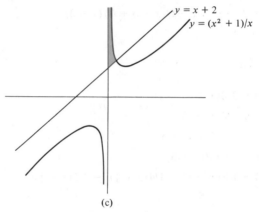

(c)

Figure C.42

EXERCISES 4.4

1. a. $x^7 + 7x^6 + 21x^5 + 35x^4 + 35x^3 + 21x^2 + 7x + 1$
   c. $16x^4 - 32x^3 + 24x^2 - 8x + 1$
   e. $(a/b)^6 - 6(a/b)^5 + 15(a/b)^4 - 20(a/b)^3 + 15(a/b)^2 - 6(a/b) + 1$
   g. $1^7 - 7(1)^6(1/10) + 21(1)^5(1/10)^2 - 35(1)^4(1/10)^3 + 35(1)^3(1/10)^4 - 21(1)^2(1/10)^5 + 7(1)^1(1/10)^6 - (1/10)^7 = 1 - .7 + .21 - .035 + .0035 - .00021 + .000007 - .0000001 = .4782969$

2. a. .608704; Upper bound for error: .004536
   c. 1076.3673744; Upper bound for error is .0000048384

3. a. $C(11,5)$ c. $4C(8,2)$ e. $-8C(9,3)$

5. $P(s,r) = (s-r)!C(s,r) = 1 \cdot 2 \cdot \ldots \cdot (s-r)C(s,r) =$
   $s(s-1) \ldots (s-r+1)$.

6. $C(s,r)$

7. 2^s

REVIEW EXERCISES 4.5

1. Monomials with odd degree and positive coefficients are strictly increasing.

3. $f(x) \leqslant f(y)$ for $x \leqslant y$. Constant monomials are increasing but not strictly increasing.

5. a. To be strictly increasing, a rational function can have no vertical asymptotes and no horizontal asymptotes. An example is $f(x) = (x-a)(x-b)/(x-a)$.
   c. Yes; Yes; No; No; No; No; Yes.

7. Yes. No. Yes. No. The only even function that can be in any of these categories is the constant function.

9. See Figure C.43.

11. a. Vertical asymptote $x = 3$; zero at $x = \frac{1}{2}$; horizontal asymptote $y = 2$.
    c. A pole at $x = -2$; no zeros and no vertical asymptotes; oblique asymptote $y = x + 2$.
    e. Zeros at $x = -3$ and $x = -2$; vertical asymptote $x = 0$; oblique asymptote $y = x + 5$.
    g. Zero at $x = 0$; vertical asymptote $x = -1$; asymptotic to $y = x^3$.

12. a. $(7, 0)$, $(-1, 8/3)$
    c. $((-1 - \sqrt{5})/2, 2/(-1 - \sqrt{5}))$, $((-1 + \sqrt{5})/2, 2/(-1 + \sqrt{5}))$

13. a. See Figure C.43. The colored segments on the axis show the values x where $f(x) \leqslant 0$.
    c. $-\frac{5}{2} < x < -2$
    e. No such x g. $x < -1$

14. a. $x \geqslant -3$; $x = -3$ is a zero.   c. all x; $x = 3$ is a zero

15. a. $x > 0$ c. $x > 0$

16. a. $x^5 - 15x^4 + 90x^3 - 270x^2 + 405x + 243$
    c. $x^3 - 6x^{5/2} + 15x^2 - 20x^{3/2} + 15x - 6x^{1/2} + 1$

17. a. $C(10, 2)/4$ c. $C(20, 4) = C(20, 16)$

19. a. 8528.998 c. 1771561

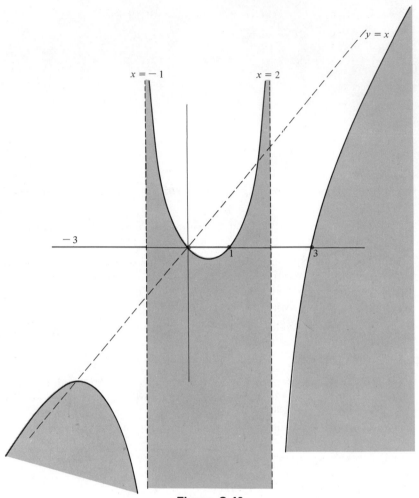

Figure C.43

QUIZ 4.6

1. a. Vertical asymptotes (and poles) $x = -8$, $x = 0$; horizontal asymptote
   $y = 0$.   b. Zero at $x = 0$; no poles; horizontal asymptote $y = 0$.
   See Figure C.44.

2. $(1/2, 4/17)$

3. Shown in color in Figure C.44.

4. $f(x) = 1/8x - 1/8(x + 8)$

5. No zeros.

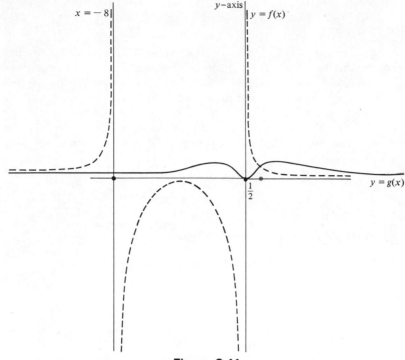

Figure C.44

6. $-C(12,5)2^7 y^5 = -128C(12,5) y^5$

7. 2163.4387

ADVANCED EXERCISES 4.7

1. a. Let $r = s = |a - b|/2$ b. Let $r = |a - b| + 1$ and examine $N_r(a)$.
    c. If A is bounded, there are m and M such that $m < a < M$ for all
    $a \in A$. Let $r = (M - m)/2$ and take $N_r((m + M)/2)$. For the converse,
    argue similarly.
    d. Suppose $N_r(a) \cap N_s(b)$ contains a point x. Let d be the minimum
    of $r - |a - x|$ and $s - |b - x|$. $N_d(x) \subset N_s(b) \cap N_r(a)$.
    f. If $b \neq a$, $|b - a| = r > 0$. Choose $1/n < r$. Then $N_{1/n}(a) \subset N_r(a)$
    with $b \notin N_r(a)$.

2. a. Pick $N_s(a) = N_r(a)$.   b. If $a = 0$, pick $s = \sqrt{r}$. If $a \neq 0$, pick s
    so that $x \in N_s(a)$ implies $|x + a| \leq 3|a|$ whereby $s \leq r/3|a|$. Then if
    $x \in N_s(a)$, $|x^2 - a^2| = |x - a| |x + a| \leq 3|a|s \leq 3|a| r/3|a| = r$. That is,
    $f(x) = x^2 \in N_r(a^2)$.
    d. Pick s to be any positive number: $f[N_s(0)] \subset N_r(1)$ for any $r > 0$.

e. Pick s_1 so that $f[N_{s_1}(a)] \subset N_{r/2}(L)$ and s_2 so that $g[N_{s_2}(a)] \subset N_{r/2}(K)$. Let s be the minimum of s_1 and s_2. Then $(f+g)[N_s(a)] \subset N_r(K+L)$.

3. a. If $f(x) = x$, $f = i_R$ is continuous. Suppose $f(x) = x^n$ represents a continuous function. Then $g(x) = x^{n+1} = i_R(x)f(x)$ and is continuous by the statements on limits in 4.7.2 e. Thus, each positive integral power of x represents a continuous function.

   b. Use a. and 4.7.2 e.

   c. Use a. and b. and induction on sums using 4.7.2 e.

   d. Rational functions are continuous at all domain points (use c. and 4.7.2 e.).

4. a. 5   b. 3/4   c. −3   d. −3/2   e. 2

5. a. 1   b. m c. $3x^2$ d. nx^{n-1} e. $-1/x^2$
   f. $1/(x+2)^2$ g. $Df(x) + Dg(x)$ h. $kDf(x)$

   i. $g(x)Df(x) + f(x)Dg(x)$ j. $\dfrac{g(x)Df(x) - f(x)Dg(x)}{g^2(x)}$

   Tangent to the curve $y = f(x)$ at $(x, f(x))$.

EXERCISES 5.1

2. If $\alpha(a)$ is in quadrant I or II, $\sin a > 0$ and conversely. Similarly, $\cos a > 0$ if and only if $\alpha(a)$ is in quadrant I or IV.

3. a. $x = \pi/6$ or $5\pi/6$ are the special values. However, $x = \pi/6 + 2n\pi$ and $x = 5\pi/6 + 2n\pi$ for $n \in Z$ are valid choices. The total set can be given by $\{x \in R : x = (-1)^n \pi/6 + n\pi, n \in Z\}$.

   c. $\{x \in R : x = \pi/3 + 2n\pi$ or
       $x = 2\pi/3 + 2n\pi\} = \{x \in R : x = (-1)^n \pi/3 + n\pi\}$

   e. $\{x \in R : x = (-1)^n \pi/4 + n\pi\}$

   g. $\{x \in R : \pi/2 + 2n\pi = x\}$ i. ϕ

4. $\alpha(a) = (x, y)$

5. If $\sin \alpha = y$ and $\cos a = x$, $\alpha(a) = (x, y)$ and $\alpha(-a) = (x, -y)$, whence $\sin(-a) = -y = -\sin a$ and $\cos(-a) = x = \cos a$. See Exercise 5.1.6.

6. $\alpha(a) = (x, y)$ and $\alpha(\pi/2 - a) = (y, x)$. Thus, $\sin(\pi/2 - a) = x = \cos a$ and $\cos(\pi/2 - a) = y = \sin a$. See Exercise 5.1.4.

7. $\left(\dfrac{x}{(x^2 + y^2)^{1/2}}, \dfrac{y}{(x^2 + y^2)^{1/2}}\right)$ is on the unit circle and there is, then, an $a \in [0, 2\pi)$ with $\cos a = x/(x^2 + y^2)^{1/2}$ and $\sin a = y/(x^2 + y^2)^{1/2}$.

EXERCISES 5.2

1. a. −4/5   c. 12/13   e. 4/5   g. −12/13   i. −4/5
    k. 5/13   m. 4/5   o. $-17/13\sqrt{2}$

2. a. $-36/85$ b. $77/85$ c. $-84/85$ d. $13/85$

3. a. $-24/25$ b. $-7/25$ i. $(3/5, -4/5)$ m. $(77/85, -36/85)$

4. a. $\cos \pi/5 = (1 - \sin^2 \pi/5)^{1/2}$
    c. $\sin \pi/10 = (1 - \cos^2 \pi/10)^{1/2}$

5. a. $\cos 3\pi/10 = \sin \pi/5 = .5878$
    c. $\cos 3\pi/8 = \sin \pi/8 = .3827$
    f. $\sin 2\pi/5 = \cos \pi/10 = .9511$

6. a. $\alpha(b) = (4/5, 3/5)$
    c. $\alpha(b) = (-1/\sqrt{2}, 1/\sqrt{2})$
    e. $\alpha(b) = (15/17, 8/17)$

7. a. $\cos (23\pi/10) = \cos 3\pi/10$
    c. $\cos 25\pi/8 = \cos (9\pi/8) = -\cos (\pi/8)$
    e. $\cos 21\pi/5 = \cos \pi/5$
    g. $\cos 83\pi/8 = \cos 3\pi/8 = \sin \pi/8$
    i. $\cos 13\pi/40 = \cos (\pi/8 + \pi/5)$
    k. $\cos \pi/20 = \cos (\pi/4 - 3\pi/10)$
            $= \cos \pi/4 \cos 3\pi/10 + \sin \pi/4 \sin 3\pi/10$
            $= \cos \pi/4 \sin \pi/5 + \sin \pi/4 \cos \pi/5$
    m. $\cos 3\pi/5 = \cos (\pi/2 + \pi/10)$

11. b. $\sin 23\pi/6 = \sin (4\pi - \pi/6) = \sin (-\pi/6) = -\sin \pi/6$
     e. $\sin 72\pi/5 = \sin (14\pi + 2\pi/5) = \sin 2\pi/5 = \cos \pi/10$
     h. $\cos 3163\pi/16 = \cos 27\pi/16 = \cos (-5\pi/16) = \sin 3\pi/16$

12. a. $y = x$ d. $y = -x$ f. $\sqrt{3}y = x$

14. Sine is odd; cosine is even.

EXERCISES 5.3

1. a. $24/25, -7/25$ c. $240/289, -161/289$
    e. $-120/169, -119/169$

2. a. $\sin a/2 = 1/\sqrt{5}, \cos a/2 = 2/\sqrt{5}$
    c. $\sin a/2 = -3/\sqrt{34}, \cos a/2 = -5/\sqrt{34}$
    e. $\sin a/2 = 2/\sqrt{13}, \cos a/2 = -3/\sqrt{13}$
    g. $\sin a/2 = 7/5\sqrt{2}, \cos a/2 = 1/5\sqrt{2}$

3. a. $(1/2)(\sin 5a - \sin a)$ c. $(1/2)(1/\sqrt{2} - \sin \pi/8)$
    e. $(1/2)(\cos 3a - \cos 7a)$

4. a. $\sqrt{2}/4$ c. $(1/2)(1/\sqrt{2} - \sin \pi/8)$

5. a. $2 \sin 3a/2 \cos a/2$
    c. $2 \cos 15a/12 \sin 5a/12$
    e. $2 \sin 4a \sin a$

6. a. $1/\sqrt{2}$ c. 1.11, approximately

7. Slope $= \sin a/\cos a$, where $(x, y) = (r \cos a, r \sin a)$.

EXERCISES 5.4

In the exercises below, the indicated solution is but one method of approach.

1. $\sin (\pi/2 + a) = \sin \pi/2 \cos a + \cos \pi/2 \sin a = \cos a = \sin (\pi/2 - a)$.
    The substitution set is R and the restriction set is ϕ.

4. $\sin a/\cos a + \cos a/\sin a = \sin^2 a/\cos a \sin a + \cos^2 a/\cos a \sin a =$
    $(\sin^2 a + \cos^2 a)/\cos a \sin a = 1/\cos a \sin a$

9. $\dfrac{1}{\cos^2 a \sin^2 a} = \dfrac{\sin^2 a + \cos^2 a}{\cos^2 a \sin^2 a} = \dfrac{\sin^2 a}{\cos^2 a \sin^2 a} + \dfrac{\cos^2 a}{\cos^2 a \sin^2 a} =$
    $1/\cos^2 a + 1/\sin^2 a$. The restriction set A is $\{x \in R: x = n\pi/2\}$, while the
    substitution set is $R - A$.

15. Let $\alpha(b) = (3/5, 4/5)$. Then $\cos b = 3/5$ and $\sin b = 4/5$, whence $3 \sin a +$
    $4 \cos a = 5[(3/5) \sin a + (4/5) \cos a] = 5(\sin a \cos b + \cos a \sin b) =$
    $5 \sin (a + b)$.

22. $\sin 2a/\cos a = 2 \sin a \cos a/\cos a = 2 \sin a$. The restriction set
    $A = \{x \in R: x = (2n + 1)\pi/2\}$ and the substitution is $R - A$.

30. $\dfrac{\sin a}{1 + \cos a} + \dfrac{1 + \cos a}{\sin a} = \dfrac{\sin^2 a + (1 + \cos a)^2}{\sin a(1 + \cos a)} =$

    $\dfrac{1 + 2 \cos a + (\sin^2 a + \cos^2 a)}{\sin a(1 + \cos a)} = \dfrac{2(1 + \cos a)}{\sin a(1 + \cos a)} = 2/\sin a$. The restric-
    tion set is $A = \{x \in R: x = n\pi\}$ and the substitution set is $R - A$.

EXERCISES 5.5

2.

$\alpha(a)$	I	II	III	IV
tan a	+	−	+	−
cot a	+	−	+	−
sec a	+	−	−	+
csc a	+	+	−	−

4. $\tan 2a = 2 \tan a/(1 - \tan^2 a)$, whence $\cot 2a = (1 - \tan^2 a)/2 \tan a =$
    $\dfrac{(1 - \tan^2 a) \cot^2 a}{2 \tan a \cot^2 a} = \dfrac{\cot^2 a - 1}{2 \cot a}$

6. $\sec a = 1/\cos a = 1/\sin (\pi/2 - a) = \csc (\pi/2 - a)$;
    $\csc a = 1/\sin a = 1/\cos (\pi/2 - a) = \sec (\pi/2 - a)$

7. a. $\tan (-a) = \sin (-a)/\cos (-a) = -\sin a/\cos a = -\tan a$

8. $\tan (a - b) = \tan (a + (-b)) = \dfrac{\tan a + \tan (-b)}{1 - \tan a \tan (-b)}$

which, by Exercise 5.5.7, gives

$$\tan (a - b) = \frac{\tan a - \tan b}{1 + \tan a \tan b}$$

9. a. $\tan (a + n\pi) = \dfrac{\sin (a + n\pi)}{\cos (a + n\pi)} = \dfrac{(-1)^n \sin a}{(-1)^n \cos a} = \dfrac{\sin a}{\cos a} = \tan a$

10. a. $\sec (\pi/2 + a) = 1/\cos (\pi/2 + a) = 1/(-\sin a) = -\csc a$

c. $\tan a + \cot a = \sin a/\cos a + \cos a/\sin a = \sin^2 a/\cos a \sin a + \cos^2 a/\cos a \sin a = 1/\cos a \sin a = \sec a \csc a$

e. $1 + \cot^2 a = 1 + \cos^2 a/\sin^2 a = (\sin^2 a + \cos^2 a)/\sin^2 a = \csc^2 a$

h. By e. of this exercise, $\cot a/(1 + \cot^2 a) = \cot a/\csc^2 a = \sin^2 a \cot a = \sin^2 a(\cos a/\sin a) = \sin a \cos a$

j. $\sin x - \sin y = \dfrac{2 \cos (x - y)/2 \ \sin (x - y)/2}{2 \cos (x - y)/2 \ \cos (x - y)/2} = \tan (1/2)(x - y)$

o. $\cos^2 a - \sin^2 a = \cos^2 a \left[1 - \dfrac{\sin^2 a}{\cos^2 a}\right] = \dfrac{1 - \tan^2 a}{\sec^2 a} = \dfrac{1 - \tan^2 a}{1 + \tan^2 a}$

[See d. of this section.]

11. a. $\dfrac{\sin a}{1 + \cos a} = \dfrac{\sin a(1 - \cos a)}{1 - \cos^2 a} = \dfrac{\sin a(1 - \cos a)}{\sin^2 a} = 1/\sin a - $
$\cos a/\sin a = \csc a - \cot a$

REVIEW EXERCISES 5.6

1. $\pi r/4, \ \pi r/3, \ \pi r/6$

3. Consult the special value list.

4. a. I   c. IV   e. II   g. III

5. a. $(1, 0)$ d. $(-1/2, \sqrt{3}/2)$ h. $(0, 1)$

8. a. $\sqrt{.2944}$ c. $1/.84$ f. $1.68 \sqrt{.2944}$ j. $\left(\dfrac{1 + \sqrt{.2944}}{2}\right)^{1/2}$

s. $\sqrt{.2944}$ w. $(\sqrt{.2944}, .84)$

9. $0 < a < \pi/2$ implies $0 < \cos a < 1$ and $0 < \sin a < 1$, whence $0 < \sin a/1 < \sin a/\cos a$ or $0 < \sin a < \tan a$

10. g. $\dfrac{\sin a + \cos a}{\sec a + \csc a} = \dfrac{(\sin a + \cos a) \sin a \cos a}{(\sec a + \csc a) \sin a \cos a} = $

$\dfrac{(\sin a + \cos a) \sin a \cos a}{\sin a + \cos a} = \sin a \cos a$

i. $2 \sin a + \sec a = \sec a(2 \sin a \cos a + 1) = \sec a(2 \sin a \cos a + \sin^2 a + \cos^2 a) = \sec a(\sin a + \cos a)^2$

p. $\dfrac{\sin a}{1 + \cos a} = \dfrac{2 \sin a/2 \cos a/2}{1 + 2 \cos^2 a/2 - 1} = \dfrac{\sin a/2}{\cos a/2} = \tan a/2$

11. a. $2 \sin 8a \cos 2a$
    c. $2 \cos 8a \cos 2a$
    h. $2 \sin (x - h/2) \sin h/2$

12. b. $(\sqrt{2} \cos \pi/4, \; \sqrt{2} \sin \pi/4)$
    d. $(\sqrt{2}x \cos \pi/4, \; \sqrt{2}x \sin \pi/4)$
    e. $(\sqrt{2} |x| \cos 5\pi/4, \; \sqrt{2} |x| \sin 5\pi/4)$
    j. $(2x \cos \pi/6, \; 2x \sin \pi/6)$

14. $\cos u = x$, $\sin a = y$, $\tan a = y/x$, $\cot a = x/y$, $\csc a = 1/y$, $\sec a = 1/x$

15. a. $b = \sqrt{3}$, $c = 2\sqrt{3}$, measure $B = \pi/6$ radians
    b. $a = 6$, $c = 6\sqrt{2}$, measure $B = \pi/4$ radians
    c. $a = 1$, $b = \sqrt{3}$, measure $B = \pi/3$ radians
    measure $B = \pi/2 - $ measure A

QUIZ 5.7

1. $\sin a = -15/17$, $\cos a = 8/17$, $\tan a = -15/8$, $\csc a = -17/15$, $\sec a = 17/8$,
   $\cot a = -8/15$

2. $\alpha(2a) = (\cos 2a, \sin 2a) = (-161/289, -240/289)$
   $\alpha(a/2) = (\cos a/2, \sin a/2) = (\pm 5/\sqrt{34}, \pm 3/\sqrt{34})$

3. a. $-y\sqrt{1-x^2} + x\sqrt{1-y^2}$ b. $xy - \sqrt{(1-y^2)(1-x^2)}$
   c. $(-1)^n \cos a = (-1)^{n+1} \sqrt{1-y^2}$ d. xy

4. $\cos \pi/8 \cos 3\pi/8 = (1/2)[\cos \pi/4 + \cos \pi/2] = (1/2) \cos \pi/4 = 1/2\sqrt{2}$

5. Sine: $R \to [-1, 1]$;   cosine: $R \to [-1, 1]$;
   tangent: $R - \{(2n+1)\pi/2 : n \in Z\} \to R$;
   cosecant: $R - \{n\pi : n \in Z\} \to R - (-1, 1)$;
   secant: $R - \{(2n+1)\pi/2 : n \in Z\} \to R - (-1, 1)$;
   cotangent: $R - \{n\pi : n \in Z\} \to R$

6. $\sin^4 a - \cos^4 a = (\sin^2 a + \cos^2 a)(\sin^2 a - \cos^2 a) = (\sin^2 a - \cos^2 a)$
   $= -(\cos^2 a - \sin^2 a) = -\cos 2a$

7. $\dfrac{\tan^4 a + \tan^2 a}{\sec^2 a - \sec^2 a \cos^2 a} = \dfrac{\tan^2 a(\tan^2 a + 1)}{\sec^2 a(1 - \cos^2 a)} = \dfrac{\tan^2 a \sec^2 a}{\sec^2 a \sin^2 a} = \dfrac{\tan^2 a}{\sin^2 a}$
   which is not identical to $\tan^2 a$. (Let $a = \pi/4$.)

8. $\dfrac{(\sin^2 a - \cos^2 a)(\sin a + \cos a)}{\sin a - \cos a}$
   $= \dfrac{(\sin a + \cos a)(\sin a - \cos a)(\sin a + \cos a)}{\sin a - \cos a} = (\sin a + \cos a)^2$
   which is not identical to 1. Use $a = \pi/4$.

9. $(\tan a + \sec a)(\sin a - 1) \sec^2 a = (\tan a + \sec a)(\tan a - \sec a)\sec a$
   $= (\tan^2 a - \sec^2 a)\sec a = -\sec a$

ADVANCED EXERCISES 5.8

2. a. $(5, 0) = (5, 0)^*$; $(0, 5) = (5, \pi/2)^*$; $(5/\sqrt{2}, 5/\sqrt{2}) = (5, \pi/4)^*$;
   $(5/2, 5\sqrt{3}/2) = (5, \pi/3)^*$; $(-5\sqrt{3}/2, 5/2) = (5, 5\pi/6)^*$; and so forth.

4. a. For $A = \{(x, y): x^2 + y^2 = 1, x \geqslant 0\}$, $p_2|A$ is $1:1$ and $\alpha|[-\pi/2, \pi/2]$
   is $1:1$ so that $\sin|[-\pi/2, \pi/2] = p_2 \circ \alpha|[-\pi/2, \pi/2]$ is $1:1$, since it is
   the composition of two $1:1$ functions.

EXERCISES 6.1

1. a. $\{x \in R: x = n\pi, n \in Z\}$
   c. $\{x \in R: x = 7\pi/6 + 2n\pi \text{ or } x = 11\pi/6 + 2n\pi, n \in Z\} =$
      $\{x \in R: x = (-1)^{n+1}\pi/6 + n\pi\}$
   e. $\{x \in R: x = 4\pi/3 + 2n\pi \text{ or } x = 5\pi/3 + 2n\pi\} =$
      $\{x \in R: x = (-1)^{n+1}\pi/3 + n\pi\}$
   g. $\{x \in R: x = 3\pi/2 + 2n\pi\}$
   i. $\{x \in R: x = 5\pi/4 + 2n\pi \text{ or } x = 7\pi/4 + 2n\pi\} =$
      $\{x \in R: x = (-1)^{n+1}\pi/4 + n\pi\}$

2. a. $\{x \in R: x = \pi/2 + n\pi\}$ c. $\{x \in R: x = \pm 4\pi/3 + 2n\pi\}$
   e. $\{x \in R: x = \pm 5\pi/6 + 2n\pi\}$ g. $\{x \in R: x = (2n + 1)\pi\}$
   i. $\{x \in R: x = \pm 3\pi/4 + 2n\pi\}$

3. a. $\{x \in R: x = n\pi\}$ c. $\{x \in R: x = 3\pi/4 + n\pi\}$
   e. $\{x \in R: x = 2\pi/3 + n\pi\}$ g. $\{x \in R: x = 5\pi/6 + n\pi\}$

4. a. $\{x \in R: x = \pi/2 + n\pi\}$ c. $\{x \in R: x = 3\pi/4 + n\pi\}$
   e. $\{x \in R: x = 5\pi/6 + n\pi\}$ g. $\{x \in R: x = 2\pi/3 + n\pi\}$

5. a. $\{x \in R: x = 2n\pi\}$ c. $\{x \in R: x = \pm\pi/4 + 2n\pi\}$
   e. $\{x \in R: x = \pm\pi/6 + 2n\pi\}$ g. $\{x \in R: x = \pm\pi/3 + 2n\pi\}$

6. a. $\{x \in R: x = \pi/2 + 2n\pi\}$
   c. $\{x \in R: x = \pi/4 + 2n\pi \text{ or } x = 3\pi/4 + 2n\pi\} =$
      $\{x \in R: x = (-1)^n \pi/4 + n\pi\}$
   e. $\{x \in R: x = \pi/3 + 2n\pi \text{ or } x = 2\pi/3 + 2n\pi\} =$
      $\{x \in R: x = (-1)^n \pi/3 + n\pi\}$
   g. $\{x \in R: x = \pi/6 + 2n\pi \text{ or } x = 5\pi/6 + 2n\pi\} =$
      $\{x \in R: x = (-1)^n \pi/3 + n\pi\}$

7. a. $\{x \in R: x = \pi/2 + 2n\pi\}$ c. ϕ
   e. $\{x \in R: x = 2\pi/3 + 2n\pi\}$
   g. $\{x \in R: x = \pi/4 + n\pi/2\} = \{x \in R: x = (2n + 1)\pi/4\}$
   i. $\{x \in R: x = n\pi/2\}$

8. a. $\{0\}$ c. $\{\pi/6\}$ e. $\{\pi/2\}$ g. $\{-\pi/6\}$ i. $\{-\pi/4\}$
   k. $\{\pi/4\}$ m. $\{\pi/6\}$ o. $\{\pi\}$ q. $\{5\pi/6\}$ s. $\{0\}$
   u. $\{\pi/4\}$ w. $\{-\pi/3\}$ y. $\{-\pi/6\}$

9. a. 3 c. $1/4$ e. $\pm 15/17$ g. $\pm 4/5$

10. a. $\{x \in R : x = \pi/6 + 2n\pi \text{ or } x = 5\pi/6 + 2n\pi\} =$
$\{x \in R : x = (-1)^n \pi/6 + n\pi\}$
    c. $\{x \in R : x = \pi/3 + n\pi\}$

11. a. $\theta \in \arcsin \sqrt{2}/2 = \{x \in R : x = (-1)^n \pi/4 + n\pi\}$
    c. $\theta \in \arccos \sqrt{3}/2 = \{x \in R : x = \pm\pi/6 + 2n\pi\}$
    e. $a \in \arccos 1 \cup \arccos (-1) = \{x \in R : x = n\pi\}$
    g. $a \in \arctan 1 \cup \arctan (-1) = \{x \in R : x = \pi/4 + n\pi/2\} =$
$\{x \in R : x = (2n + 1) \pi/4\}$
    i. $2 \sin^2 a - \sin a - 1 = 0$, $(2 \sin a + 1)(\sin a - 1) = 0$, $\sin a = -1/2, 1$,
$a \in \arcsin (-1/2) \cup \arcsin 1 = \{x \in R : x = (-1)^{n+1} \pi/6 + n\pi \text{ or }$
$x = \pi/2 + 2n\pi\}$
    k. $\sin^2 a - \cos^2 a = 0$, $2 \sin^2 a - 1 = 0$, $\sin a = \pm 1/\sqrt{2}$,
$a \in \arcsin 1/\sqrt{2} \cup \arcsin (-1/\sqrt{2}) = \{x \in R : x = \pi/4 + n\pi/2\}$
    l. $\cos a (\sin^2 a - 1) = 0$, $\cos a = 0$ or $\sin^2 a - 1 = 0$, whence $\sin a = \pm 1$. Thus $a \in \arccos 0 \cup \arcsin 1 \cup \arcsin (-1) = \arccos 0 =$
$\{x \in R : x = (2n + 1) \pi/2\}$.
    n. $\tan a = \cos a$, $\sin a = \cos^2 a = 1 - \sin^2 a$, $\sin^2 a + \sin a - 1 = 0$,
$\sin a = \dfrac{-1 \pm \sqrt{1+4}}{2} = \dfrac{-1 \pm \sqrt{5}}{2}$; $a \in \arcsin \dfrac{-1 + \sqrt{5}}{2} \cup \arcsin$

$\dfrac{-1 - \sqrt{5}}{2}$. However, $\arcsin \dfrac{-1 - \sqrt{5}}{2} = \phi$ so that $a \in \arcsin \dfrac{-1 + \sqrt{5}}{2}$.

    r. $\sin a \tan a = 0$ implies $\sin a = 0$ or $\tan a = 0$. Thus, $a \in \arcsin 0 = \arctan 0 = \{x \in R : x = n\pi\}$.
    x. $\sin 2\theta = \cos \theta$, $2 \sin \theta \cos \theta = \cos \theta$, $(2 \sin \theta - 1) \cos \theta = 0$, $\cos \theta = 0$ or $\sin \theta = 1/2$, $\theta \in \arccos 0 \cup \arcsin 1/2 = \{x \in R : x = (2n + 1)\pi/2$ or $x = (-1)^n \pi/6 + n\pi\}$
    b'. $\theta \cos \theta - \theta - \cos \theta + 1 = 0$, $\theta(\cos \theta - 1) - (\cos \theta - 1) = 0$,
$(\theta - 1)(\cos \theta - 1) = 0$, $\theta = 1$ or $\theta \in \arccos 1$.
Thus $\theta = 1$ or $\theta \in \{x \in R : x = 2n\pi\}$.

12. a. $(0, 1)$
    c. $(-1/2, \sqrt{3}/2)$, $(1/2, -\sqrt{3}/2)$
    e. $(\sqrt{3}/2, 1/2)$, $(\sqrt{3}/2, -1/2)$
    g. $(-5/13, 12/13)$, $(-5/13, -12/13)$
    i. $(24/25, -7/25)$, $(-24/25, -7/25)$

EXERCISES 6.2

1. a. 0 c. $\pi/4$ e. $\pi/6$ g. 0 i. $\pi/6$
   k. $-\pi/4$ m. $-\pi/3$ o. $5\pi/6$ q. $3\pi/4$ s. $-\pi/3$
2. a. $1/4$ c. 6 e. $3/5$ g. $-\pi/7$ i. $\pi/5$ k. $\pi/10$
4. a. $(\sqrt{3}/2, 1/2)$ c. $(1/\sqrt{2}, -1/\sqrt{2})$ e. $(-15/17, -8/17)$
   g. $(1/2, -\sqrt{3}/2)$ i. $(\sqrt{3}/2, 1/2)$

EXERCISES 6.3

1. a. 1/2   c. 65   e. $\pi/12$ g. Not possible   i. Not possible
2. a. $\pi/4$ c. $\pi/3$ e. $\sqrt{3}/2$ g. $1/\sqrt{2}$ i. $2/\sqrt{3}$
    k. 4/5   m. 12/13   o. $-7/25$ q. 12/13
3. a. $(\sqrt{3}-1)/2\sqrt{2}$ c. 36/85   e. $-323/36$ g. 24/25
    i. $[(\sqrt{2}+1)/2\sqrt{2}]^{1/2}$ k. $-336/527$ m. 1/3

4. a. $0 < \tan^{-1} 6 < \pi/2$ and $0 < \tan^{-1} 1/6 < \pi/2$ so that $0 < \tan^{-1} 6 +$
        $\tan^{-1} 1/6 < \pi$, $\cos(\tan^{-1} 6 + \tan^{-1} 1/6) = 6/\sqrt{37} - 1/\sqrt{37} \cdot 6/\sqrt{37} = 0$,
        whence $\tan^{-1} 6 + \tan^{-1} 1/6 = \pi/2$. The problem could be worked
        by noting that $\tan^{-1} 1/6 = \cot^{-1} 6$, whereby $\tan^{-1} 6$ and $\cot^{-1} 6$
        (or $\tan^{-1} 1/6$) are complementary numbers. Hence, they add to $\pi/2$.
    e. $0 < \cos^{-1} 12/13 + \cos^{-1} 24/25 < \pi$, so that, if $x = \cos(\cos^{-1} 12/13 +$
        $\cos^{-1} 24/25)$, $\cos^{-1} x = \cos^{-1} 12/13 + \cos^{-1} 24/25$. $\cos(\cos^{-1} 12/13 +$
        $\cos^{-1} 24/25) = 12/13 \cdot 24/25 - 5/13 \cdot 7/25 = 253/325$.
        The problem follows.
    f. Since $\sin^{-1} 4/5 = \cos^{-1} 3/5$, $\sin^{-1} 4/5$ and $\sin^{-1} 3/5$ are complemen-
        tary numbers [both lie in $(0, \pi/2)$] and necessarily $\sin^{-1} 3/5 +$
        $\sin^{-1} 4/5 = \pi/2$.

EXERCISES 6.4

1. a. period $= 2\pi/5$, phase $= 0$, amplitude $= 3$.
    c. period $= \pi$, phase $= \pi/6$, amplitude $= 1/2$.
    e. period $= 2\pi$, phase $= \pi/2$, amplitude $= 1$.
    g. period $= 6\pi/5$, phase $= 3\pi/5$, amplitude $= 5/3$.
    i. period $= \pi$, phase $= (1/2) \sin^{-1} 12/13$, amplitude $= 13$.
2. a. $5/2\pi$ c. $1/\pi$ e. $1/2\pi$ g. $5/6\pi$ i. $1/\pi$
3. See Figures C.45, C.46, C.47, C.48, and C.49 for a., c., e., g., and i.,
    respectively.
4. a. 24π c. 12π

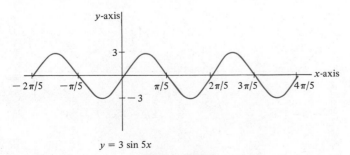

$y = 3 \sin 5x$

Figure C.45

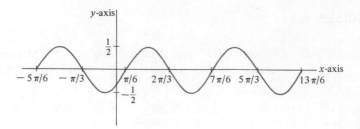

$$y = \frac{1}{2} \cos 2(x - \pi/6)$$

Figure C.46

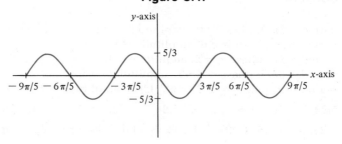

$$y = \sin (x - \pi/2)$$

Figure C.47

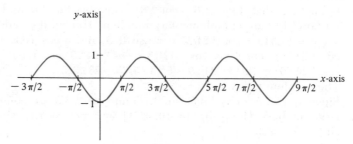

$$y = (5/3) \sin (5x/3 + \pi)$$

Figure C.48

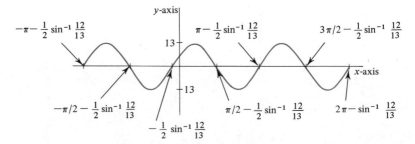

$$y = 5 \sin 2x + 12 \cos 2x$$

Figure C.49

REVIEW EXERCISES 6.5

1. a. $\arcsin 0$ c. $\{x \in R: x = \pi/4 + 2n\pi\}$ e. ϕ g. ϕ
   i. $\{x \in R: x = -\pi/4 + 2n\pi\}$
   k. **arctan** $1 = \{x \in R: x = \pi/4 + n\pi\}$
   m. $\{x \in R: x = n\pi/2\}$
   o. $\{x \in R: x = \pi/4 + n\pi \text{ or } x = 3\pi/4 + 2n\pi\}$
   q. $\{x \in R: x = \pi/4 + 2n\pi\}$
   s. $\{x \in R: x = -\pi/4 + 2n\pi\}$

2. a. π c. $5\pi/6$ e. $\pi/2$ g. $7\pi/6$ i. $5\pi/4$
   k. $-\pi/4$ m. $-\pi/6$ o. $-\pi$ q. $-5\pi/6$ s. π
   u. $5\pi/4$ w. $2\pi/3$ y. $5\pi/6$

4. a. $x \in \{n\pi/2: n \in Z\}$
   c. $x \in \{\pi/4 + n\pi: n \in Z\}$
   e. No solution

5. a. a if $-1 \leqslant a \leqslant 1$ c. a e. $-\pi/6$ g. $\pi/4$
   i. $\pi/7$ k. $15/17$ m. $5/12$ o. $15/17$ q. π

7. a. amplitude $= 3$, phase $= 1/7$, period $= 2\pi/7$
   c. amplitude $= 3/2$, phase $= 1/3$, period $= 2\pi/5$
   e. amplitude $= \sqrt{10}$, phase $= \sin^{-1} 1/\sqrt{10}$, period $= 2\pi$
   g. amplitude $= 5$, phase $= (\sin^{-1} 4/5)/2$, period $= \pi$

8. a. $7/2\pi$ c. $5/2\pi$ e. $1/2\pi$ g. $1/\pi$

10. a. $c = 5/3$, $b = 4/3$, $B \sin^{-1} 4/5$
    c. $a = 16/17$, $b = 30/17$, $B = \tan^{-1} 15/8$
    e. $c = 10$, $A = \tan^{-1} 3/4$, $B = \tan^{-1} 4/3$
    g. $a = \sqrt{3}$, $A = \pi/3$, $B = \pi/6$

QUIZ 6.6

1. a. $\{a: a = \tan^{-1} (1 \pm \sqrt{21})/2 + n\pi, n \in Z\}$
   b. $\{a: a = (2n + 1) \pi/2, n \in Z\}$
   c. ϕ

2. See Figure C.50.

3. a. $.61$ b. $3\pi/10$ c. $-\pi/3$ d. $\pi/4$

4. a. $(1/2, -\sqrt{3}/2)$ b. $(-\sqrt{3}/2, 1/2)$ c. $(1/\sqrt{2}, 1/\sqrt{2})$

5. a. Since $0 < \tan^{-1} 2 < \tan^{-1} 3 < \pi/2$, $\tan^{-1} 3 + \tan^{-1} (-2) > 0$ whence
   $\tan^{-1} 3 + \tan^{-1} (-2) \neq \tan^{-1} (-1/7)$ $[\tan^{-1} (-1/7) < 0]$.
   b. False, $\tan^{-1} 2 + \tan^{-1} 3 = 3\pi/4$.

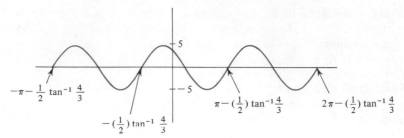

Figure C.50

c. $0 < \sin^{-1} 3/5 < \pi/2, 0 < \cos^{-1} 12/13 < \pi/2$ so that $\sin^{-1} 3/5 = \tan^{-1} 3/4$
and $\cos^{-1} 12/13 = \tan^{-1} 5/12$. Then $\tan\ [\sin^{-1} 3/5 + \cos^{-1} 12/13] =$
$\dfrac{3/4 + 5/12}{1 - 3/4 \cdot 5/12} = 56/33$. Thus, $\sin^{-1} 3/5 + \cos^{-1} 12/13 = \tan^{-1} 56/33$.

ADVANCED EXERCISES 6.7

1. $[\pi/2, 3\pi/2]$ is one such interval.

4. $\tan \pi/2$ and $\tan (-\pi/2)$ are not defined.

5. $\cos|[-\pi/2, \pi/2]$ is not $1:1$.

6. $2\pi s/mp$, where s is the least common multiple of np and qm.

9. a. $\sqrt{3}y = x$ c. $x = y$ e. $y = 0$

13. G is the diagonal of $[-\pi/2, \pi/2] \times [-\pi/2, \pi/2]$, that is,
$G = \{(x, x) : x \in [-\pi/2, \pi/2]\}$.

EXERCISES 7.1

1. a. $(3, 0)^*, (-3, \pi)^*$
    c. $(1, \pi)^*, (-1, 0)^*$
    e. $(5, \sin^{-1} 4/5)^*, (-5, \sin^{-1} 4/5 + \pi)^*$
    g. $(17, \sin^{-1} 15/17)^*, (-17, \sin^{-1} 15/17 + \pi)^*$
    i. $(13, \pi - \sin^{-1} 12/13)^*, (-13, 2\pi - \sin^{-1} 12/13)^*$
    k. $(\sqrt{61}, \tan^{-1} 5/6)^*, (-\sqrt{61}, \pi + \tan^{-1} 5/6)^*$
    m. $(\sqrt{13}, \cos^{-1} (-2/\sqrt{13}))^*, (-\sqrt{13}, \pi + \cos^{-1} (-2/\sqrt{13}))^*$
    o. $(10, 5\pi/6)^*, (-10, 11\pi/6)^*$

3. a. The rectangular form of $(r, a)^*$ is $(r \cos a, r \sin a)$ and
    $p_1((r \cos a, r \sin a)) = r \cos a$.

4. y/x

5. a. $(3, 0)$ d. $(-5, 0)$ g. $(-\sqrt{2}, \sqrt{2})$ j. $(0, -3)$
    m. $(-120/13, -50/13)$

6. If $y = mx$, $y/x = m$, (x, y) is given as $(r \cos a, r \sin a)$ for some $a \in R$
    and $r^2 = x^2 + y^2$. Thus $y/x = \tan a = m$, whence, if a is chosen as $\tan^{-1} m$,
    the equation is $a = \tan^{-1} m$.

EXERCISES 7.2

1. a. $(4, 7)$ f. $(5, -2)$ k. $(0, 5)$ u. $(1/2, 1/2)$

2. a. (a, b) b. (a, b) c. (a, b) d. (a, b) e. $(0, 0)$
    i. $(-1, 0)$ k. $(0, -1)$ m. $(1, 0)$

5. $(0, 0)$

6. $(1, 0)$

9. a. $(-bk, ak)$ c. $(2a, 0)$ d. $(0, 2b)$ e. $(a^2 + b^2, 0)$

10. If $(a, b) = \alpha(p)$, $\alpha(-p) = (a, -b)$.

11. a. $(rs \cos (a + b), rs \sin (a + b))$
     b. $((r/s) \cos (a - b), (r/s) \sin (a - b))$
     c. $(r^2 \cos 2a, r^2 \sin 2a)$
     d. $(r^3 \cos 3a, r^3 \sin 3a)$

EXERCISES 7.3

1. If $a \neq b$, $(a, 0) \neq (b, 0)$.

2. $i(a + b) = (a + b, 0) = (a, 0) + (b, 0) = i(a) + i(b)$, $i(ab) = (ab, 0) =$
    $a(b, 0) = ai(b) = (a, 0)(b, 0) = i(a)i(b)$, $i(0) = (0, 0)$ and, since i is
    $1:1$, $i(b) = (0, 0)$ implies $b = 0$.

6. $p_1(a, b)$ is the real part of (a, b) while $p_2(a, b)$ is the imaginary part.

7. $(a^2 + b^2)^{1/2}$

EXERCISES 7.4

2. a. $(0, -5832)$ c. $(-1, 0)$ e. $(16, 0)$ g. $(0, 216)$

5. b. $re^{i\theta} se^{i\varphi} = r(\cos \theta, \sin \theta) s (\cos \varphi, \sin \varphi) = rs(\cos (\theta + \varphi),$
     $\sin (\theta + \varphi)) = rse^{i(\theta + \varphi)}$

6. a. $3i$, $(0, 3)$ c. $5\sqrt{3}/2 + 5i/2$, $(5\sqrt{3}/2, 5/2)$ e. -2, $(-2, 0)$

7. a. $(1/\sqrt{2}, 1/\sqrt{2})$ c. $(\sqrt{3}/2, 1/2)$

EXERCISES 7.5

1. $(4, 0)$, $(-4, 0)$

3. $(3\sqrt{3}/2, 3/2)$, $(-3\sqrt{3}/2, 3/2)$, $(0,-3)$

5. $10(\cos \pi/8, \sin \pi/8)$, $10(\cos 9\pi/8, \sin 9\pi/8)$

7. $(\sqrt{2}/2)(\sqrt{3}, 1)$, $(\sqrt{2}/2)(-\sqrt{3}, -1) = (-\sqrt{3}/2)(3, 1)$

9. $3e^{3\pi i/8}$, $3e^{7\pi i/8}$, $3e^{11\pi i/8}$, $3e^{15\pi i/8}$

15. Show that $(-b \pm \sqrt{b^2 - 4ac})/2a$ represents complex numbers if $a \neq 0$.

16. a. The four fourth roots of 1.
    c. $-2 \pm \sqrt{3}$ e. $\dfrac{-1 \pm \sqrt{11}\,i}{2}$

REVIEW EXERCISES 7.6

1. a. $(3/\sqrt{2}, -3/\sqrt{2})$ c. $(-13, 0)$ e. $(64/5, 48/5)$
   g. $(-96/17, 180/17)$ i. $(r\cos\theta, r\sin\theta)$

2. a. $(65, \sin^{-1} 12/13)^\circ$ c. $(85, \sin^{-1} 4/5)^\circ$ e. $(30, 0)^\circ$
   g. $(102, 2\pi/3)^\circ$

3. a. Line given by $y = x$ in rectangular coordinates.
   c. Circle $(x - 32)^2 + y^2 = 32^2$
   e. Circle $x^2 + y^2 = 36$
   g. Spiral

4. a. $3\sqrt{2}, \pi/4, 3, 3$ c. $4, \pi/3, 2, 2\sqrt{3}$ e. $16, \pi, -16, 0$
   g. $\sqrt{5}, \tan^{-1}(-1/2), 2, -1$ i. $3\sqrt{2}, -\pi/4, 3, -3$

5. a. $8 - i$ c. $3 - 2i$ e. 2 f. $-1 + i$ i. 1
   k. $-8/13 + 27i/13$

6. a. $18i$ c. $-8 + 8\sqrt{3}i$ e. 256 g. $3 - 4i$ i. $-18i$

7. c. -64 e. -4096

8. c. $\sqrt{3} + i, -\sqrt{3} - i$ e. $4i, -4i$

9. a. $\sqrt{2} + \sqrt{2}i, -\sqrt{2} + \sqrt{2}i, -\sqrt{2} - \sqrt{2}i, \sqrt{2} - \sqrt{2}i$

QUIZ 7.7

1. a. $(5, \tan^{-1} 4/3)^\circ$ b. $(5\sqrt{2}, -\pi/4)^\circ$ c. $(6, a)^\circ$

2. a. $(36/13, 15/13)$ b. $(-40/17, 75/17)$ c. $(1, \sqrt{3})$

3. a. $8 + i$ b. $3 + 4i$ c. $26i$ d. $23/5 - 11i/5$
   e. 24 f. $5 + 12i$

4. a. The real part is 8 and the imaginary part is 1.
   d. The real part is 23/5 while the imaginary part is $-11/5$.

5. a. -4 b. 3^0

6. $3e^{\pi i/10}$, $3e^{\pi i/2}$, $3e^{9\pi i/10}$, $3e^{13\pi i/10}$, $3e^{17\pi i/10}$

7. $x^3 - x^2 + x - 6 = (x - 2)(x^2 + x + 3)$. The zeros are 2, $(-1 \pm \sqrt{11}\,i)/2$.

ADVANCED EXERCISES 7.8

1. Write $(r, \theta)^*$, $(s, \Phi)^*$ in rectangular coordinates and calculate the distance.

2. See Figures C.51, C.52, and C.53 for graphs of a., c., and e., respectively.

3. If $k = 0$, the problem is trivial. If $k \neq 0$, the two intersect only at points (s, φ), where $2\varphi \in$ arccos $1/4$. Thus, since $r = k$ is a circle of radius $k(\neq 0)$ about the origin, the points of intersection must be $(k/2, (1/2)$ $\cos^{-1} 1/4)^*$ and $(k/2, (-1/2) \cos^{-1} 1/4)^*$. It is also possible to solve for the points of intersection by rewriting the equations in "rectangular" form.

5. a. $z = (8\sqrt{2} + 8\sqrt{2}i)^{1/2} - 1$ for both square roots of $8\sqrt{2} + 8\sqrt{2}i$.
    c. $z = 3 + 4i$.

6. a. Circle of center a and radius 2.
    c. Interior of the circle from a.
    e. Line given by $x = 2$.
    g. "Left half plane."
    i. Line $y = x$ except for $(0, 0)$.
    k. The ray from $(0, 0)$ through $(1/2, \sqrt{3}/2)$ except for the point $(0, 0)$.

EXERCISES 8.1

1. If $x = a + m_1 t$ and $y = b + m_2 t$, $\dfrac{(y - b)}{m_2} = t$. Thus, $m_1 \left(\dfrac{y - b}{m_2} \right) + a = x$

    or $(y - b) = \dfrac{m_2}{m_1}(x - a)$. This represents a line of slope m_2/m_1 passing through (a, b). We have assumed that neither m_2 nor m_1 is zero. If $m_1 = 0$, the line represented is vertical and if $m_2 = 0$, the line is horizontal. Now suppose l is given by $(y - b) = m(x - a)$. If we write $x = a + m_1 t$, $(y - b) = mm_1 t$ or $y = mm_1 t + b$. Let $m_2 = mm_1$. We now have the parametric form.

3. For the segment from (a, b) to (c, d), t takes on all values of $[0, 1]$. For the ray from (a, b) through (c, d), t takes on all values of $[0, \infty)$.

4. a. $x = t$, $y = 3t$. Alternatively, $x = kt$, $y = 3kt$, $k \neq 0$.
    c. $x = t - 1$, $y = 2t + 2$

5. See Figure C.54 for a. and c.

6. See Figure C.55 for a. and c.

7. See Figure C.56 for a. and b.

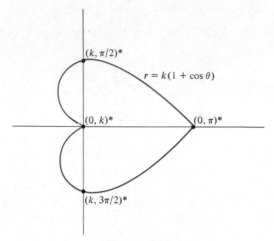

Figure C.51

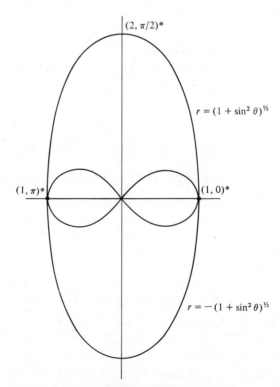

Figure C.52

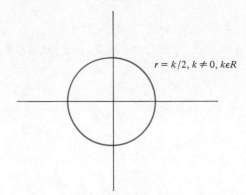

$r = k/2, k \neq 0, k\epsilon R$

Figure C.53

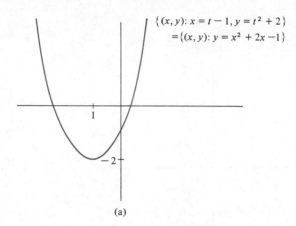

$\{(x,y): x = t - 1, y = t^2 + 2\}$
$= \{(x,y): y = x^2 + 2x - 1\}$

(a)

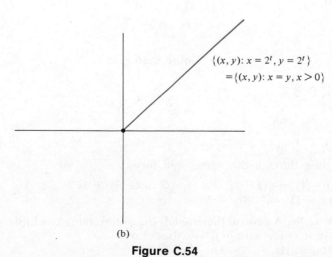

$\{(x,y): x = 2^t, y = 2^t\}$
$= \{(x,y): x = y, x > 0\}$

(b)

Figure C.54

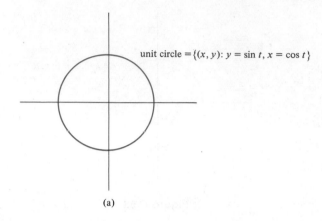

unit circle $= \{(x, y): y = \sin t, x = \cos t\}$

(a)

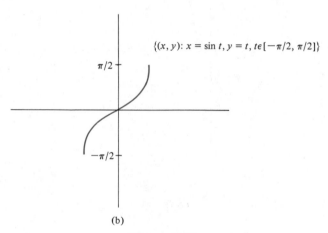

$\{(x, y): x = \sin t, y = t, t\epsilon[-\pi/2, \pi/2]\}$

(b)

Figure C.55

EXERCISES 8.2

1. $p > 0; p < 0$

3. $y - b = 4p(x - a); (a, b)$

4. The line through the vertex and focus.

5. a. $(y - 4)^2 = 8(x - 7)$ c. $(x - 5)^2 = 4y$
    e. $(x - 1)^2 = -16(y - 5)$

6. a. Yes; No. A vertical (horizontal) line intersecting an ellipse intersects
        the curve twice with only two exceptions.
    b. Major axis.

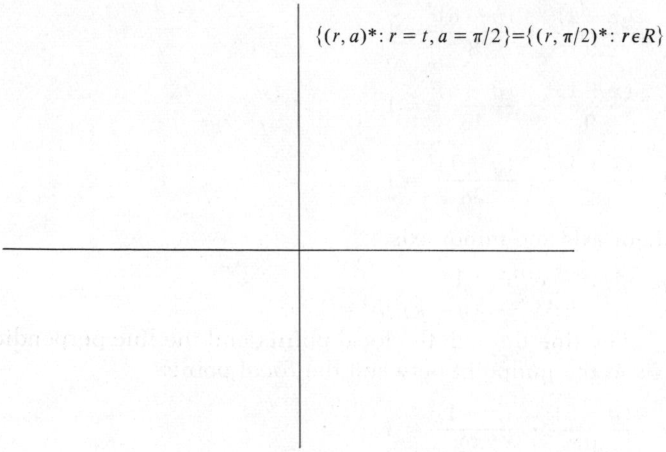

$\{(r, a)^*: r = t, a = \pi/2\} = \{(r, \pi/2)^*: r \epsilon R\}$

(a)

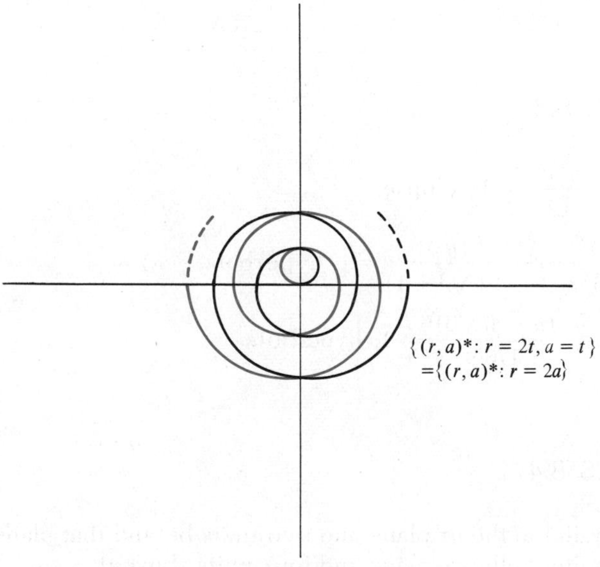

$\{(r, a)^*: r = 2t, a = t\}$
$= \{(r, a)^*: r = 2a\}$

(b)

Figure C.56

7. $x^2/a^2 + y^2/(a^2 - c^2) = 1$

8. a. $\dfrac{(x - h)^2}{a^2} + \dfrac{(y - k)^2}{a^2 - c^2} = 1$. The geometric center.

    b. $\dfrac{(x - h)^2}{a^2 - c^2} + \dfrac{(y - k)^2}{a^2} = 1$

9. a. $\dfrac{(x-2)^2}{9} + \dfrac{(y-6)^2}{8} = 1$

    b. $\dfrac{(x+1)^2}{9} + \dfrac{(y-6)^2}{13} = 1$

    d. $\dfrac{(x+1)^2}{25} + \dfrac{(y+1)^2}{36} = 1$

10. Major axis and minor axis.

11. b. $y^2/a^2 - x^2/b^2 = 1$
     c. $(x-h)^2/a^2 - (y-k)^2/b^2 = 1$
     e. The line through the focal points and the line perpendicular to that axis at the midpoint between the focal points.

12. a. $\dfrac{(y-3)^2}{16} - \dfrac{(x-1)^2}{12} = 1$

     b. $\dfrac{(x-6)^2}{25} - \dfrac{(y+2)^2}{11} = 1$

EXERCISES 8.3

1. $\dfrac{(x')^2}{5} + \dfrac{(y')^2}{4} = 1$; ellipse.

3. $\dfrac{(x')^2}{8/(2\sqrt{3}-1)} - \dfrac{(y')^2}{8/(\sqrt{3}+1)} = 1$; hyperbola.

5. $\dfrac{(x-4)^2}{176/9} - \dfrac{(y-4/\sqrt{3})^2}{176/9} = 1$; hyperbola.

EXERCISES 8.4

1. a. Parallel to the yz plane and two units behind that plane.
    c. Parallel to the xy plane and four units above it.

3. a. $4x + 4y + z - 9 = 0$ b. $z = 1$ c. $y = 1$ e. $y = 1$

4. a. $x + y - 5z + 3 = 0$ c. $5x + y - 2z - 5 = 0$

5. a. $x = a(1-t),\ y = b(1-t),\ z = c(1-t)$
    c. $x = 5 - 3t,\ y = 3 - 6t,\ z = -3 - 4t$
    e. $x = 4t - 3,\ y = 5t - 2,\ z = 1 - 6t$

6. a. $x = 2 - 2t,\ y = 2t,\ z = 1$
    c. $x = -2,\ y = 4 - 4t,\ z = 1 + 16t$
    e. $y = 1,\ z = 4$

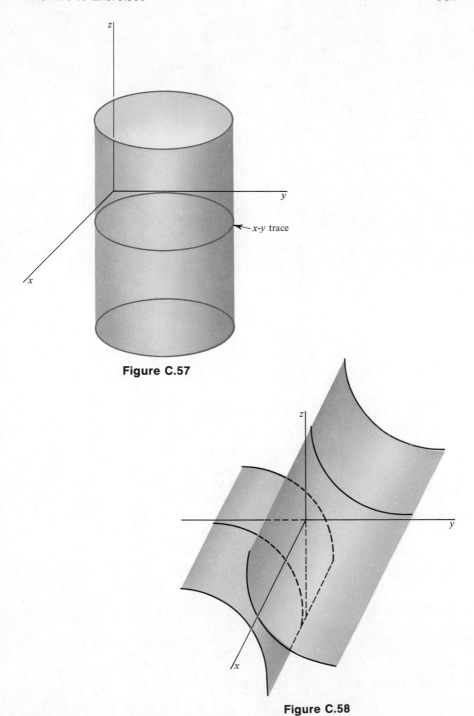

Figure C.57

Figure C.58

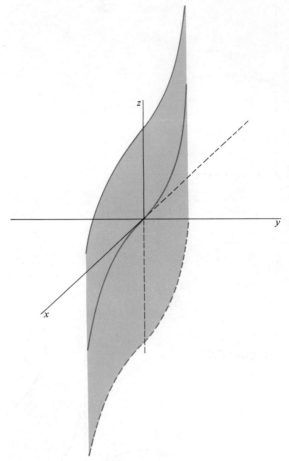

Figure C.59

EXERCISES 8.5

1. See Figure C.57.   6. See Figure C.59.
3. See Figure C.58.   9. See Figure C.60.

EXERCISES 8.6

1. See Figure C.61.   13. 1   18. 0
4. See Figure C.62.   14. 2   19. 2
8. See Figure C.63.   15. 2   21. 11
11. 1   16. 1
12. 5   17. 1

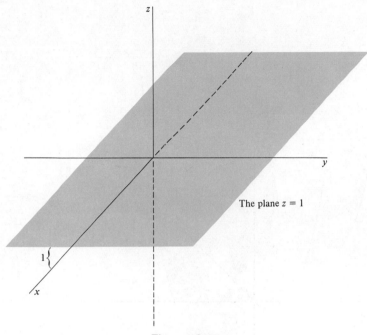

The plane $z = 1$

Figure C.60

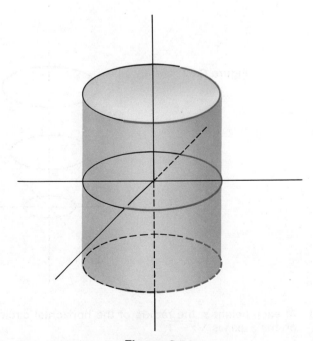

Figure C.61

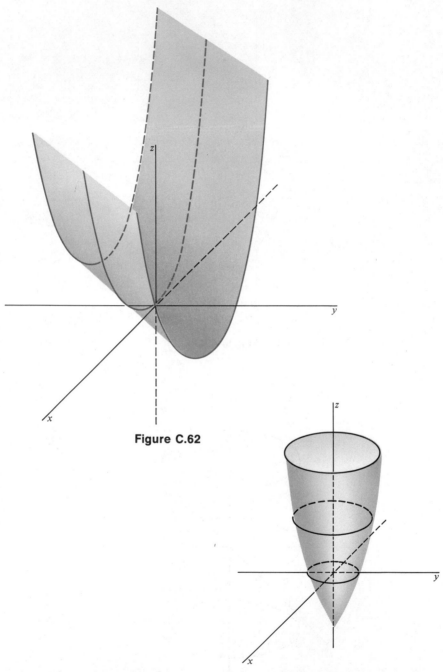

Figure C.62

Figure C.63 At each height z, the radius of the horizontal circle centered on the z axis is $\sqrt{z - 1}$.

REVIEW EXERCISES 8.7

3. a. $y = 1$ c. $x + y - 1 = 0$

4. a. $x = 1 - t$, $y = 1 - t$, $z = 1 - t$
    c. $x = 2 - 3t$, $y = 4$, $z = 3t - 1$
    d. $y = 1$, $x = -1$

6. a. See Figure C.64.

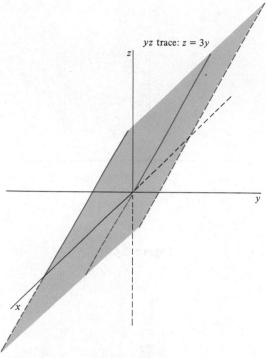

yz trace: $z = 3y$

Figure C.64

7. a. Parabola with vertex at $(-1, 3)$ and $p = -4$ opening downward.
    c. Ellipse of center $(-1, 1)$ with $a = 12$ and $b = 5$, and with the major axis vertical.
    e. Hyperbola with center at $(-2, 1)$, traverse axis (axis containing the focal points) horizontal, $a = b = 1$.

8. a. $x + 1 = 24(y - 7)^2$

    c. $\dfrac{(x - 1)^2}{36} + \dfrac{(y - 8)^2}{40} = 1$

    e. $\dfrac{(x - 15)^2}{49} - \dfrac{(y - 2)^2}{51} = 1$

QUIZ 8.8

1. $x - 8y + z - 5 = 0$.
2. $x = 0,\ y = 1 - 2t,\ z = 13 - 16t$
3. See Figure C.65.
4. See Figure C.66. $R \times R - (0, 0)$

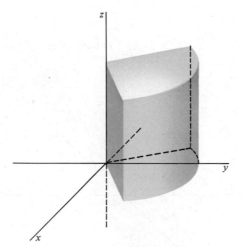

Figure C.65

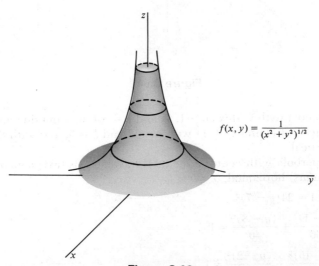

$f(x, y) = \dfrac{1}{(x^2 + y^2)^{1/2}}$

Figure C.66

5. A parabola opening to the right having vertex $(-5, 3)$, focus $(1, 3)$, and directrix $x = -11$.

6. $\dfrac{(x-7)^2}{91} + \dfrac{(y-13)^2}{100} = 1$

7. See Figure C.67. Rotate the axes $(1/2)\cot^{-1}(7/24)$ and the equation becomes $\dfrac{(y'-1)^2}{6} - \dfrac{(x'-1)^2}{3} = 1$.

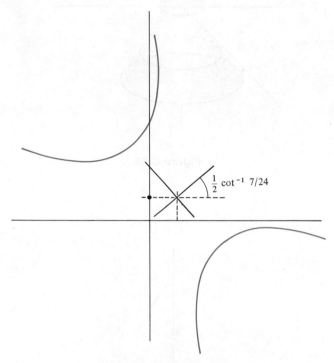

Figure C.67

ADVANCED EXERCISES 8.9

1. a. See Figure C.68. c. See Figure C.69.

2. a. The graph is similar to that in Figure C.69.

EXERCISES A.1

1. a. $x \in A$ or $x \in B$ k. $x \notin A$ and $x \notin B$
   d. $x \in A$ or $x \notin B$ o. If $x \notin A$, then $x \notin B$.
   g. If $x \in B$, then $x \in A$.   s. If $x \notin A$, then $x \notin B$.

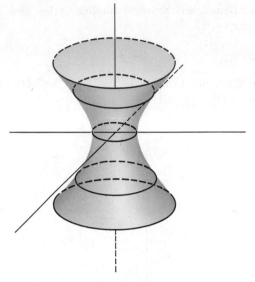

Figure C.68

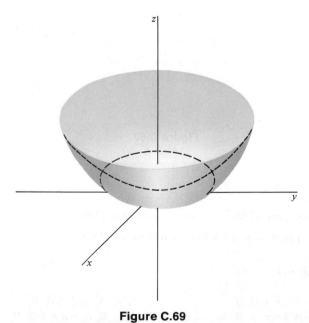

Figure C.69

2. Expression a. is the only self-contradiction.

EXERCISES A.2

1. The universal set ϕ.

2. a. $P \cap Q$ d. $P \cup Q'$

3. a. $P \cap Q = \phi$ c. $P \cap Q \subset R$ e. $P = Q$

5. a. The universal set   b. The universal set

EXERCISES B.1

2. a. -2 c. 0 e. 21

3. 0; 1 and -1

4. c. $1, 3, 5, 15$
   e. $1, 2, 3, 5, 13, 6, 10, 26, 15, 39, 65, 30, 78, 195, 390$

5. c. $3, 5$ e. $2, 3, 5, 13$ g. $2, 3, 4, 23$

6. 2

8. $p = p + 0$ (B.1.5); $p + 0 = 0 + p$ (B.1.2); $0 + p = 0$ (stated property of p);
   $p = 0$ (repeated use of the transitive property of "$=$").

15. If $b \neq 0$, $0 = 0b^{-1} = (ab)b^{-1} = a(bb^{-1}) = a$.

EXERCISES B.2

1. True

2. No

3. a. $1/6$ c. $3/2$ e. $10/9$

5. a. -1 c. -2 e. 11 g. -7 j. $143/27$

10. a. $1/8$ c. $5/3$ e. $230/999$ g. $2/165$ i. $1/4$

EXERCISES B.3

1. a. $7 + 3 = 10$ c. $-20 + 22 = 2$

2. a. 5 c. 0 e. 0

5. If $a < b$ and $b < c$ there are positive numbers d and e with $a + d = b$
   and $b + e = c$. Thus $a + (d + e) = c$ and since $d + e$ is positive, $a < c$.

11. a. $\{x : x < 3\}$ c. $\{x : x < 3\}$ e. $\{x : x > 0\}$

12. If $x \geqslant 0$, $|x| = x \geqslant 0$. If $x < 0$, $|x| = -x$. Since $x < 0$, $-x > 0$ whence $|x| > 0$. In any case, $|x| \geqslant 0$.

16. a. $\{x: -2 < x < 4\}$ c. $\{x: -4 < x < 6\}$

17. $x^+ + x^- = x$; $x^+ - x^- = |x|$

EXERCISES B.4

1. a. Let $n = 1$. $x^m x^n = x^m x = x^{m+1}$ by definition. Thus for $n = 1$, $x^m x^n = x^{m+n}$. Now assume that $x^m x^k = x^{m+k}$. Then $x^m x^{k+1} = x^m (x^k x) = (x^m x^k) x = x^{m+k} x = x^{(m+k)+1} = x^{m+(k+1)}$. The theorem holds for $n = k + 1$ whenever it holds for k. By induction, $x^m x^n = x^{m+n}$ for all m and n in N.

4. a. $f(1) = 5$, $f(2) = 7$, $f(3) = 9$, $f(4) = 11$, $f(5) = 13$
 e. If $n = 1$, $f(1) = f(1) + (1 - 1)d = f(1) + (n - 1)d$. Suppose that $f(k) = f(1) + (k - 1)d$. $f(k + 1) = f(k) + d = f(1) + (k - 1)d + d = f(1) + kd = f(1) + [(k + 1) - 1]d$. Therefore, $f(n) = f(1) + (n - 1)d$ for all $n \in N$.

5. a. $f(1) = 6$, $f(2) = 12$, $f(3) = 24$, $f(4) = 48$, $f(5) = 96$

6. a. $f(n) = 1$ for all n. $g(n) = 0$ for all n.
 b. $f(n) = 1/n$ c. $f(n) = 6 + 1/n$ d. $f(n) = n$

MISCELLANEOUS EXERCISES B.5

1. We know that a_1, $a_1 + a_2$, and $a_1 + a_2 + a_3$ are real numbers. Suppose that for any collection $a_1, a_2, \ldots, a_k$ their sum $a_1 + a_2 + \cdots + a_k$ is a real number. Examine the sum $b_1 + b_2 + \cdots + b_k + b_{k+1}$ of $k + 1$ real numbers. $(b_1 + b_2 + \cdots + b_k) + b_{k+1}$ is a real number since $b_1 + b_2 + \cdots + b_k$ is a real number (our induction hypothesis). Thus, the sum of n real numbers is a real number for any n.

3. Either $x - 1 = 0$ or $x - 2 = 0$ whereby $x = 1$ or $x = 2$.

4. No.

6. $ab > 0$ means $a > 0$ and $b > 0$ or $a < 0$ and $b < 0$. The numbers a and b must have the same polarity.
 a. $x > 1$ or $x < -1$ b. $-1 < x < 1$

7. a. 2   c. 24   e. $n + 1$

8. a. 1   c. 5   e. 10   g. n

9. a. $4, -4$ c. -3 e. $\frac{1}{2}, -\frac{1}{2}$

10. a. 3   c. 2

11. a. 1/16   c. 125

12. a. 4   c. $\frac{1}{4}$

14. a. 21   c. 252   e. 36

Index

absolute value
  function, 74
  of a complex number, 254
  of a real number, 74, 317
addition
  of complex numbers, 244, 250, 253, 259
  of ordered pairs, 244
  of real numbers, 70, 312
  of vectors, 246
additive
  identity, 312
  inverse, 73, 312–313
algebraic
  structure, 244, 250, 310
algorithm
  division, 171
amplitude, 231–232, 235
angle, 216
arc
  of a circle, 78
  length, 79, 114, 155
    function, 84
  cosecant, 219
  cosine, 219
  cotangent, 219
  secant, 219
  sine, 219, 224
  tangent, 219
argument
  of a complex number, 254
arithmetic
  mean, 321
  progression, 321
associative
  operation, 72, 312
asymptote
  horizontal, 161
  oblique, 164
  vertical, 159

axis
  horizontal, 55
  minor, 272
  major, 272
  of symmetry, 274–275, 277
  vertical, 55

between, 155
bijective
  function, 48
binary
  operation, 70
binomial
  coefficients, 179
  theorem, 177
bisector
  perpendicular, 154
bound
  greatest lower, 54
  least upper, 54
  lower, 53
  upper, 53
bounded
  above, 53
  below, 53
  function, 68, 160

Cartesian
  product, 12, 37
chord, 79
circle, 76, 190, 277
  arc of, 78
  circumference, 78
  equation, 76, 78, 244, 268
  unit, 78, 190
circumference
  of a circle, 78, 112
closed, 70, 311
codomain
  of a function, 29, 37, 39–40

Date Due